CISM International Centre for Mechanical Sciences

Courses and Lectures

Volume 566

The series presents lecture notes, monographs, edited works and proceedings in the field of Mechanics, Engineering, Computer Science and Applied Mathematics. Purpose of the series is to make known in the international scientific and technical community results obtained in some of the activities organized by CISM, the International Centre for Mechanical Sciences.

More information about this series at http://www.springer.com/series/76

Jörg Schröder · Peter Wriggers
Editors

Advanced Finite Element Technologies

Springer

Editors
Jörg Schröder
Institut für Mechanik
Universität Duisburg-Essen
Essen
Germany

Peter Wriggers
Institut für Kontinuumsmechanik
Leibniz Universität Hannover
Hannover
Germany

ISSN 0254-1971 ISSN 2309-3706 (electronic)
CISM International Centre for Mechanical Sciences
ISBN 978-3-319-31923-0 ISBN 978-3-319-31925-4 (eBook)
DOI 10.1007/978-3-319-31925-4

Library of Congress Control Number: 2016936423

Printed on acid-free paper

This Springer imprint is published by Springer Nature
The registered company is Springer International Publishing AG Switzerland

Preface

Advanced Finite Element Technologies are essential for the solution of almost all problems in computational mechanics. One of the great attractions of the finite element method is its enormous range of applicability, which varies from classical subjects like mechanical, aerospace, automotive, and civil engineering, to new scientific disciplines like information technology, applied physics, or biomechanics. Due to the substantial developments in several fields, as for instance materials science, production methods or optimization processes, many engineering and mathematical approaches for novel finite elements were developed during the last decades. The growing demand for reliable, accurate, and highly efficient finite elements particularly in the field of nonlinearities has led to a number of interesting finite element formulations.

The CISM course on "Advanced Finite Element Technologies", held in Udine from October 6 to 10, 2014, was addressed to master students, doctoral students, postdocs, and experienced researchers in engineering, applied mathematics, and materials science who wished to broaden their knowledge in e.g. advanced mixed Galerkin and least-squares FEM, discontinuous Galerkin methods as well as the related mathematical analysis.

It is our pleasure to thank the lecturers of the CISM course: Ferdinando Auricchio (Pavia, Italy), Antonio Huerta (Barcelona, Spain), Daya Reddy (Cape Town, South Africa), Gerhard Starke (Essen, Germany), as well as the additional contributors to these CISM lecture notes Adrien Lefieux (Atlanta, USA), Benjamin Müller (Essen, Germany), Alessandro Reali (München, Germany), Alexander Schwarz (Essen, Germany), Ruben Sevilla (Swansea, Wales), and Karl Steeger (Essen, Germany). We furthermore thank the 55 participants from 13 countries who made the course a success. Finally, we extend our thanks to the Rectors, the Board, and the staff of CISM for the excellent support and kind help.

Jörg Schröder
Peter Wriggers

Contents

Functional Analysis, Boundary Value Problems and Finite Elements

Batmanathan Dayanand Reddy

Abstract This chapter presents, first, an overview of the mathematical tools required to undertake studies of the well-posedness of linear boundary value problems and their approximations by finite elements. In the remainder of this work, these tools are used to examine the existence and uniqueness of solutions to weak boundary value problems, and convergence of finite element approximations. The emphasis is on second-order partial differential equations, with the governing equations for linear elasticity being the key model problem.

1 Introduction

We will be concerned with boundary value problems that arise in solid mechanics. These typically take the form of a single partial differential equation (PDE) or, often, a system of PDEs. In applications such as plasticity and contact, the problem comprises a set of equations as well as inequalities.

The purpose of this chapter is to present an overview of mathematical fundamentals that are essential to qualitative studies of boundary value problems as well as their approximations by the finite element method. The aim of a qualitative study is to glean information about a problem and its solution in the absence of a closed-form solution, which is generally the case in complex problems of continuum mechanics. When approximate approaches such as the finite element method are used, such studies are able to tell us about the quality of an approximation and its rate of convergence to the actual solution, again in the absence of knowledge about the exact solution.

We focus in this chapter on linear problems, in which the governing equation takes the form $Au = f$. Here A is a linear operator and the right-hand side f is given. A simple example is the problem of an Euler–Bernoulli beam. The governing equation is a fourth-order differential equation

B.D. Reddy (✉)
Centre for Research in Computational and Applied Mechanics,
University of Cape Town, Rondebosch, South Africa
e-mail: daya.reddy@uct.ac.za

J. Schröder and P. Wriggers (eds.), *Advanced Finite Element Technologies*,
CISM International Centre for Mechanical Sciences 566,
DOI 10.1007/978-3-319-31925-4_1

$$Au = EI\frac{d^4u}{dx^4} = f \tag{1}$$

in which u denotes the transverse displacement, f the applied loading, and E and I are, respectively, Young's modulus and the second moment of area of the beam cross-section. From a qualitative point of view, we wish to know: (a) whether the problem has a solution, that is, whether $f \in R(A)$, the range of A; and (b) whether that solution is unique, that is, whether the null space of A, denoted $N(A)$, consists of only the zero element. These terms have yet to be defined. It will be seen that the answer to the two questions will depend to some extent on the specification of the boundary conditions. The range of A comprises all continuous functions. Suppose, for example, that the load f is constant and that the boundary conditions are $u(0) = u''(L) = u'''(L) = 0$, for a beam of length L: that is, zero displacement at one end and zero moment and shear force at the other. This gives the solution

$$u(x) = \frac{f}{EI}\left(\frac{1}{24}x^4 - \frac{1}{6}Lx^3 + \frac{1}{4}L^2x^2 + Cx\right). \tag{2}$$

There remains one constant C to be found, which is as it should be because there is a further boundary condition that needs to be specified at $x = 0$. If we assume that the outstanding boundary condition is $u'(0) = 0$, so that the end $x = 0$ is clamped, then we find that $C = 0$. The null space of A comprises all solutions u such that $Au = 0$, and here the only such solution is $u = 0$, so that $N(A) = \{0\}$. The solution (2) with $C = 0$ is thus unique.

On the other hand, assume that the boundary condition is $u''(0)$, so that we have zero moment at the end $x = 0$. Physically it is clear that the system cannot be in equilibrium. This is confirmed by the fact that C is now undetermined: $N(A) = \{u \mid u(x) = Cx\}$ which gives an infinite number of further solutions corresponding to the beam rotating about the end $x = 0$.

The above simple example illustrates for a somewhat obvious case the kind of information that can be obtained by seeking qualitative information about the solution. We shall formalize this process in the following sections.

A further example of a boundary value problem is the system of equations for equilibrium of isotropic linear elastic bodies. The governing equations of the problem are as follows:

Equilibrium: $$-\text{div}\, \boldsymbol{\sigma} = \boldsymbol{f}, \tag{3a}$$

Hooke's law: $$\boldsymbol{\sigma} = \mathbb{C}\boldsymbol{\varepsilon}(\boldsymbol{u}) = \lambda\, \text{div}\ \boldsymbol{u} + 2\mu\boldsymbol{\varepsilon}(\boldsymbol{u}), \tag{3b}$$

Strain-displacement: $$\boldsymbol{\varepsilon}(\boldsymbol{u}) = \frac{1}{2}(\nabla\boldsymbol{u} + [\nabla\boldsymbol{u}]^T). \tag{3c}$$

Here λ and μ are the Lamé constants, $\boldsymbol{\sigma}$ is the stress and $\boldsymbol{\varepsilon}(\boldsymbol{u})$ is the strain tensor. Substitution in the equilibrium equation yields the *Lamé equation*

$$A\boldsymbol{u} := -\ \text{div}\ [\mathbb{C}\boldsymbol{\varepsilon}(\boldsymbol{u})] = -(\lambda + \mu)\nabla\ \text{div}\ \boldsymbol{u} - \mu\nabla^2\boldsymbol{u} = \boldsymbol{f} \tag{4}$$

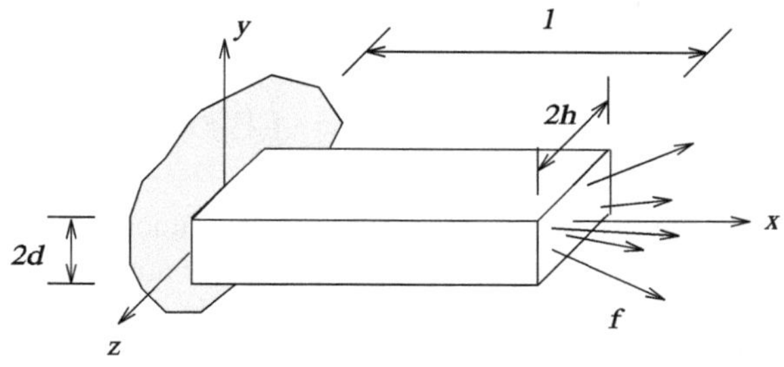

Fig. 1 Deformation of an elastic bar

for a given body force $\boldsymbol{f}$. To this system of equations a set of boundary conditions must be added. For the domain shown in Fig. 1 , the displacement $\boldsymbol{u} = \boldsymbol{0}$ on the end $x = 0$ while the traction $\boldsymbol{t} = \boldsymbol{\sigma n}$ is prescribed over the rest of the boundary.

A PDE of order $2m$ requires m boundary conditions at each point on the boundary, with derivatives of order no greater than $2m - 1$. Thus the Euler–Bernoulli beam equation is an equation of the fourth order and this requires two boundary conditions at each end of the beam. The elasticity problem leads to a system of PDEs of second order, and so requires *one vector-valued* boundary condition at each point on the boundary. For a second-order PDE or system of PDEs a boundary condition involving the displacement is called a *Dirichlet* condition, while that involving the first derivatives of the displacement, through the traction for example, is called a *Neumann* boundary condition.

1.1 Weak Formulations

Questions around the existence of solutions and their uniqueness may be effectively approached not only by studying the original PDE and boundary conditions, but also by reformulating the problem in a *weak* or *variational* form. We take the governing equations as a model problem for equilibrium of a linear elastic body as set out in (4) and use these to show how the weak formulation is constructed. Suppose that the body occupies the domain $\Omega \subset \mathbb{R}^d$ ($d = 2, 3$) with boundary Γ comprising two nonoverlapping parts Γ_u and Γ_t. Suppose further that the boundary conditions are

$$\boldsymbol{u} = \boldsymbol{0} \quad \text{on } \Gamma_u\,, \quad \boldsymbol{t} = \boldsymbol{\sigma n} = \bar{\boldsymbol{t}} \quad \text{on } \Gamma_t. \tag{5}$$

We start by introducing a *test function* $\boldsymbol{v}$, which is a function smooth enough to be differentiated, and which satisfies the homogeneous Dirichlet boundary condition $\boldsymbol{v} = \boldsymbol{0}$ on Γ_u. Next, take the inner product of both sides of Eq. (4) with $\boldsymbol{v}$ and integrate over the domain Ω:

$$-\int_\Omega \operatorname{div}\, \mathbb{C}\varepsilon(\boldsymbol{u}) \cdot \boldsymbol{v}\, dx = \int_\Omega \boldsymbol{f} \cdot \boldsymbol{v}\, dx. \tag{6}$$

Now using the Green–Gauss theorem, the term on the left-hand side becomes

$$-\int_\Omega \text{div } \mathbb{C}\varepsilon(\boldsymbol{u}) \cdot \boldsymbol{v}\, dx = -\int_\Gamma [\mathbb{C}\varepsilon(\boldsymbol{u})]\boldsymbol{n} \cdot \boldsymbol{v}\, dx + \int_\Omega \mathbb{C}\varepsilon(\boldsymbol{u}) : \varepsilon(\boldsymbol{v})\, dx. \quad (7)$$

Finally, we use the property of the test function $\boldsymbol{v}$ that $\boldsymbol{v} = \boldsymbol{0}$ on Γ_u, and also the boundary condition (5b): in this way, we arrive at the weak formulation for linear elasticity: find $\boldsymbol{u}$ which satisfies (5a) and

$$\int_\Omega \mathbb{C}\varepsilon(\boldsymbol{u}) : \varepsilon(\boldsymbol{v})\, dx = \int_\Omega \boldsymbol{f} \cdot \boldsymbol{v}\, dx + \int_{\Gamma_t} \bar{\boldsymbol{t}} \cdot \boldsymbol{v}\, dx \quad \text{for all test functions } \boldsymbol{v}. \quad (8)$$

That any solution of the strong formulation, that is, the PDE (4) and boundary conditions (5), solves the weak problem (8) is clear. Furthermore, we can show by a reverse process that a solution of the weak formulation is also a solution of the PDE and boundary conditions, provided that the weak solution is smooth enough for derivatives to be taken where required. Thus in the weak formulation, it is only necessary that the first derivatives of $\boldsymbol{u}$ and $\boldsymbol{v}$ in (8) make sense. On the other hand, the strong formulation requires that the second derivatives of the displacement be defined.

Later, we will see that the weak formulation is also important as the basis for finite element approximations. If we denote such an approximation by $\boldsymbol{u}_h$, then we would be interested in estimating the error $\boldsymbol{u} - \boldsymbol{u}_h$ of the approximation, and in determining the rate of convergence of $\boldsymbol{u}_h$ to $\boldsymbol{u}$ as the mesh size h goes to zero.

We proceed to build the mathematical tools required to study well-posedness and convergence of finite element approximations. More detailed accounts of the relevant background may be found in the works of Atkinson and Han (2001), Ciarlet (2002), Reddy (1998), for example.

2 Function Spaces and Operators

Much motivation for the construction of function spaces and their properties derives from our understanding of $\mathbb{R}^n$, or equivalently of vectors in n-dimensional space. Indeed, the properties of vectors in $\mathbb{R}^2$ include the following: there is a zero vector $\boldsymbol{0}$; the addition of multiples $\alpha\boldsymbol{a} + \beta\boldsymbol{b}$ of vectors $\boldsymbol{a}$ and $\boldsymbol{b}$ is again a vector; the inner product or dot product of vectors $\boldsymbol{a}$ and $\boldsymbol{b}$ is defined by $\boldsymbol{a} \cdot \boldsymbol{b} = a_1b_1 + a_2b_2$, where a_i and b_i are components of the vectors relative to an orthonormal basis; has the properties $\boldsymbol{a} \cdot \boldsymbol{b} = \boldsymbol{b} \cdot \boldsymbol{a}$, and $\boldsymbol{a} \cdot \boldsymbol{a} \geq 0$; and is equal to zero if and only if $\boldsymbol{a} = \boldsymbol{0}$. In this way, we define the length or norm $\|\boldsymbol{a}\|$ of $\boldsymbol{a}$ by $\|\boldsymbol{a}\| = (\boldsymbol{a} \cdot \boldsymbol{a})^{1/2}$. A further property of the norm is the *triangle inequality*: $\|\boldsymbol{a} + \boldsymbol{b}\| \leq \|\boldsymbol{a}\| + \|\boldsymbol{b}\|$. We generalize these features so that they apply to sets or spaces of *functions*.

Continuous functions Roughly, a continuous function $f(x)$ of a single variable x is one that can be sketched without picking up one's pen. The rigorous definition, for functions of any number of variables, is as follows: suppose that a function f is

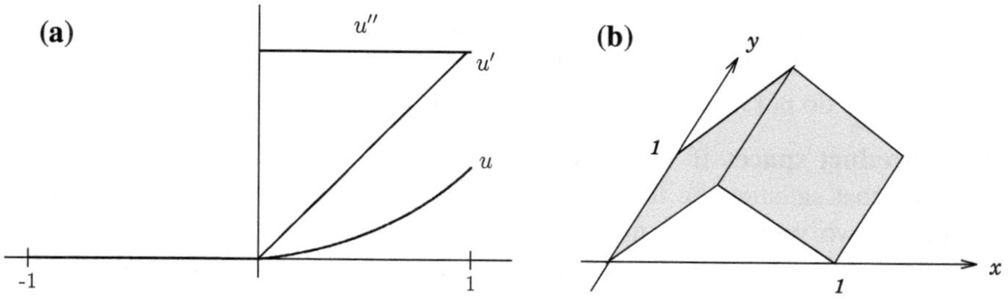

Fig. 2 **a** An example of a function $u \in C^1(-1, 1)$; **b** a function $u \in C(\Omega)$ where Ω is the square domain $(0, 1) \times (0, 1)$

defined on a bounded domain[1] Ω (for example, the interval $(-1, 1)$ in one dimension or the unit square in two). Then f is continuous if, for any given number $\epsilon > 0$, it is possible to find a number $\delta > 0$ (depending on ϵ) such that

$$|f(\boldsymbol{x}) - f(\boldsymbol{y})| < \epsilon \quad \text{whenever} \quad |\boldsymbol{x} - \boldsymbol{y}| < \delta \text{ for any } \boldsymbol{x}, \boldsymbol{y} \in \Omega.$$

Examples of continuous functions are shown in Fig. 2. We denote the set of continuous functions on Ω by $C(\Omega)$. More generally, the set of functions which together with their derivatives up to and including those of order m is denoted by $C^m(\Omega)$; thus $C^0(\Omega) = C(\Omega)$. Finally, the set of functions all of whose derivatives are continuous is denoted by $C^\infty(\Omega)$.

The Lebesgue space $L^2(\Omega)$ We are also interested in functions that may possibly not be differentiable but which can be integrated. We define the space $L^2(\Omega)$ to be the set of functions which are square-integrable on Ω; that is, the integral

$$\int_\Omega [f(\boldsymbol{x})]^2 \, dx \quad \text{is finite.}$$

The functions in Fig. 2 are members of L^2, and so is the function $f(x) = x^{-1/3}$ on $(0, 1)$. Though it blows up at $x = 0$ we find that $\int_0^1 f^2 \, dx = 3$.

Vector spaces The spaces $C^m(\Omega)$ and $L^2(\Omega)$ are examples of *vector spaces*. That is, they mimic the properties of $\mathbb{R}^n$: in particular, multiplication of a function in $L^2(\Omega)$ by a scalar α gives a function in $L^2(\Omega)$, as does the sum of two functions u, v, in that $u + v$ also belongs to $L^2(\Omega)$. Furthermore, there exists a zero member, viz. the zero function, and a negative function $-f$ corresponding to any function f.

What about subsets of these spaces? If a subset of a vector space V is itself a vector space, then it is referred to as a *subspace* of V. So, for example, the set of all points in a plane passing through the origin is a subspace of $\mathbb{R}^3$. Likewise, the set $P_k(0, 1)$ of polynomials of degree not exceeding k is a subspace of $C(0, 1)$ and

[1] A connected set Ω is one for which every pair of points can be connected by a curve that lies entirely in Ω; Ω is open if it comprises only interior points; and a domain is an open connected set.

in fact of $L^2(0, 1)$; but the set X of all nonnegative functions $f \in L^2(\Omega)$, that is, functions satisfying $f(\boldsymbol{x}) \geq 0$ for all $\boldsymbol{x} \in \Omega$, is *not* a subspace because for $f \in X$, functions $-f$ do not belong to X (except for the trivial case $f(\boldsymbol{x}) \equiv 0$).

Inner product spaces If V is a vector space then the inner product $(\cdot, \cdot)$ is an operation that satisfies the following axioms (following the properties of the dot product of two vectors): for members u and v of V, (u, v) is a real number; the inner product is *symmetric* in that $(v, u) = (u, v)$; it is *linear* in each slot, in that $(\alpha u + \beta w, v) = \alpha(u, v) + \beta(w, v)$ for $u, v, w \in V$ and real numbers α and β; and finally it is *positive-definite* in that $(v, v) \geq 0$ and $(v, v) = 0$ if and only if $v = 0$. A vector space with an inner product defined on it is called an *inner product space*.

An important example of an inner product space is $L^2(\Omega)$, for which the inner product is defined by

$$(f, g)_{L^2} = \int_\Omega f(\boldsymbol{x})g(\boldsymbol{x})\, dx. \tag{9}$$

The property of *orthogonality* in inner product spaces carries over in an obvious way from the vectorial definition: two members u, v of an inner product space V are orthogonal if $(u, v)_V = 0$. For example, the functions $u(x) = \sin \pi x$ and $v(x) = \cos \pi x$ are orthogonal in $L^2(0, 1)$ since

$$(u, v)_{L^2} = \int_0^1 (\sin \pi x)\,(\cos \pi x)\, dx = 0.$$

Normed spaces The final property that we extend from sets of vectors is that of a norm, which is a quantity that measures 'length' or 'magnitude'. For a vector space V, a real-valued operation $\|\cdot\|$ is called a norm if it satisfies the following: first, it is positive-definite in that $\|v\| \geq 0$ and $\|v\| = 0$ if and only if $v = 0$; second, it is positively homogeneous in that $\|\alpha v\| = |\alpha|\|v\|$ for all real numbers α; and finally it satisfies the triangle inequality: $\|u + v\| \leq \|u\| + \|v\|$. A vector space with a norm defined on it is called a *normed space*.

When dealing with vectors in two-dimensional space, for example, we know that the length or norm may be defined through the scalar product: that is, $\|\boldsymbol{a}\| = (\boldsymbol{a}\cdot\boldsymbol{a})^{1/2}$. This feature carries over to the general case: every inner product space is a normed space, with norm $\|\cdot\|$ defined by $\|u\| = (u, u)^{1/2}$. But it is possible to define a norm independent of the existence of an inner product: for example, in $\mathbb{R}^2$ the quantity $\|\cdot\|_1$ defined by $\|\boldsymbol{a}\|_1 = |a_1| + |a_2|$ is a norm, as can be easily verified. The case $\|\boldsymbol{a}\|_2 = (|a_1|^2 + |a_2|^2)^{1/2}$ corresponds to the Euclidean norm, which is generated by the standard scalar product on $\mathbb{R}^2$. Every norm may be generated by an inner product, and so *every inner produce space is a normed space, while the converse is not true.*

The relation $|\boldsymbol{a} \cdot \boldsymbol{b}| = |\boldsymbol{a}|\,|\boldsymbol{b}| \cos(\boldsymbol{a}, \boldsymbol{b}) \leq |\boldsymbol{a}|\,|\boldsymbol{b}|$ has a counterpart in the context of inner product spaces, known as the *Cauchy–Schwarz* inequality: for X an inner product space, the inequality states that

$$|(u, v)_X| \leq \|u\|_X \|v\|_X \quad \text{for all } u, v \in X. \tag{10}$$

Hilbert and Banach spaces The final important property that we introduce is *completeness*, which can be defined roughly as follows. We start with a normed space X and consider a sequence $u_1, u_2, \ldots$ of members of a subset Y of X. This is called a *Cauchy sequence* if successive members are progressively closer: that is, if the 'distance' $\|u_n - u_m\|_X$ between members u_n and u_m of X goes to zero as m and n become progressively larger. We write this succinctly as $\lim_{m,n\to\infty} \|u_m - u_n\|_X = 0$. The question as to whether the Cauchy sequence is *convergent* is a separate question, which may have a positive or negative answer. For example, take $X = \mathbb{R}$ and $Y =$ the set of rational numbers. The sequence $\{1, 1.4, 1.41, 1.414, \ldots\}$ is a Cauchy sequence in Y, but its limit $\sqrt{2}$ is irrational and does *not* belong to Y.

As another example, consider the space $C[0, 1]$ of continuous functions with the norm $\|u\| = [\int_0^1 [u(x)]^2\, dx]^{1/2}$. The sequence of functions shown in Fig. 3 is a Cauchy sequence in $C[0, 1]$ with the norm $\|\cdot\|$, but its limit, the *discontinuous function* u shown in the figure, is not a member of $C[0, 1]$. This is an example of an incomplete normed space: we say that a normed space is *complete* if every Cauchy sequence in the space converges to a member of that space. The same definition applies to subsets. Important examples of complete spaces or sets are the closed interval $[a, b]$ in $\mathbb{R}$, the space of continuous functions $C(\overline{\Omega})$ *with the max-norm* $\|u\|_\infty = \max_{x\in\Omega} |u(\boldsymbol{x})|$, *and the space* $L^2(\Omega)$ *with the standard* L^2-norm.

A complete normed space is called a *Banach space*, while a complete inner product space is called a *Hilbert space*.

Finite-dimensional spaces This topic will probably be familiar from undergraduate courses in linear algebra. Given a vector space X, an expression of the form $I = \alpha_1 u_1 + \alpha_2 u_2 + \cdots + \alpha_n u_n$, where α_i $(i = 1, \ldots, n)$ are real numbers and $Y = \{u_1, \ldots, u_n\}$ a set of members of X, is said to be a *linear combination* of $u_1, \ldots, u_n$. The set Y is linearly dependent if we can find α_i, not all zero, such that the linear combination $I = 0$. Otherwise, it is linearly independent.

The set Y *spans* the space X if every member of X can be written as a linear combination of members of Y. Finally, the set Y is a *basis* if it is both linearly independent and spans X. If Y has n members, then X is said to have dimension n.

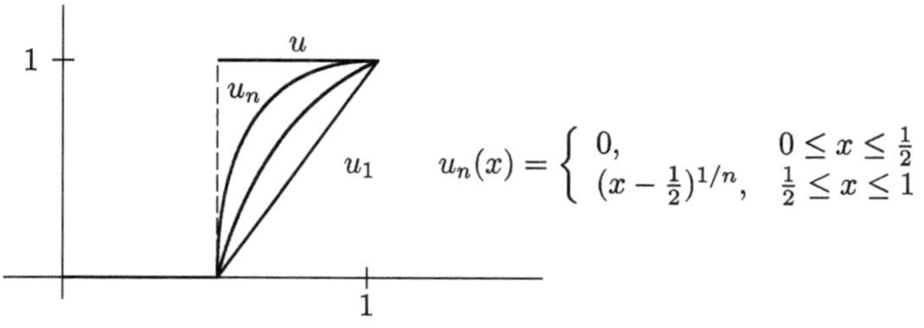

Fig. 3 A Cauchy sequence of continuous functions with 'limit' a discontinuous function

Some simple examples will help to fix ideas. The space $\mathbb{R}^n$ of n-tuples is a vector space of dimension n. So for $\mathbb{R}^2$, the standard basis is $\{\boldsymbol{e}_1 = (1,0), \boldsymbol{e}_2 = (0,1)\}$ and every vector may be written as a linear combination of $\boldsymbol{e}_1$ and $\boldsymbol{e}_2$. The space $P_3[0,1]$ of polynomials of degree at most 3 on the interval $[0,1]$ (that is, of the form $p(x) = a_0 + a_1x + a_2x^2 + a_3x^3$) is a vector space of dimension 4; the set of monomials $\{1, x, x^2, x^3\}$ forms a basis for $P_3[0,1]$.

The Sobolev spaces $H^m(\Omega)$ We introduce here a family of spaces that are central to the study of weak boundary value problems and their approximation by finite elements. For convenience we confine attention to problems in $\mathbb{R}^2$; the extension to three dimensions is immediate. We introduce the very convenient multi-index notation as a precursor to defining the Sobolev spaces. Let $\mathbb{Z}^+$ denote the set of all nonnegative integers $0, 1, 2, \ldots$. A multi-index α is a pair of nonnegative integers: $\alpha = (\alpha_1, \alpha_2)$ where $\alpha_i \in \mathbb{Z}^+$. We define $|\alpha| = \alpha_1 + \alpha_2$, and a partial derivative of order k by $D^\alpha v$, for $|\alpha| = k$. So for example $\sum_{|\alpha|=2} D^\alpha v = \frac{\partial^2 v}{\partial x^2} + 2\frac{\partial^2 v}{\partial x \partial y} + \frac{\partial^2 v}{\partial y^2}$, viz. the sum of all second derivatives of v. Then $H^m(\Omega)$ is defined to be the space of functions which are square-integrable (that is, in $L^2(\Omega)$), and furthermore all of whose derivatives[2] of order $1, 2, \ldots, m$ are also square-integrable. Thus,

$$H^m(\Omega) = \{v \,|D^\alpha v \in L^2(\Omega), |\alpha| \le m\}. \tag{11}$$

It can be shown that $H^m(\Omega)$ is a *Hilbert space* with inner product $(\cdot,\cdot)_{H^m}$ and norm $\|\cdot\|_{H^m}$ defined by

$$(u, v)_{H^m} = \int_\Omega \sum_{|\alpha| \le m} D^\alpha u \, D^\alpha v \, dx\,, \quad \|v\|_{H^m} = (v, v)_{H^m}^{1/2}.$$

For example, for the case $m = 2$ and $u \in H^2(\Omega)$,

$$\|u\|_{H^2}^2 = \int_\Omega \left[u^2 + \left(\frac{\partial u}{\partial x}\right)^2 + \left(\frac{\partial u}{\partial y}\right)^2 + \left(\frac{\partial^2 u}{\partial x^2}\right)^2 + 2\left(\frac{\partial^2 u}{\partial x \partial y}\right)^2 + \left(\frac{\partial^2 u}{\partial y^2}\right)^2 \right] dx.$$

As a further example consider $u(x)$ defined on $\Omega = (0, 2) \in \mathbb{R}$ by

$$u(x) = \begin{cases} x^2, & 0 < x \le 1, \\ 2x^2 - 2x + 1, & 1 < x < 2. \end{cases}$$

Then

$$u'(x) = \begin{cases} 2x, & 0 < x \le 1, \\ 4x - 2, & 1 < x < 2, \end{cases} \qquad u''(x) = \begin{cases} 2, & 0 < x \le 1, \\ 4, & 1 < x < 2, \end{cases}$$

---

[2]Strictly speaking, we should define these as weak derivatives: see Reddy (1998).

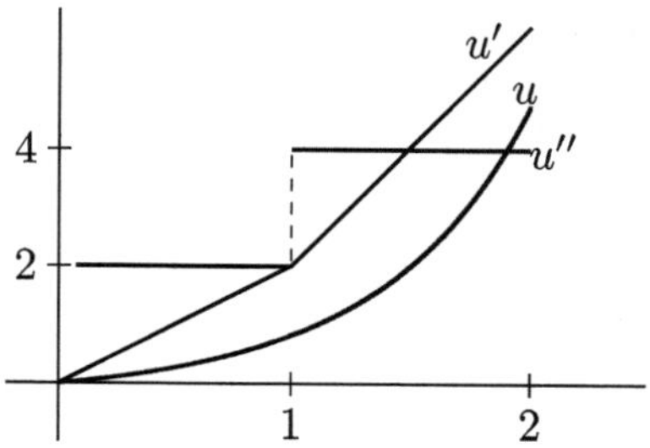

Fig. 4 An example of a function $u \in H^2(0, 2)$

so that u, u', and u'' all belong to $L^2(0, 2)$, but $u''' = 2\delta(x - 1)$, where δ is the Dirac delta, and so is not a member of $L^2(0, 2)$. Hence u is a member of $H^2(0, 2)$ (see Fig. 4), and $\|u\|_{H^2}^2 = \int_0^2 [u^2 + (u')^2 + (u'')^2]\, dx = 71.37$.

Boundary values of functions in $H^m(\Omega)$ It is not possible in general to define the boundary values of a function that is merely in $L^2(\Omega)$. But according to an important result known as the trace theorem it can be shown that any function $u \in H^1(\Omega)$ has a well-defined boundary value that belongs to $L^2(\Gamma)$. Furthermore, this coincides with the conventional definition $u|_\Gamma$ if the function u is continuous. Similar considerations apply to boundary values of a function u on a part Γ_u of the boundary. The space $H_0^1(\Omega)$ is defined by

$$H_0^1(\Omega) = \{u \in H^1(\Omega) \mid u = 0 \text{ on } \Gamma\}.$$

Linear operators An operator or map T from a set X to a set Y is a rule whereby an element u of X is mapped or transformed to an element v of Y. We write $T : X \to Y, \;\; Tu = v \;\; (\text{or } T(u) = v), \;\; u \in X, \; v \in Y$. Here X is called the domain of T, written $D(T)$; $R(T)$, the range of T comprises those elements of Y that are images of members of X, and the null space $N(T)$ of T comprises those members of X which map to $0 \in Y$.

A *linear operator* is an operator whose domain is a vector space X, and for which $T(\alpha u + \beta v) = \alpha T u + \beta T v$ for all $u, v \in X$ and real numbers α, β. For example, a 2×3 matrix M is a linear operator from $\mathbb{R}^3$ to $\mathbb{R}^2$. If its elements are $M_{11} = M_{21} = 1$ and all other $M_{ij} = 0$ then $\mathsf{Mx} = (x_1, x_1)$ and so $R(\mathsf{M})$ is the straight line going through the origin at 45°. Its null space is $N(\mathsf{M}) = \{\boldsymbol{x} \mid x_1 = 0\}$, that is, the $x_2 - x_3$ plane.

The differential operators A defined in (1) and (4) are linear operators, as is easily verified.

An operator of some importance, for example in finite element error analysis, is the *projection* P from a vector space X into itself, defined by $P^2 = P$; that is, $P(Pu) = Pu \;\; \text{for all } \; u \in X$. The motivation for such an operator may be found in $\mathbb{R}^2$: the map defined by $P\boldsymbol{x} = (x_1, 0)$ projects a point in the plane onto the x_1 axis and is a projection. As a further example, the operator $P : C[0, 1] \to C[0, 1]$ defined by $Pu = u(0)(1 - x) + u(1)x$ is a projection, called the linear interpolate of u (see Fig. 5).

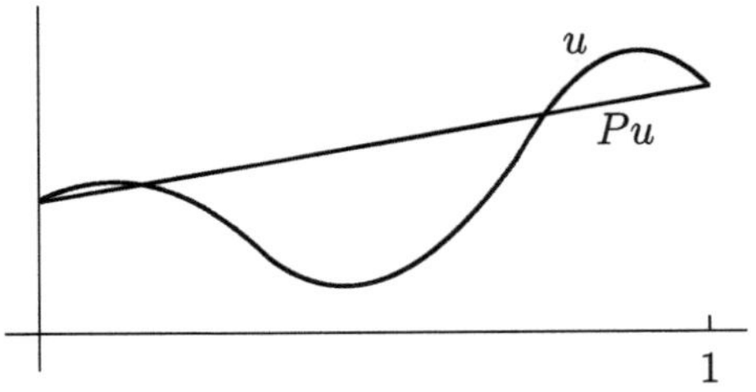

Fig. 5 The linear interpolate Pu of a function u as a projection

Bilinear forms For vector spaces X and Y, a *bilinear form* $a : X \times Y \to \mathbb{R}$ is an operator which is linear in both slots: that is,

$$\left.\begin{aligned} a(\alpha u + \beta v, w) &= \alpha a(u, w) + \beta a(v, w) \\ a(u, \alpha w + \beta z) &= \alpha a(u, w) + \beta a(u, z) \end{aligned}\right\} \quad \text{for all } u, v \in X, w, z \in Y.$$

The bilinear form is *symmetric* if $a(u, v) = a(v, u)$.

An inner product is a simple example of a bilinear form. Another example is for the case in which $X = Y = C[0, 1]$, and

$$a(u, v) = \int_0^1 [u(x)v(x) + u'(x)v'(x)]\, dx.$$

Bounded linear operators and continuous bilinear forms A linear operator $T : X \to Y$, where X and Y are normed spaces, is *bounded* or *continuous* if there is a constant $C > 0$ such that

$$\|Tu\|_Y \le \|u\|_X \quad \text{for all } u \in X.$$

If $Y = \mathbb{R}$ then the operator is called a *functional*. The set of all bounded linear functionals $\ell : X \to \mathbb{R}$ is called the *dual space* of X, and is denoted by X'. Similarly, a bilinear operator $a : X \times Y \to \mathbb{R}$, is *continuous* if there is a constant $K > 0$ such that

$$|a(u, v) \le K\|u\|_X\|v\|_Y \quad \text{for all } u, v \in X.$$

Furthermore, the bilinear form $a : X \times X \to \mathbb{R}$ is X-*elliptic* or *coercive* if there exists a constant $\alpha > 0$ such that

$$a(v, v) \ge \alpha\|v\|_X^2 \quad \text{for all } v \in X.$$

For example, define $a : H^1(0, 1) \times H^1(0, 1) \to \mathbb{R}$, $a(u, v) = \int_0^1 [u'v' + \kappa uv]\, dx$, where $\kappa(x)$ is a bounded continuous function satisfying $\kappa_1 \ge \kappa(x) \ge \kappa_2 > 0$. Then by repeated use of the Cauchy–Schwarz inequality and noting that $\|u\|_{L^2} \le \|u\|_{H^1}$, we can show that $|a(u, v)| \le (1+\kappa_1)\|u\|_{H^1}\|v\|_{H^1}$. Likewise, $a(\cdot, \cdot)$ is coercive since $a(v, v) \ge (1 + \kappa_2)\|v\|_{H^1}^2$.

Weak formulations of boundary value problems We return to the material in Sect. 1.1. The ingredients are as follows: a space V, which will generally be a subspace of a Hilbert space; a *symmetric* bilinear form $a : V \times V \to \mathbb{R}$; and a linear functional $\ell : V \to \mathbb{R}$. Then the weak problem is as follows: find $u \in V$ such that

$$a(u, v) = \ell(v) \quad \text{for all } \ v \in V. \tag{12}$$

For example, for the elasticity problem (8) define the space of admissible displacements or test functions V, the bilinear form a and linear functional ℓ by

$$\begin{aligned} &V = \{\boldsymbol{v} \mid v_i \in H^1(\Omega),\ \boldsymbol{v} = \boldsymbol{0} \text{ on } \Gamma_u\}, \\ &a : V \times V \to \mathbb{R},\ a(\boldsymbol{u}, \boldsymbol{v}) = \int_\Omega \mathbb{C}\varepsilon(\boldsymbol{u}) : \varepsilon(\boldsymbol{v})\, dx, \\ &\ell : V \to \mathbb{R},\ \ell(\boldsymbol{v}) = \int_\Omega \boldsymbol{f} \cdot \boldsymbol{v}\, dx + \int_{\Gamma_t} \bar{\boldsymbol{t}} \cdot \boldsymbol{v}\, ds. \end{aligned} \tag{13}$$

Then the problem (8) can be rewritten as follows: find $\boldsymbol{u} \in V$ that satisfies

$$a(\boldsymbol{u}, \boldsymbol{v}) = \ell(\boldsymbol{v}) \quad \text{for all } \ \boldsymbol{v} \in V. \tag{14}$$

We can now formulate the set of conditions that will guarantee existence of a unique solution.

Theorem 2.1 *Let V be a Hilbert space and assume that*

(i) $a(\cdot,\cdot) : V \times V \to \mathbb{R}$ is continuous;
(ii) $a(\cdot,\cdot)$ is V-elliptic;
(iii) $\ell : V \to \mathbb{R}$ is a bounded linear functional on V.

Then the problem of finding $u \in V$ that satisfies (12) *has one and only one solution, which depends continuously on the data in the sense that $\|u\|_V \le \frac{1}{\alpha}\|\ell\|_{V'}$.*

Returning to the elasticity problem, assume that $\boldsymbol{u} = \boldsymbol{0}$ on the entire boundary, so that $\Gamma_u = \Gamma$; then $V = [H_0^1(\Omega)]^d$ and

$$\begin{aligned} |a(\boldsymbol{u}, \boldsymbol{v})| &= \left| \int_\Omega [\lambda(\operatorname{div} \boldsymbol{u})(\operatorname{div} \boldsymbol{v}) + 2\mu\varepsilon(\boldsymbol{u}) : \varepsilon(\boldsymbol{v})]\, dx \right| \\ &\le \lambda \|\operatorname{div} \boldsymbol{u}\|_{L^2} \|\operatorname{div} \boldsymbol{v}\|_{L^2} + 2\mu\|\varepsilon(\boldsymbol{u})\|_{L^2} \|\varepsilon(\boldsymbol{v})\|_{L^2} \le M\|\boldsymbol{u}\|_{H^1} \|\boldsymbol{v}\|_{H^1}. \end{aligned}$$

The elasticity tensor is said to be *pointwise stable* if there exists a constant $C_0 > 0$ such that $C_{ijkl}M_{ij}M_{kl} \ge C_0 \sum_{i,j=1}^n M_{ij}M_{ij}$ for all matrices $\boldsymbol{M}$. For an isotropic elastic material, pointwise stability is equivalent to requiring $\mu > 0$ and $3\lambda + 2\mu > 0$. Assuming this property to hold, we have

$$\int_\Omega C_{ijkl}\varepsilon_{ij}(\boldsymbol{u})\varepsilon_{kl}(\boldsymbol{u})\, dx \ge C_0 \int_\Omega |\varepsilon(\boldsymbol{u})|^2\, dx, \tag{15}$$

where $|\boldsymbol{\varepsilon}|^2 = \sum_{i,j=1}^{n} \varepsilon_{ij}\varepsilon_{ij}$. This is not quite the final result because we need the norm $\|\boldsymbol{v}\|_V$ on the right-hand side of (15). This is achieved using Korn's inequality, an important and nontrivial result which states that there exists a constant $C_2 > 0$ such that

$$\|\boldsymbol{\varepsilon}(\boldsymbol{v})\|_{L^2}^2 \geq C_2\|\boldsymbol{v}\|_V \quad \text{for all } \boldsymbol{v} \in V.$$

Thus the weak problem (14) possesses a unique solution $\boldsymbol{u} \in V$.

Consider next a pure *natural* or *Neumann* boundary condition in which the traction is prescribed on the entire boundary. Then $V = [H^1(\Omega)]^d$; as before we have

$$a(\boldsymbol{v}, \boldsymbol{v}) \geq C_0 \int_\Omega |\boldsymbol{\varepsilon}(\boldsymbol{v})|^2 \, dx,$$

but this time we cannot show V-ellipticity because $\boldsymbol{\varepsilon}(\boldsymbol{v}) = \boldsymbol{0}$ for a rigid-body displacement $\boldsymbol{v}(\boldsymbol{x}) = \boldsymbol{a} + \boldsymbol{b} \times \boldsymbol{x}$. So we do not have a unique solution. Physically this is clear: without a displacement constraint on the boundary the body does not occupy a unique location. Furthermore, by choosing $\boldsymbol{v}$ in (14) to be a rigid-body displacement and noting that $\boldsymbol{a}$ and $\boldsymbol{b}$ are arbitrary, we obtain the familiar conditions for force and moment equilibrium of the applied loads, viz.

$$\int_\Omega \boldsymbol{f} \, dx + \int_\Gamma \bar{\boldsymbol{t}} \, ds = \boldsymbol{0} \quad \text{and} \quad \int_\Omega \boldsymbol{x} \times \boldsymbol{f} \, dx + \int \boldsymbol{x} \times \bar{\boldsymbol{t}} \, ds = \boldsymbol{0}.$$

3 The Finite Element Method

We now consider approximate solutions of the weak formulations using the finite element method. The first step is to partition the domain Ω, assumed here to be polygonal for convenience, and to construct a finite element mesh, with the following properties (see Fig. 6a):

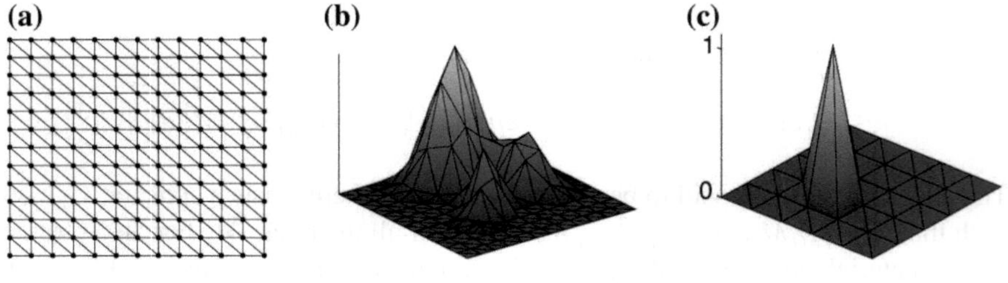

Fig. 6 **a** A finite element mesh comprising triangles in two dimensions; **b** a piecewise-linear and continuous solution; **c** a global basis function

1. The subsets Ω_e $(e = 1, \ldots, E)$ of Ω are called elements and satisfy $\Omega_e \cap \Omega_f = \varnothing$, $e \neq f$, and $\cup_{e=1}^{E} \overline{\Omega}_e = \overline{\Omega}$.
2. *Nodes* or *nodal points* $\boldsymbol{x}_i$ $(i = 1, \ldots, G)$ are identified at least at vertices of elements.
3. The set of elements and nodes forms the *finite element mesh*.

Next, we construct a finite-dimensional space X^h of functions on Ω by defining the basis or shape functions N_i of X^h to have the following properties:

(i) N_i are *bounded and continuous*: $N_i \in C(\overline{\Omega})$;
(ii) there is a total of G basis functions (that is, one for each node) and $N_i(\boldsymbol{x})|_{\Omega_e} \equiv 0$ if $\boldsymbol{x} \notin \overline{\Omega}_e$;
(iii) $N_i(\boldsymbol{x}_j) = \begin{cases} 1 & \text{if } i = j, \\ 0 & \text{otherwise} \end{cases}$;
(iv) the *restriction* $N_i^{(e)}$ of N_i to Ω_e is a polynomial: $N_i|_{\Omega_e} \equiv N_i^{(e)}$, $N_i^{(e)} \in P_k(\Omega_e)$ for $k \geq 1$.

The functions N_i in Fig. 6 are constructed from local polynomial basis functions $N_i^{(e)}$ on elements Ω_e connected to node i. These span the set X^h of piecewise-linear and continuous functions on Ω: that is, $v_h \in X^h$ can be written $v_h = \sum_{i=1}^{G} \mathsf{v}_i N_i(\boldsymbol{x})$, and furthermore $v_h(\boldsymbol{x}_j) = \sum_{i=1}^{G} \mathsf{v}_i N_i(\boldsymbol{x}_j) = \mathsf{v}_j$: that is, v_j is the value of v_h at node i.

We require the space V^h of approximations to be a subspace of V and define $V^h = X^h \cap V$; that is, V^h is the finite-dimensional space spanned by shape functions N_i that satisfy the boundary conditions of V, for example, homogeneous Dirichlet boundary conditions.

In order to have a convergent theory, we place some conditions on the shape of the individual elements Ω_e. We define on a single element its diameter h_e to be the largest length in Ω, and ρ_e to be the diameter of the largest circle or disk inscribed in Ω_e (see Fig. 7a). That is, elements should not become too 'thin' or degenerate. Then a family $\{\Omega_1, \Omega_2, \ldots, \Omega_E\}$ of E elements is said to be *regular* if: (i) there exists a constant δ such that $h_e/\rho_e \leq \delta$ for all elements and (ii) the diameters h_e approach zero. Furthermore, we set $h = \max_{e=1,\ldots E} h_e$; h is called the *mesh size* and quantifies the degree of refinement of the mesh as well as the dimension of V^h. We return now to the weak problem (12), and construct a Galerkin finite elementapproximation

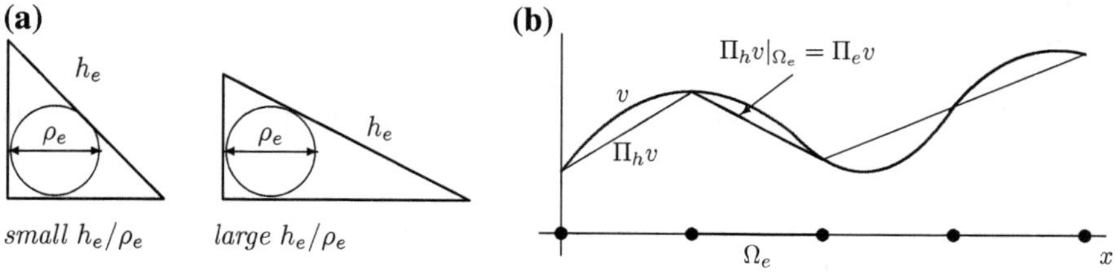

Fig. 7 **a** The measure h_e/ρ_e for a regular element; **b** local interpolate $\Pi_e v$ and global interpolate $\Pi_h v$ of a function $v \in V$

$u_h \in V^h$ that satisfies

$$a(u_h, v_h) = \ell(v_h) \quad \text{for all} \;\; v_h \in V^h. \tag{16}$$

The idea is that the approximate solution u_h will approach the exact solution u as V^h 'approaches' V. With the machinery of normed spaces at our disposal, we would like to show that

$$\|u - u_h\|_V \to 0 \quad \text{as} \;\; h \to 0.$$

This is made precise as follows. Denote the error by $e = u - u_h$; note that $e \in V$. Next, choose $v = v_h$ in (12)—this is legitimate since $v_h \in V$. This gives the equation

$$a(u, v_h) = \ell(v_h). \tag{17}$$

Subtracting (16) from (17) we find that

$$a(u - u_h, v_h) = 0 \quad \text{or} \quad a(e, v_h) = 0. \tag{18}$$

Now if $a(\cdot, \cdot)$ is continuous and V-elliptic, which we assume to be the case, then the entity $a(\cdot, \cdot) \equiv (\cdot, \cdot)_a$ satisfies all the conditions for an inner product. Thus (18) indicates that the error is *orthogonal* to the space V^h, in the inner product $(\cdot, \cdot)_a$. The following is a central result in finite element analysis.

Céa's Lemma. Let V be a Hilbert space, and let $a(\cdot, \cdot)$ and $\ell(\cdot)$ be, respectively, a continuous V-elliptic bilinear form and bounded linear functional on V. Then there exists a constant C, independent of h, such that

$$\|u - u_h\|_V \leq C \inf_{v_h \in V_h} \|u - v_h\|_V. \tag{19}$$

Thus the problem of estimating the *approximation error* becomes one of estimating distance between the exact solution u and the approximation space V^h. We estimate the error by choosing as a candidate in V^h the *interpolate* $\Pi_h u \in V^h$ of u: this is the function in V^h which satisfies $\Pi_h u(\boldsymbol{x}_i) = u(\boldsymbol{x}_i)$ (Fig. 7b). Thus we will set $v_h = \Pi_h u$ in (19); a concrete estimate for the interpolation error $u - \Pi_h u$ will then supply an estimate for the approximation error (see Ciarlet 2002; Reddy 1998 for detailed accounts).

Now consider the global interpolant $\Pi_h v(\boldsymbol{x}) = \sum_{i=1}^{N} \mathsf{v}(\boldsymbol{x}_i) N_i(\boldsymbol{x})$, so that $\Pi_h v = v$ at nodal points. Furthermore, $\Pi_h v|_{\Omega_e} = \Pi_e v$ (see Fig. 7b). Then we have the following results.

Theorem 3.1 *(a) There exists a constant c independent of h such that for any $v \in H^2(\Omega)$ and for piecewise-linear approximations,*

$$\|v - \Pi_h v\|_{H^1(\Omega)} \leq ch|v|_{H^2(\Omega)}. \tag{20}$$

(b) *For the problem (12) assume that $a(\cdot, \cdot)$ is continuous and V-elliptic and ℓ is continuous on V. If u_h is the finite element approximation of the solution in V^h, then there exists a constant C independent of h such that*

$$\|u - u_h\|_{H^1(\Omega)} \leq Ch|u|_{H^2(\Omega)}. \tag{21}$$

Here $|\cdot|_{H^2(\Omega}$ is the H^2 seminorm: $|v|^2_{H^2} = \sum_{|\alpha|=2} \int_\Omega (D^\alpha v)^2 \, dx$. Note that the estimate applies to functions that are smooth enough to belong to $H^2(\Omega)$, and the error estimate is given in terms of the H^1-norm, so that it measures the error $u - u_h$ as well as of its first derivatives.

The proof follows directly from Céa's Lemma and the estimate (20).

For the elasticity problem the unknown variable is vector-valued, and the results carry over virtually unchanged. If $V = [H^1_0(\Omega)]^2$ for example, then using, for example, piecewise-linear basis functions defined on a mesh of three-noded triangular elements as in Fig. 6 we have

$$\|\boldsymbol{u} - \boldsymbol{u}_h\|_{H^1(\Omega)} \leq Ch|\boldsymbol{u}|_{H^2(\Omega)}. \tag{22}$$

Acknowledgments The support of the South African Department of Science and Technology and National Research Foundation through the South African Research Chair in Computational Mechanics is gratefully acknowledged.

References

Atkinson, K., & Han, W. (2001). *Theoretical numerical analysis: A functional analysis framework*. Heidelberg: Springer.

Ciarlet, P. G. (2002). *The finite element method for elliptic problems*. Classics in applied mathematics. Philadelphia: Society for Industrial and Applied Mathematics.

Reddy, B. D. (1998). *Introductory functional analysis: With applications to boundary value problems and finite elements*. Heidelberg: Springer.

Discretization Methods for Solids Undergoing Finite Deformations

Peter Wriggers

Abstract Finite element methods for solving engineering problems are used since decades in industrial applications. This market is still growing and the underlying methodologies, formulations, and algorithms seem to be settled. But still there are open questions and problems when applying the finite element method to situations where finite strains occur. Another problem area is the incorporation of constraints into the formulations, such as incompressibility, contact, and directional constraints needed to formulate anisotropic material behavior. In this section, we present the basic continuum formulation and different discretization techniques that can be used to overcome the problems mentioned above. Additionally, a set of test problems is presented that can be applied to test new finite element formulations.

1 Basic Equations of Continuum Mechanics

A short summary of continuum mechanics is provided in order to be able to formulate the basis for finite element discretizations of finite strain problems. This summary includes kinematical relations, balance laws with their weak forms, and a selection of constitutive equations.

The kinematical relations and associated strain measures used within discretizations are described. Variational formulations are derived based on balance laws that are basis for nonlinear finite element methods. Isotropic hyperelastic material behavior is discussed as an example for nonlinear constitutive equations. Furthermore, all relations will be provided for small strains, leading to the well-known linear relations.

Since this chapter is devoted to nonlinear finite element formulations the underlying theory of continuum mechanics cannot be treated in depths. For this we refer to standard books, e.g., Truesdell and Toupin (1960), Truesdell and Noll (1965), Eringen (1967), Malvern (1969), Becker and Bürger (1975),

P. Wriggers (✉)
Institute of Continuum Mechanics, Gottfried Wilhelm Leibniz Universität, Hannover, Germany
e-mail: wriggers@ikm.uni-hannover.de

J. Schröder and P. Wriggers (eds.), *Advanced Finite Element Technologies*, CISM International Centre for Mechanical Sciences 566,
DOI 10.1007/978-3-319-31925-4_2

Altenbach and Altenbach (1994), Chadwick (1999) or Holzapfel (2000), Ogden (1984), Marsden and Hughes (1983) and Ciarlet (1988) which give background in continnum theory and its mathematical background.

1.1 Kinematics

The kinematical relations concern in continuum mechanics the description of the deformation and motion of a body. They are basis for the derivation of strain measures.

Motion, deformation gradient The motion of a body can be described with respect to the initial configuration or the current configuration of a body. In general, the motion of body B is given as a series of configurations

$$\mathbf{x} = \varphi\,(\mathbf{X}, t), \tag{1}$$

where $\mathbf{X}$ is the position vector defining a point in the initial configuration B. The map $\varphi\,(\mathbf{X}, t)$ maps this point to the current/spatial configuration, described by the position vector $\mathbf{x}$. The position vector can be written in component form as $\mathbf{X} = X_A\,\mathbf{E}_A$ where $\mathbf{E}_A$ defines an orthogonal base system in the initial configuration. Hence, one can write (1) in terms of components of the vectors as $x_i = \varphi_i\,(X_A, t)$. Using, instead of $\mathbf{X}$, the spatial coordinates $\mathbf{x}$ leads to the called current, spatial or Eulerian description of motion.[1]

By introducing a displacement vector $\mathbf{u}(\mathbf{X}, t)$ as difference between the position vectors of current and initial configuration

$$\mathbf{u}(\mathbf{X}, t) = \varphi(\mathbf{X}, t) - \mathbf{X} \tag{2}$$

the current coordinates $\mathbf{x} = \varphi\,(\mathbf{X}, t)$ can be expressed by $\mathbf{x} = \mathbf{X} + \mathbf{u}(\mathbf{X}, t)$.

Local deformations can be investigated by introducing the deformation gradient $\mathbf{F}$. The deformation gradient maps a material line element of the initial configuration $\mathbf{dX}$ in B, to a line element $\mathbf{dx}$ of the current configuration $\varphi(B)$.

$$\mathbf{dx} = \mathbf{F}\,\mathbf{dX}. \tag{3}$$

The components form of the deformation gradient can be written as

$$\mathbf{F} = \text{Grad}\,\varphi(\mathbf{X}, t) = \frac{\partial x_i}{\partial X_A}\,\mathbf{e}_i \otimes \mathbf{E}_A. \tag{4}$$

The connection within B during the deformation process the mapping (3) has to be one-to-one which excludes singularity of $\mathbf{F}$ and further to exclude self-penetration one obtains the restriction

[1] Small letters are used for indices of vectors and tensors which are related to the basis $\mathbf{e}_i$ of the current or spatial configuration. The quantities x_i are the spatial coordinates of X.

$$J = \det \mathbf{F} > 0, \tag{5}$$

where J defines the Jacobi determinant. Hence $\mathbf{F}$ is regular and the inverse $\mathbf{F}^{-1}$ exists.

With the deformation gradient the transformation of surface area elements between B and $\varphi(B)$ can be expressed by the formula of Nanson,

$$\mathbf{da} = \mathbf{n}\, da = J\, \mathbf{F}^{-T}\, \mathbf{N}\, dA = J\, \mathbf{F}^{-T}\, \mathbf{dA}. \tag{6}$$

In this equation $\mathbf{n}$ is the normal vector of the surface of the deformed body $\varphi(B)$ and $\mathbf{N}$ is the normal vector in the initial configuration B. J is the Jacobi determinant and da and dA are the area elements of the associated configurations, respectively.

The transformation between volume elements of the initial and current configuration is given by

$$dv = J\, dV. \tag{7}$$

The deformation gradient can be written in terms of the displacements as

$$\mathbf{F} = \text{Grad}\,[\,\mathbf{X} + \mathbf{u}(\mathbf{X}, t)\,] = \mathbf{1} + \text{Grad}\,\mathbf{u} = \mathbf{1} + \mathbf{H}. \tag{8}$$

with the displacement gradient $\mathbf{H} = \text{Grad}\,\mathbf{u}$.

Strain measures Different strain measures are introduced that are used in forthcoming formulations. The Green-Lagrange strain tensor is referred to the initial configuration B. It is defined by

$$\mathbf{E} = \frac{1}{2}\,(\,\mathbf{F}^T\mathbf{F} - \mathbf{1}\,) = \frac{1}{2}\,(\,\mathbf{C} - \mathbf{1}\,) \tag{9}$$

The tensor $\mathbf{C} = \mathbf{F}^T\,\mathbf{F}$ is the right Cauchy-Green tensor. In component form the strain tensor $\mathbf{E}$ can be written as $\mathbf{E} = E_{AB}\,\mathbf{E}_A \otimes \mathbf{E}_B$ with $E_{AB} = \frac{1}{2}\,(\,F_{iA}\,F_{iB} - \delta_{AB}\,)$. The Kronecker symbol δ_{AB} denotes the components of the unit tensor $\mathbf{1}$. The Green-Lagrange strain measure describes arbitrary rigid body motions correctly. Often the Green-Lagrange strain tensor $\mathbf{E}$ is expressed in terms of the displacement gradient. This yields with (8) $\mathbf{E} = \frac{1}{2}\,(\,\mathbf{H} + \mathbf{H}^T + \mathbf{H}^T\mathbf{H}\,)$. The higher order term $\mathbf{H}^T\mathbf{H}$ depicts the nonlinear part of the Green-Lagrange strain tensor. Within the geometrically linear theory this term is neglected, hence $\mathbf{E}$ reduces to the linear strain measure ε

$$\varepsilon = \frac{1}{2}\,(\,\mathbf{H} + \mathbf{H}^T\,) = \frac{1}{2}\,(\,u_{A,B} + u_{B,A}\,)\,\mathbf{E}_A \otimes \mathbf{E}_B. \tag{10}$$

The poplar decomposition of the deformation gradient splits the deformation gradient in a multiplicative way in a proper orthogonal rotation tensor $\mathbf{R}$ (with $\mathbf{R}^{-1} = \mathbf{R}^T$) and the symmetrical stretch tensors $\mathbf{U}, \mathbf{V}$, see e.g. Ogden (1984)

$$\mathbf{F} = \mathbf{R}\,\mathbf{U} = \mathbf{V}\,\mathbf{R}. \tag{11}$$

Due to the orthogonality of $\mathbf{R}$ one can write the right Cauchy-Green tensor in terms of $\mathbf{U}$: $\mathbf{C} = \mathbf{F}^T\,\mathbf{F} = \mathbf{U}^T\,\mathbf{R}^T\,\mathbf{R}\,\mathbf{U} = \mathbf{U}^T\,\mathbf{U} = \mathbf{U}^2$. The last result follows from the symmetry of $\mathbf{U}$. Since $\mathbf{C} = \mathbf{U}^2$ it can be easily shown that the spectral decomposition of the right Cauchy-Green tensor is given by

$$\mathbf{C} = \sum_{i=1}^{3} \lambda_i^2\, \mathbf{N}_i \otimes \mathbf{N}_i . \tag{12}$$

The stretch tensors can be computed via spectral decomposition

$$\mathbf{U} = \sum_{i=1}^{3} \lambda_i\, \mathbf{N}_i \otimes \mathbf{N}_i \qquad \mathbf{V} = \sum_{i=1}^{3} \lambda_i\, \mathbf{n}_i \otimes \mathbf{n}_i \tag{13}$$

The principal values λ_i of the stretch tensors are called principal stretches. They are equal for $\mathbf{U}$ and $\mathbf{V}$. The eigenvectors $\mathbf{N}_i$ of $\mathbf{U}$ are related to the reference configuration. The eigenvectors $\mathbf{n}_i$ of $\mathbf{V}$ are referred to the spatial configuration.

With respect to the spatial configuration $\varphi(B)$ the left Cauchy-Green tensor can be defined

$$\mathbf{b} = \mathbf{F}\,\mathbf{F}^T = \mathbf{V}\,\mathbf{R}\,\mathbf{R}^T\,\mathbf{V}^T = \mathbf{V}^2 . \tag{14}$$

In case of incompressibility, that plays a prominent role in rubber materials and metal plasticity, the constraint condition $\det \mathbf{F} = J = 1$ holds. To account for this kinematic constraint the multiplicative decomposition of the deformation gradient

$$\mathbf{F} = J^{\frac{1}{3}}\,\widehat{\mathbf{F}} \qquad \widehat{\mathbf{F}} = J^{-\frac{1}{3}}\,\mathbf{F} \tag{15}$$

was introduced in Flory (1961). It preserves a priori the volume of $\widehat{\mathbf{F}}$ (isochoric motion), since $\det \widehat{\mathbf{F}} \equiv 1$.

By inserting (15) in (9) one obtains a relation between the isochoric part of the right Cauchy-Green deformation tensor $\widehat{\mathbf{C}}$ and $\mathbf{C}$

$$\widehat{\mathbf{C}} = \widehat{\mathbf{F}}^T\,\widehat{\mathbf{F}} = J^{-\frac{2}{3}}\,\mathbf{F}^T\,\mathbf{F} = J^{-\frac{2}{3}}\,\mathbf{C} . \tag{16}$$

This multiplicative split corresponds to an additive decomposition of the strain tensor in the geometrically linear theory

$$\boldsymbol{\epsilon} = \mathbf{e}_D + \frac{1}{3}\,\mathrm{tr}\,\boldsymbol{\epsilon}\,\mathbf{1} . \tag{17}$$

1.2 Balance Equations

Differential equations that describe the local balance equations are introduced in this section. These are the balance of mass, balance of linear and angular momentum.

Balance of mass For processes in which the mass of the system is conserved the change one has ($\dot{m} = 0$). Hence an infinitesimal mass element in initial and current configuration has to be equal which leads to

$$\rho \, dv = \rho_0 \, dV. \tag{18}$$

Here ρ_0 and ρ are the densities in initial and current configuration, respectively. With Eq. (7) the volume elements dV and dv can be transformed leading to the Lagrangian description of the mass balance $\rho_0 = J \, \rho$.

Balance of linear and angular momentum The balance of linear momentum can be expressed in its local form by

$$\operatorname{div} \boldsymbol{\sigma} + \rho \, \bar{\mathbf{b}} = \rho \dot{\mathbf{v}}, \quad \sigma_{ik,i} + \rho \, \bar{b}_k = \rho \, \dot{v}_k. \tag{19}$$

The stress tensor $\boldsymbol{\sigma}$ is called Cauchy stress tensor. $\rho \dot{\mathbf{v}}$ describes the inertial forces that can be neglected in case of purely static investigations. The stress vector $\mathbf{t}$ can be obtained using the Cauchy's theorem

$$\mathbf{t} = \boldsymbol{\sigma} \, \mathbf{n}, \quad t_i = \sigma_{ik} \, n_k. \tag{20}$$

The angular momentum yields after some manipulations the local balance of angular momentum which simply demands the symmetry of the Cauchy stress tensor

$$\boldsymbol{\sigma} = \boldsymbol{\sigma}^T, \qquad \sigma_{ik} = \sigma_{ki}. \tag{21}$$

Introduction of different stress tensors Equations (19) and (21) are referred to the current configuration. In finite element formulations it can be desirable to relate these equations to the initial configuration B. This leads with Nanson's formula (6) to

$$\int\limits_{\partial\varphi(B)} \boldsymbol{\sigma} \, \mathbf{n} \, da = \int\limits_{\partial B} \boldsymbol{\sigma} \, J \, \mathbf{F}^{-T} \, \mathbf{N} \, dA = \int\limits_{\partial B} \mathbf{P} \, \mathbf{N} \, dA. \tag{22}$$

The relation defines the first Piola-Kirchhoff stress tensor $\mathbf{P}$ that can be written in terms of the Cauchy stress

$$\mathbf{P} = J \, \boldsymbol{\sigma} \mathbf{F}^{-T} \quad P_{Ak} = J \, \sigma_{ik} (F_{iA})^{-1}. \tag{23}$$

$\mathbf{P}$ is a two-field tensor with one basis referred to the current and the other to the initial configuration.

Often it is convenient to work totally in the initial configuration. For this purpose the second Piola-Kirchhoff stress tensor

$$\mathbf{S} = \mathbf{F}^{-1} \, \mathbf{P} = J \, \mathbf{F}^{-1} \, \boldsymbol{\sigma} \mathbf{F}^{-T}, \tag{24}$$

$$S_{AB} = (F_{Ai})^{-1} \, P_{Bi} = J \, (F_{Ai})^{-1} \, \sigma_{ik} (F_{kB})^{-1}. \tag{25}$$

was introduced. It is symmetric, but a pure mathematical quantity that generally cannot be interpreted as a physical measure for stress.

Besides the Cauchy stress tensor $\boldsymbol{\sigma}$ the Kirchhoff stress tensor

$$\boldsymbol{\tau} = \mathbf{F}\,\mathbf{S}\,\mathbf{F}^T, \quad \boldsymbol{\tau} = J\,\boldsymbol{\sigma}. \tag{26}$$

can be used. For some formulations the Biot stress tensor can be applied which is defined by

$$\mathbf{T}_B = \mathbf{R}^T\mathbf{P}. \tag{27}$$

Balance equations with respect to the initial configuration By using the first Piola-Kirchhoff stress tensor the local balance of linear momentum (19) can be recast in the initial configuration

$$\text{DIV}\,\mathbf{P} + \rho_0\,\bar{\mathbf{b}} = \rho_0\,\dot{\mathbf{v}} \tag{28}$$

where DIV denotes the divergence operation with respect to the coordinates of the initial configuration. The balance of angular momentum (21) yields with (23)

$$\mathbf{P}\,\mathbf{F}^T = \mathbf{F}\,\mathbf{P}^T. \tag{29}$$

This relation depicts that the first Piola-Kirchhoff stress tensor is nonsymmetric. The balance of angular momentum in terms of the second Piola-Kirchhoff stress tensor leads to $\mathbf{S} = \mathbf{S}^T$ which shows the symmetry of $\mathbf{S}$.

1.3 Constitutive Equations

For the solution of boundary or initial value problems in solid mechanics equations are needed that characterize the material response of a body. Here only isotropic elastic materials will be considered under the assumption of hyperelastic behavior, see e.g. Ogden (1984). This description is valid for materials that undergo finite deformations. In case of small strains these constitutive relations reduce to the classical law of Hooke.

The constitutive equation for the second Piola-Kirchhoff stress tensor can be derived from the potential ψ in case of a hyper elastic material. ψ describes the strain energy stored in the body. The derivative of ψ with respect to the right Cauchy-Green tensor yields

$$\mathbf{S} = 2\,\rho_0\,\frac{\partial\psi(\mathbf{C})}{\partial\mathbf{C}}; \qquad S_{AB} = 2\,\rho_0\,\frac{\partial\psi(C_{CD})}{\partial C_{AB}}. \tag{30}$$

For isotropic material behavior a strain energy function can be introduced that only depends on the invariants of the strain tensors

$$\psi(\mathbf{C}) = \psi\,(I_C,\, II_C,\, III_C\,). \tag{31}$$

The invariants I_C, II_C, III_C are defined in the standard way:

$$I_C = \mathrm{tr}\mathbf{C}, \quad II_C = \frac{1}{2}\left[(\mathrm{tr}\mathbf{C})^2 - \mathrm{tr}(\mathbf{C})^2\right] \quad \text{and} \quad III_C = \det \mathbf{C}.$$

With the relations

$$\begin{aligned} I_C &= \lambda_1^2 + \lambda_2^2 + \lambda_3^2 \\ II_C &= \lambda_1^2\,\lambda_2^2 + \lambda_2^2\,\lambda_3^2 + \lambda_3^2\,\lambda_1^2 \\ III_C &= \lambda_1^2\,\lambda_2^2\,\lambda_3^2 \end{aligned} \tag{32}$$

one can express the invariants in terms of the principal stretches λ_i^2 of $\mathbf{C}$. Then the strain energy function assumes the form

$$\psi(\mathbf{C}) = \psi\,\left(\lambda_1^2,\, \lambda_2^2,\, \lambda_3^2\right). \tag{33}$$

The isotropic constitutive Eq. (30) can also be expressed in quantities related to the current configuration. With Eq. (25) one obtains for the Cauchy stress $\boldsymbol{\sigma} = J^{-1}\,\mathbf{F}\,\mathbf{S}\,\mathbf{F}^T$ and hence by considering (18)

$$\boldsymbol{\sigma} = 2\,\rho\,\mathbf{F}\,\frac{\partial\psi(\mathbf{C})}{\partial\mathbf{C}}\,\mathbf{F}^T.$$

It can be shown that this equation is equivalent to

$$\boldsymbol{\sigma} = 2\,\rho\,\mathbf{b}\,\frac{\partial\psi(\mathbf{b})}{\partial\mathbf{b}}. \tag{34}$$

Constitutive equations of the form (31) are still very complex because ψ can be an arbitrary function of the invariants. It is desirable to formulate material functions in nonlinear elasticity with a minimum number of constitutive parameters. To shorten notation $W = \rho_0\,\psi$ is introduced in the following to define the strain energy function. The choice of

$$W(I_C) = g(J) + \frac{1}{2}\,\mu\,(I_C - 3) \quad \text{with} \qquad J = \sqrt{\det \mathbf{C}} \tag{35}$$

yields the special case of a compressible Neo-Hooke material. A special ansatz can be selected for $g(J)$ in (35)

$$g(J) = c\,(J^2 - 1) - d\ln J - \mu\ln J \quad \text{with} \quad c > 0, d > 0. \tag{36}$$

The constitutive parameters Λ, μ are known as the Lamé constants.

Other strain energy functions for incompressible rubber materials are the so called Mooney–Rivlin materials

$$W(I_C, II_C) = c_1\,(I_C - 3) + c_2\,(II_C - 3). \tag{37}$$

For a complete formulation of an incompressible problem the incompressibility constraint ($J = 1$) has to be considered.

In case of geometrically linear materials a linear relation between the stresses and the linear strains is proposed. This is given by

$$\boldsymbol{\sigma} = \Lambda\,\mathrm{tr}\,\boldsymbol{\epsilon}\,\mathbf{1} + 2\,\mu\,\boldsymbol{\epsilon} \tag{38}$$

in terms of the two Lamé constants Λ and μ. Since in the linear case we do not have to distinguish different stress measures, we use here for convenience the Cauchy stress.

In finite elasticity of rubber materials one splits of the deformation in a volumetric part represented by J and the isochoric part described by $\widehat{\mathbf{C}}$, see (16), since both part can depict different material behavior. This split is useful in mixed finite element formulations when quasi-incompressible materials are described by special numerical formulations since the split permits a different treatment of the incompressible part. One possibility for the formulation of the constitutive equation is given by an additive split of the strain energy function in its volumetric and isochoric parts: $W(\widehat{\mathbf{C}}, J) = \hat{W}(\widehat{\mathbf{C}}) + U(J)$. This leads for a strain energy function that is analogous to the one introduced in (35)

$$W(\widehat{\mathbf{C}}, J) = U(J) + \frac{1}{2}\,\mu\,(I_{\widehat{C}} - 3). \tag{39}$$

Here the term $U(J)$ is different from $g(J)$: $U(J) = \frac{K}{4}\,(J^2 - 1) - \frac{K}{2}\ln J$ where K denotes the modulus of compression.[2]

For the special choice of the strain energy function (39) one specifies

$$\mathbf{S}_{ISO} = \mu\,J^{-\frac{2}{3}}\left(\mathbf{1} - \frac{1}{3}\,\mathrm{tr}\mathbf{C}\,\mathbf{C}^{-1}\right), \qquad \mathbf{S}_{VOL} = \frac{K}{2}\,(J^2 - 1)\,\mathbf{C}^{-1}. \tag{40}$$

The transformation to the current configuration yields for the Kirchhoff stress tensor introduced in (27) together with the operator $\mathrm{dev}(\bullet) = (\bullet) - \frac{1}{3}\,\mathrm{tr}(\bullet)\,\mathbf{1}$ for (40)

$$\boldsymbol{\tau} = J\,p\,\mathbf{1} + \mathrm{dev}\,\widehat{\boldsymbol{\tau}} = \tau_{vol}\,\mathbf{1} + \boldsymbol{\tau}_{iso}. \tag{41}$$

[2]Note that this split represents physically a different strain energy function than (35).

This depicts the split of the stress tensor into a volumetric and an isochoric part clearly. The following definitions were used in (41)

$$p = \frac{\partial U}{\partial J} \qquad \text{and} \qquad \widehat{\boldsymbol{\tau}} = \widehat{\mathbf{F}}\, 2\, \frac{\partial \hat{W}}{\partial \widehat{\mathbf{C}}}\, \widehat{\mathbf{F}}^T. \tag{42}$$

In case of small strains, we can use (17) to represent the linear elastic constitutive equation by

$$\boldsymbol{\sigma} = \Lambda \operatorname{tr} \boldsymbol{\epsilon}\, \mathbf{1} + 2\,\mu \left(\mathbf{e}_D + \frac{1}{3} \operatorname{tr} \boldsymbol{\epsilon} \right). \tag{43}$$

This equation can be rewritten with the modulus of compression $K = \Lambda + \frac{2}{3}\,\mu$ as

$$\boldsymbol{\sigma} = K \operatorname{tr} \boldsymbol{\epsilon}\, \mathbf{1} + 2\,\mu\, \mathbf{e}_D. \tag{44}$$

With the introduction of the pressure

$$p = K \operatorname{tr} \boldsymbol{\epsilon} \tag{45}$$

and the deviatoric stress

$$\mathbf{s} = 2\,\mu\, \mathbf{e}_D \tag{46}$$

the stress can be decomposed in a deviatoric and volumetric part

$$\boldsymbol{\sigma} = p\, \mathbf{1} + \mathbf{s}. \tag{47}$$

1.4 Weak Forms of Balance of Momentum

The principle of virtual work is an equivalent formulation of the balance of linear and angular momentum (19) or (28). In the mathematical literature it is called weak form of the partial differential equation. Since no further assumptions, like existence of a potential, are made, the weak form is applicable to general problems like inelastic materials, friction, etc. The derivation of the weak form starts from the equation of linear momentum ($\operatorname{Div} \mathbf{P} + \rho_0\, \bar{\mathbf{b}} = \rho_0\, \dot{\mathbf{v}}$) which is multiplied scalar by a vector valued function $\boldsymbol{\eta} = \{\boldsymbol{\eta} \mid \boldsymbol{\eta} = \mathbf{0} \text{ auf } \partial B_u\}$—often called virtual displacement or test function. The following integration over the volume of the solid yields, together with a partial integration of the first term, application of the divergence theorem and introduction of the traction boundary condition the weak form of linear momentum

$$\int_B \mathbf{P} \cdot \operatorname{Grad} \boldsymbol{\eta}\, dV - \int_B \rho_0\, (\bar{\mathbf{b}} - \dot{\mathbf{v}}) \cdot \boldsymbol{\eta}\, dV - \int_{\partial B_\sigma} \bar{\mathbf{t}} \cdot \boldsymbol{\eta}\, dA = 0. \tag{48}$$

The gradient of the test function $\boldsymbol{\eta}$ can also be interpreted as variation $\delta\,\mathbf{F}$ of the deformation gradient. In the weak form (48) the first Piola-Kirchhoff stress tensor can be replaced by the second Piola-Kirchhoff stress tensor leading to

$$\mathbf{P}\cdot \mathrm{Grad}\,\boldsymbol{\eta} = \mathbf{S}\cdot \mathbf{F}^T\,\mathrm{Grad}\,\boldsymbol{\eta} = \mathbf{S}\cdot \frac{1}{2}\,(\mathbf{F}^T\,\mathrm{Grad}\,\boldsymbol{\eta} + \mathrm{Grad}^T\boldsymbol{\eta}\,\mathbf{F}) = \mathbf{S}\cdot \delta\mathbf{E}, \qquad (49)$$

where the fact has been used that the scalar product of a symmetrical tensor (here $\mathbf{S}$) with a antisymmetrical part of a tensor is zero. $\delta\,\mathbf{E}$ denotes the variation of the Green-Lagrange strain tensor. Using (49) equation (48) can be rewritten as

$$\int\limits_B \mathbf{S}\cdot\delta\mathbf{E}\,dV - \int\limits_B \rho_0\,(\bar{\mathbf{b}} - \dot{\mathbf{v}})\cdot\boldsymbol{\eta}\,dV - \int\limits_{\partial B_\sigma} \bar{\mathbf{t}}\cdot\boldsymbol{\eta}\,dA = 0. \qquad (50)$$

The first term in (50) denotes the internal virtual work also called stress divergence term. The last two terms describe the virtual work of the applied loading and the inertia term.

The transformation of the weak form (48) to the current or spatial configuration is performed by using the transformation of the first Piola-Kirchhoff stress tensor to the Cauchy stress tensor, see (23): $\boldsymbol{\sigma} = \frac{1}{J}\,\mathbf{P}\,\mathbf{F}^T$. This leads to

$$\mathbf{P}\cdot\mathrm{Grad}\,\boldsymbol{\eta} = J\,\boldsymbol{\sigma}\,\mathbf{F}^{-T}\cdot\mathrm{Grad}\,\boldsymbol{\eta} = J\,\boldsymbol{\sigma}\cdot\mathrm{Grad}\,\boldsymbol{\eta}\,\mathbf{F}^{-1} = J\,\boldsymbol{\sigma}\cdot\mathrm{Grad}\,\boldsymbol{\eta}.$$

Furthermore we have with (7) $dv = J\,dV$ which is equivalent to $\rho = \rho_0\,J$. With these relations the weak form (48) can be written in terms of the current configuration

$$\int\limits_{\varphi(B)} \boldsymbol{\sigma}\cdot\mathrm{Grad}\,\boldsymbol{\eta}\,dv - \int\limits_{\varphi(B)} \rho\,(\bar{\mathbf{b}} - \dot{\mathbf{v}})\cdot\boldsymbol{\eta}\,dv - \int\limits_{\varphi(\partial B_\sigma)} \hat{\mathbf{t}}\cdot\boldsymbol{\eta}\,da = 0. \qquad (51)$$

The symmetry of the Cauchy stress tensor facilitates the replacement of the spatial gradient by its symmetric part. Hence with the definition

$$\nabla^S\,\boldsymbol{\eta} = \frac{1}{2}\,(\mathrm{Grad}\,\boldsymbol{\eta} + \mathrm{Grad}^T\,\boldsymbol{\eta}) \qquad (52)$$

the weak form with respect to the spatial configuration follows

$$\int\limits_{\varphi(B)} \boldsymbol{\sigma}\cdot\nabla^S\,\boldsymbol{\eta}\,dv - \int\limits_{\varphi(B)} \rho\,(\bar{\mathbf{b}} - \dot{\mathbf{v}})\cdot\boldsymbol{\eta}\,dv - \int\limits_{\varphi(\partial B_\sigma)} \hat{\mathbf{t}}\cdot\boldsymbol{\eta}\,da = 0. \qquad (53)$$

For the geometrical linear theory the weak form is formally equivalent to (53). Only the operator defined in (52) has to be evaluated with respect to the coordinates of the initial configuration as well as the integrals

$$g^{lin}(\mathbf{u}, \boldsymbol{\eta}) = \int_B \boldsymbol{\sigma} \cdot \nabla_X^S \boldsymbol{\eta} \, dv - \int_B \rho \bar{\mathbf{b}} \cdot \boldsymbol{\eta} \, dv - \int_{\partial B_\sigma} \hat{\mathbf{t}} \cdot \boldsymbol{\eta} \, da = 0 \tag{54}$$

with $\nabla_X^S \boldsymbol{\eta} = \frac{1}{2} (\mathrm{Grad}\, \boldsymbol{\eta} + \mathrm{Grad}^T \boldsymbol{\eta})$.

1.5 Variational Functionals

In this section two variational functionals will be discussed which can alternatively be applied within the discretization process of the finite element method.

Based on this strain energy W the classical principle of minimum of potential energy can be formulated for the finite deformation theory.[3] Here one has in general consider that deformations can occur which are unstable. Due to that only a stationary value of the potential can be reached. We assume further that the applied loads are conservative which means path independent. Then one can state for static problems

$$\Pi(\boldsymbol{\varphi}) = \int_B [W(\mathbf{C}) - \rho_0 \, \bar{\mathbf{b}} \cdot \boldsymbol{\varphi}] \, dV - \int_{\partial B_\sigma} \bar{\mathbf{t}} \cdot \boldsymbol{\varphi} \, dA \Longrightarrow STAT. \tag{55}$$

Of all possible deformations $\boldsymbol{\varphi}$ the ones which make Π stationary fulfill the equilibrium equation. The stationary value of (55) can be computed by the variation of Π with respect to the deformation. This is achieved by the directional derivative

$$\delta \Pi = D\,\Pi(\boldsymbol{\varphi}) \cdot \boldsymbol{\eta} = \frac{d}{d\alpha} \Pi(\boldsymbol{\varphi} + \alpha \, \boldsymbol{\eta}) \Big|_{\alpha=0}, \tag{56}$$

which is also called first variation of Π. The associated mathematical operation yields

$$D\,\Pi(\boldsymbol{\varphi}) \cdot \boldsymbol{\eta} = \int_B \left[\frac{\partial W}{\partial \mathbf{C}} \cdot D\,\mathbf{C} \cdot \boldsymbol{\eta} - \rho_0 \, \bar{\mathbf{b}} \cdot \boldsymbol{\eta} \right] dV - \int_{\partial B_\sigma} \bar{\mathbf{t}} \cdot \boldsymbol{\eta} \, dA = G(\mathbf{u}, \boldsymbol{\eta}) = 0. \tag{57}$$

The directional derivative of the right Cauchy-Green strain tensor can be written in terms of the Green-Lagrange strain tensor: $D\,\mathbf{C} \cdot \boldsymbol{\eta} = 2\,D\,\mathbf{E} \cdot \boldsymbol{\eta}$.[4]

[3]The construction of such principle has advantages. One of them is that the development of efficient algorithms for the solution of the nonlinear equations can be based on optimization strategies.

[4]This result corresponds to the variation $\delta \mathbf{E}$, defined already (49). The partial derivative of W with respect to $\mathbf{C}$ leads to the second Piola-Kirchhoff stress tensor $\mathbf{S}$, see (30): $\mathbf{S} = 2\,\partial W / \partial \mathbf{C}$. Hence Eq. (57) is equivalent to the weak form (50) for a hyperelastic material.

The stationary value of (55) can also be computed by the variation of Π with respect to the deformation. For this purpose the directional derivative (56) is applied. The application of this mathematical operation yields

$$D\,\Pi(\varphi)\cdot\eta = \int_B \left[\frac{\partial W}{\partial \varphi}\cdot\eta - \rho_0\,\bar{\mathbf{b}}\cdot\eta\right] dV - \int_{\partial B_\sigma} \bar{\mathbf{t}}\cdot\eta\, dA = 0. \tag{58}$$

Note that weak form (58) is obtained directly by the variation of (55). Hence equation (58) is equivalent to the weak form (57) for hyperelastic materials.

Another more general variational principle is the Hu–Washizu principle, see Washizu (1975). It has gained significance during the last years for the construction of finite elements. This principles can be derived by writing the weak formulation with additional constraints that contain kinematics and constitutive equations. Hence deformations, strains and stresses occur as independent variables. Once this principle is constructed its variation with respect to all variables yields the static equilibrium equations, the kinematical relations and the constitutive equation. The nonlinear version of the Hu–Washizu principle can formulated by using any set of work conjugated variables. Here, we will state it in terms of the deformation gradient $\mathbf{F}$, the first Piola-Kirchhoff stress tensor $\mathbf{P}$ and the Deformation φ

$$\begin{aligned}\Pi(\varphi,\mathbf{F},\mathbf{P}) = &\int_B [W(\mathbf{F}) + \mathbf{P}\cdot(\operatorname{Grad}\varphi - \mathbf{F})]\, dV \\ &- \int_B \varphi\cdot\rho_0\,\bar{\mathbf{b}}\, dV - \int_{\partial B_\sigma} \varphi\cdot\hat{\mathbf{t}}\, dA.\end{aligned} \tag{59}$$

The variation, according to the definition of the directional derivative given above, yields now three independent equations

$$\begin{aligned} D\Pi(\varphi,\mathbf{F},\mathbf{P})\cdot\eta &= \int_B (\mathbf{P}\cdot\operatorname{Grad}\eta - \eta\cdot\rho_0\,\bar{\mathbf{b}})\, dV - \int_{\partial B_\sigma} \eta\cdot\hat{\mathbf{t}}\, dA = 0,\\ D\Pi(\varphi,\mathbf{F},\mathbf{P})\cdot\mathbf{Q} &= \int_B \mathbf{Q}\cdot(\operatorname{Grad}\varphi - \mathbf{F})\, dV = 0,\\ D\Pi(\varphi,\mathbf{F},\mathbf{P})\cdot\mathbf{A} &= \int_B \mathbf{A}\cdot\left(\frac{\partial W}{\partial \mathbf{F}} - \mathbf{P}\right) dV = 0 \end{aligned} \tag{60}$$

where η, $\mathbf{Q}$ and $\mathbf{A}$ are the virtual displacements, the virtual stresses and the virtual strains, respectively. Observe that they represent the weak form (48), the kinematical relation (4) and a hyperelastic constitutive equation for $\mathbf{P}$. Note further that the last two terms can be interpreted as constraint terms which have to be fulfilled additionally together with the linear momentum equation. In that case $\mathbf{Q}$ and $\mathbf{A}$ are Lagrangian multipliers.

The Hu–Washizu principle was originally derived in Washizu (1975) for small elastic deformations. It then is stated in terms of the strain tensor $\boldsymbol{\epsilon}$, the first stress tensor $\boldsymbol{\sigma}$ and the displacement $\mathbf{u}$

$$\Pi(\mathbf{u}, \boldsymbol{\epsilon}, \boldsymbol{\sigma}) = \int_B \left[\frac{\Lambda}{2} (\mathrm{tr}\boldsymbol{\epsilon})^2 + \mu\, \boldsymbol{\epsilon} \cdot \boldsymbol{\epsilon} + \boldsymbol{\sigma} \cdot (\nabla_X^S \mathbf{u} - \boldsymbol{\epsilon}) \right] dV$$
$$- \int_B \boldsymbol{\varphi} \cdot \rho_0\, \bar{\mathbf{b}}\, dV - \int_{\partial B_\sigma} \boldsymbol{\varphi} \cdot \hat{\mathbf{t}}\, dA. \tag{61}$$

The variation, according to the definition of the directional derivative given above, yields again three independent equations.

A special form of the Hu–Washizu variational principle can be applied for the construction of finite elements which have to represent nearly incompressible material behavior. Since incompressibility is associated with a constraint for the volumetric deformation ($J \equiv 1$), one can use the split (15) to distinguish volumetric and isochoric parts of the deformation. Based on this idea Simo et al. (1985) formulated a three-field functional which is only defined for the volumetric part of the deformation. Hence the independent variable are now the deformation $\boldsymbol{\varphi}$, the pressure p and a strain variable θ which is equivalent to J. The last variable has to fulfill the constraint condition $\theta = J$. With the multiplicative split of the deformation gradient one obtains the split into volumetric and deviatoric parts. Note that we have in relation (15) $\widehat{\mathbf{F}} = J^{-\frac{1}{3}}\, \mathrm{Grad}\, \boldsymbol{\varphi}$. Furthermore $\bar{\mathbf{C}} = \theta^{\frac{2}{3}} J^{-\frac{2}{3}}\, \mathbf{C} = \theta^{\frac{2}{3}}\, \widehat{\mathbf{C}}$ holds with (16). Also the strain energy function has to be defined on the basis of the new variables: $W(\theta^{\frac{2}{3}}\, \widehat{\mathbf{C}})$. Using the additive split $W = W(\theta) + W(\widehat{\mathbf{C}})$, see (39), the following three-field variational functional can be defined

$$\Pi(\boldsymbol{\varphi}, p, \theta) = \int_B [\, W(\widehat{\mathbf{C}}) + W(\theta) + p(J - \theta)\,]\, dV$$
$$- \int_B \boldsymbol{\varphi} \cdot \rho_0\, \bar{\mathbf{b}}\, dV - \int_{\partial B_\sigma} \boldsymbol{\varphi} \cdot \hat{\mathbf{t}}\, dA. \tag{62}$$

This form is actually a mixture of a Hu–Washizu principle for the pressure p and the volumetric deformation J and a deformation based functional like (55) for $\widehat{\mathbf{C}}$.

The Hu–Washizu principle is the most general principle which depends on three independent fields. Based on the fulfillment of a constraints we can eliminate such constraints directly. This yields two-field formulation with displacements and stresses as primary variables. The adequate Hellinger–Reissner functional can be written for small strains as

$$\Pi_{HR}(\boldsymbol{\sigma}, \mathbf{u}) = \int_B \boldsymbol{\sigma} \cdot \nabla_X^S \mathbf{u}\, dV - \frac{1}{2} \int_B \boldsymbol{\sigma} \cdot \mathbb{C}^{-1}[\boldsymbol{\sigma}]\, dV - \int_B \hat{\mathbf{b}} \cdot \mathbf{u}\, dV - \int_{\partial B_\sigma} \hat{\mathbf{t}} \cdot \mathbf{u}\, dA. \tag{63}$$

Here $\mathbb{C}$ is the constitutive tensor which relates strains and stresses in case of linear elasticity, see (38),

$$\boldsymbol{\sigma} = \mathbb{C}\,[\,\boldsymbol{\epsilon}(\mathbf{u})\,] = \mathbb{C}\,[\,\nabla_X^S \mathbf{u}\,]. \tag{64}$$

$\hat{\mathbf{b}}$ and $\hat{\mathbf{t}}$ are the applied body forces and boundary tractions, respectively.

The variation of this variational functional yields

$$\begin{aligned} D\Pi \cdot \boldsymbol{\eta} &= \int_B \boldsymbol{\sigma} \cdot \nabla_X^S \boldsymbol{\eta}\, dV - \int_B \hat{\mathbf{b}} \cdot \boldsymbol{\eta}\, dV - \int_{\partial B_\sigma} \hat{\mathbf{t}} \cdot \boldsymbol{\eta}\, dA = 0, \\ D\Pi \cdot \boldsymbol{\tau} &= \int_B \boldsymbol{\tau} \cdot \nabla_X^S \mathbf{u}\, dV - \int_B \boldsymbol{\tau} \cdot \mathbb{C}^{-1}\,[\boldsymbol{\sigma}]\, dA = 0. \end{aligned} \tag{65}$$

The first equation represents the weak form of equilibrium. The second equation is the weak statement of the constitutive equation. Hence the two-field Hellinger–Reissner principle has besides the weak form of equilibrium one additional constraint equation, which is the inverse of constitutive Eq. (64).

More general, we have the relation stemming from the Legendre transformation

$$W(\boldsymbol{\epsilon}) = \boldsymbol{\sigma} \cdot \boldsymbol{\epsilon} - U(\boldsymbol{\sigma}) \tag{66}$$

in linear elasticity we have $W(\boldsymbol{\epsilon}) = \frac{1}{2}\,\boldsymbol{\varepsilon} \cdot \mathbb{C}\,[\,\boldsymbol{\epsilon}\,]$ which yields with (64)

$$U(\boldsymbol{\sigma}) = \frac{1}{2}\,\boldsymbol{\sigma} \cdot \mathbb{C}^{-1}[\boldsymbol{\sigma}]. \tag{67}$$

Hence it is easy to see that the first two terms in (63) are equivalent to the integral of the strain energy $W(\boldsymbol{\epsilon})$.

Another version of a functional has as only unknowns the stresses. This is based on the dual formulation

$$U(\boldsymbol{\sigma}) = \frac{1}{2} \int_B \boldsymbol{\sigma} \cdot \mathbb{C}^{-1}\,[\boldsymbol{\sigma}]\, dV \tag{68}$$

in which the complementary energy $U(\boldsymbol{\sigma})$ has to be minimized by a stress field which fulfills the strong form of linear and angular momentum (19) and (21) in B for the case of statics and the stress boundary conditions (Neumann boundary conditions) on ∂B_σ. It is of course complicated to find a stress field which fulfills these local conditions. Hence one can add the linear momentum to (68) by using the Lagrangian multiplier method

$$\frac{1}{2} \int_B \boldsymbol{\sigma} \cdot \mathbb{C}^{-1}\,[\boldsymbol{\sigma}]\, dV + \int_B (\text{div}\boldsymbol{\sigma} + \rho\,\mathbf{b}) \cdot \mathbf{u}\, dV \rightarrow \mathit{EXTR}. \tag{69}$$

Now the stresses have to be computed for fields which fulfill the divergence operator and the angular momentum (21) in B and the Neumann boundary conditions on ∂B_σ. The Lagrange multiplier $\mathbf{u}$ can be interpreted as a displacement. Since it is not simple to find stable finite element discretizations for (69), see Brezzi and Fortin (1991), one can relax the strong symmetry condition (21) of the stress field , see e.g. Fraeijs de Veubeke (1975), Arnold et al. (1984) and Stenberg (1988). This yields the modified functional

$$\frac{1}{2}\int_B \boldsymbol{\sigma}\cdot\mathbb{C}^{-1}[\boldsymbol{\sigma}]\,dV + \int_B (\mathrm{div}\boldsymbol{\sigma} + \rho\,\mathbf{b})\cdot\mathbf{u}\,dV + \int_B \boldsymbol{\sigma}\cdot\boldsymbol{\beta}\,dV \to EXTR. \tag{70}$$

Here the stress field has to fulfill the divergence operator and the Neumann boundary conditions. The additional Lagrange multiplier $\boldsymbol{\gamma}$ is associated with the linear rotation field. At the extremum it has the value $\boldsymbol{\beta} = \frac{1}{2}\,(\,\nabla\mathbf{u} - \nabla^T\mathbf{u}\,)$. Variation of (70) leads to the weak form for the three independent fields $\boldsymbol{\sigma}$, $\mathbf{u}$ and $\boldsymbol{\beta}$

$$\begin{aligned}
\int_B \boldsymbol{\tau}\cdot\mathbb{C}^{-1}[\boldsymbol{\sigma}]\,dV + \int_B \mathrm{div}\boldsymbol{\tau}\cdot\mathbf{u} + \int_B \boldsymbol{\tau}\cdot\boldsymbol{\beta}\,dV &= 0,\\
\int_B (\mathrm{div}\boldsymbol{\sigma} + \rho\,\mathbf{b})\cdot\boldsymbol{\eta}\,dV &= 0,\\
\int_B \boldsymbol{\sigma}\cdot\boldsymbol{\gamma}\,dV &= 0.
\end{aligned} \tag{71}$$

Here $\boldsymbol{\tau}$, $\boldsymbol{\eta}$ and $\boldsymbol{\gamma}$ are the test functions.

1.6 Linearization of Variational Formulations

The solutions of nonlinear initial or boundary value problems in solid mechanics can be obtained in general only by employing approximate solution techniques. Since many of these methods—like also the finite element method—rely on a variational formulation of the field equations, the basis for numerical methods are provided by the weak forms of the associated field equations.

For the solution of the set of nonlinear equations many different algorithms are known. Often Newton's method is applied since it possesses the advantage of a quadratic convergence close to the solution point. In case of Newton's method, an improved solution is obtained from the Taylor series expansion of the nonlinear equation at the already computed approximate solution. This Taylor expansion corresponds in finite element applications to the linearization of the weak form, or in solid mechanics to the linearization of the virtual work or variational principles and can be obtained by the directional derivative.

In this contribution, the automatic tool *AceGen* is employed to generate the finite element formulations. This tool which is based on *Mathemaica* has the ability to automatically derive the linearizations of weak form expressions, thus there is no need to derive explicit expressions for the linearization of the weak forms used within the finite element method.

2 Finite Element Methods

Many formulation exists for the construction of finite elements. They are based on different interpolations and ansatz functions and different mathematical models related to the weak forms. The element formulations can be basically classified as follows.

- Finite elements with ansatz functions for the primary variables, in solid mechanics these are the deformations. Classically, the interpolation is provided by polynomials,
- Formulations based on mixed ansatz functions for different variables, like deformations, stresses, strains. Here polynomials of different order are applied to interpolate the field variables.
- Elements using Bezier interpolations, like isogeometric ansatz functions, in primal or mixed form.

In this contribution, we will discuss elements related to the first two points, for the third point see, e.g., Cottrell et al. (2009).

2.1 Requirements for Solid Elements

The ideal case for application of the finite method would be that one interpolation scheme could be used for a large class of problems. Due to several reasons that are discussed below, this is not possible. However, the search for such an element family has led to numerous scientific contributions in this area, for overviews, see, e.g., Zienkiewicz and Taylor (2000) and Wriggers (2008). The basic problem classes that arise in solid mechanics problems undergoing finite deformations are

- elastic deformations,
- inelastic deformations,
- deformations with volume constraints, like incompressibility or anisotropic unidirectional constraints,
- deformations, constraint by contact and interface conditions and
- deformations of thin solids, like beams, plates, and shells.

For these problem classes one would like to have elements that can predict deformation and stress fields in an accurate and efficient manner. Furthermore, the formulations should be robust in the sense that they converge with respect to the necessary equilibrium iterations when applied in the large deformation range.

Based on these demands specific targets for the element development can be specified. Some of them are listed below.

- No sensitivity against mesh distortions, means that the element, undergoing large strain states, might deform very much and thus can change its shape. Thus, an element with convex geometry can become nonconvex. These types of distorsions should not affect the overall solution behavior. But also mesh generation tools can create element geometries that are severely distorted.
- Good coarse mesh accuracy is needed when available computer resources are limited. In such cases, especially in three-dimensional applications the element should provide results with sufficient accuracy even for relative coarse meshes. This demand is in line with efficiency of the element formulation which means that computing effort at local element level should be a low as possible. Here, a small number of integrations points, no solution of extra equation systems provide efficient elements. An example for such element formulation can be found, e.g. in Reese (2005).
- Finite elements should be formulated such that the implementation of nonlinear constitutive equations is relatively simple. In that way, an interface can be created that allows even nonspecialists to apply a certain finite element to another class of constitutive equations.
- When constraint equations have to be considered within the finite element formulation, standard approaches with low-order interpolation tend to lock. This occurs for incompressible and anisotropic materials, but also certain elasto-plastic material equations and for thin, elements that of undergo bending deformations for which standard interpolations lock, see e.g. Zienkiewicz and Taylor (1989) oder Hughes (1987) and Braess (1992).

Element ansatz functions that interpolate the deformation or displacement field within an element with first-order shape functions (bi- or tri-linear interpolation) do not convergence properly when applied to bending problems and to problems where incompressible material is present. In the past different variational formulations have been explored in order to construct finite elements that can be used for this problem class. Some approaches are listed below.

- Reduced integration and stabilization is the most simple method to overcome locking behavior, see, e.g., Zienkiewicz et al. (1971). Many variants were discussed in the literature. It was shown the reduced integration on works together with stabilization, see e.g. Belytschko et al. (1984) and Reese and Wriggers (2000). These elements are in general *locking* free for incompressible deformations. They are not sensitive against mesh distortion and can be used for arbitrary constitutive equations. Due to the reduced integration these elements are very efficient (e.g., for an eight-node brick element only one Gauss point is needed). On the down side, the reduced integrated and stabilized elements rely often on artificial stabilization parameters.Hence, there exists cases, like some bending problems were the finite element solution depends on the stabilization parameter.

- Finite elements based on the mixed variational principle of Hu–Washizu type were successfully developed by Simo and co-workers who introduced the *enhanced strain* elements first for the geometrical linear theory, Simo and Rifai (1990) and then for large deformations, Simo and Armero (1992) and Simo et al. (1993)

2.2 Nonlinear Finite Element Formulations

In this contribution the tool *AceGen*, see Korelc (2011), is used to produce efficient finite element code with a minimum of effort. This package is a toolbox that is based on Mathematica (2011). It has the following advantages for developers of finite elements:

- automatic generation of finite element source code for different code environments, see Korelc (2002),
- short generation times, see Korelc (1997),
- produces residual vectors and tangent matrices automatically for complex applications,
- all relevant equations can be checked in debug mode.

The workflow when using *AceGen* can be beasically described as follows

- initialize *AceGen*,
- initialize a template (type of finite element, e.g., a six-node triangle or a hexahedral element with eight nodes). This template defines automatically all essential variables related to such element, like number of nodes, etc.
- define the problem (mechanical, thermal,..):
  - define variables,
  - define shape functions and symbolically compute derivatives with respect to the selected coordinate systems,
  - compute symbolically kinematical variables and stresses (if needed, like, e.g., in the weak form),
  - define strain energy function or weak form,
- derive residual vector and tangent matrix symbolically,
- generate code and export code for the selected environment.

2.3 Three-Dimensional Finite Strain Element

To show the general procedure that has to be followed in order to derive finite elements, three-dimensional finite elements for finite strains are developed for a Neo-Hooke material based on the strain energy (55) and on the weak form (50).

Formulation based on the strain energy function In this approach, the finite element is developed based on the strain energy function with respect to the initial configuration. The strain energy can be written for a hyperelastic Neo-Hooke material as

$$W(\mathbf{u}) = \int_V \left\{ \frac{\lambda}{2} \left(J(\mathbf{u}) - 1 \right)^2 + \frac{\mu}{2} \left[\text{tr}\mathbf{C}(\mathbf{u}) - 3 - 2 \log J(\mathbf{u}) \right] \right\} dV. \tag{72}$$

with the Lamé constants λ and μ, the right Cauchy-Green tensor $\mathbf{C}$, see (9), and the Jacobian $J = \det \mathbf{F}$, see (8). All kinematic variables depend on the displacement field $\mathbf{u}$.

A finite element formulation will now be developed for the hexahedral element with tri-linear shape functions based on the isoparametric concept. In this formulation, the displacement field and the coordinated within the element are approximated by the same ansatz

$$\mathbf{u}_e = \sum_{I=1}^{8} N_I(\xi, \eta, \zeta)\, \mathbf{u}_I \qquad \mathbf{X}_e = \sum_{I=1}^{8} N_I(\xi, \eta, \zeta)\, \mathbf{X}_I. \tag{73}$$

For the 8 noded brick element the shape functions N_I can be expressed as follows

$$N_I(\xi, \eta, \zeta) = \frac{1}{8} (1 + \xi_I \xi)(1 + \eta_I \eta)(1 + \zeta_I \zeta), \quad I = 1, \ldots, 8 \tag{74}$$

where ξ_I, η_I and ζ_I are the nodal points in the reference element of the isoparametric formulation, for details see, e.g., Wriggers (2008).

Note that differentiation within the isoparametric formulation with respect to the physical coordinates $\mathbf{X}$ has to be done by using the chain rule. By defining $\Xi = \{\xi, \eta, \zeta\}$ the differentiation of the displacement field with respect to $\mathbf{X}$, leading to the displacement gradient $\mathbf{H}$, see (8), can be written within the element e as

$$\mathbf{H}_e = \text{Grad}\, \mathbf{u} = \frac{\partial \mathbf{u}}{\partial \mathbf{X}} = \frac{\partial \mathbf{u}}{\partial \Xi} \frac{\partial \Xi}{\partial \mathbf{X}} = \mathbf{J}_e^{-T} \frac{\partial \mathbf{u}}{\partial \Xi} \tag{75}$$

where $\mathbf{J}_e = \frac{\partial \mathbf{X}}{\partial \Xi}$ is the Jacobi matrix of the transformation from the coordinates Ξ reference element to the actual physical coordinates $\mathbf{X}$ of element e. After that the deformation gradient $\mathbf{F}$, its detrminant J and the right Cauchy-Green tensor $\mathbf{C}$ follow directly for the element e

$$\mathbf{F}_e = \mathbf{1} + \mathbf{H}_e, \quad J_e = \det \mathbf{F}_e \quad \text{and}\ \mathbf{C}_e = \mathbf{F}_e^T \mathbf{F}_e. \tag{76}$$

Now all basic variable needed in (72) are defined and *AceGen* can be applied to derive the nonlinear hexahedral element. Here the basic steps are shown. For details regarding the use of this tool, see Korelc (2011).

Fig. 1 Initiation of *AceGen*

```
<<AceGen`;
SMSInitialize["H1element","Environment"->"AceFEM"];
```

Fig. 2 Basic template of the element

```
SMSTemplate[
   "SMSTopology"→"H1"
   ,"SMSDomainDataNames"→
    {"E -elastic modulus","ν -poisson ratio",…}
   ,"SMSDefaultData"->{21000,0.3,…}
   ,"SMSSymmetricTangent"→True
  ];
nen=SMSNoNodes;ndim=SMSNoDimensions;
np=SMSNoDOFGlobal;
```

The toolbox has to be started, see first line in Fig. 1. The second line defines the name of the generated element "H1element" and the finite element environment for which the element will be generated, in this case the program *AceFEM* associated with *AceGen*. However also elements can be generated for the commercial program ABAQUS and the research code FEAP, see Taylor (2011).

Next comes the selection of the template, here "H1" is related to the eight-noded hexahedral element, see Fig. 2. With this *AceGen* knows the dimension, the number of nodes and the number of unknowns. Furthermore input data related to the material model, loading,... are provided in this template.

In Fig. 3 the first line retrieves the material data in form of the Young's modulus E and the Poisson ratio ν which are converted by an inbuilt function to the Lamé constants λ and μ in the next line. In the third and forth line the nodal values of the coordinates, $\mathbf{X}_i$ and the displacements, $\mathbf{u}_I$ are defined. The fifth line convert the 3×8 matrix of the nodal variables into a 1×24 vector $\mathbf{p}_e$ which is related to the final size of the residual vector (1×24) and the tangent stiffness matrix (24×24), to be generated. The sixth line defines now the loop over all integration points used to integrate residual and tangent stiffness over the element volume. Due to the selected template 'H1' *AceGen* assumes that a Gauss point integration with $2 \times 2 \times 2$ point will be sufficient for the eight-noded brick.[5] The in the seventh line the Gauss point coordinates $\boldsymbol{\Xi}$ applied in the integration loop are defined and in the last line the weight related to the Gauss point.

The shape functions of the eight-noded hexahedral element, see (74) are defined in the first four lines of Fig. 4. The fifth line computed in a very condensed way the coordinates $\mathbf{X}_e$ and displacements $\mathbf{u}_e$ in the element as given in (73). Furthermore in this line the Jacobi matrix $\mathbf{J}_e$, see (75), and its determinant are computed In the sixth line the displacement gradient is computed using the chain rule provided in (75). The seventh line is the equivalent to (76), it computed deformation gradient $\mathbf{F}_e$, its determinant J_e and right Cauchy-Green tensor $\mathbf{C}_e$. In the sixth line the strain energy function, as given in (72) is defined. Once all this is done, the residual $\mathbf{R}$ follows by

[5]The user can overrule the automatic selection of the integration rule, but this is only necessary when special shape functions are used.

Fig. 3 Basic definitions of material data, nodal values, and Gauss points and

```
{Em,ν}⊢SMSReal[Table[es$$["Data",i],{i,2}]];
{λ,μ}⊨SMSHookeToLame[Em,ν];
XIO⊢Table[SMSReal[nd$$[i,"X",j]],{i,nen},{j,ndim}];
uIO⊢SMSReal[Table[nd$$[i,"at",j],{i,nen},{j,ndim}]];
pe=Flatten[uIO];
SMSDo[Ig,1,SMSInteger[es$$["id","NoIntPoints"]]];
Ξ={ξ,η,ζ}⊢Table[SMSReal[es$$["IntPoints",i,Ig]],{i,3}];
wgp⊢SMSReal[es$$["IntPoints",4,Ig]];
```

Fig. 4 Definition of shape functions, derivatives, and generation of residual and tangent matrix

```
Ξn={{-1,-1,-1},{1,-1,-1},{1,1,-1},{-1,1,-1},{-1,-1,1},
   {1,-1,1},{1,1,1},{-1,1,1}};
Nh⊨Table[1/8 (1+ξ Ξn⟦i,1⟧) (1+η Ξn⟦i,2⟧) (1+ζ Ξn⟦i,3⟧),
   {i,1,nen}];
X⊢SMSFreeze[Nh.XIO];u⊨Nh.uIO;Je⊨SMSD[X,Ξ];Jed⊨Det[Je];
H⊨SMSD[u,X,"Dependency"→{Ξ,X,SMSInverse[Je]}];
F⊨IdentityMatrix[3]+H;Ce⊨Fᵀ.F;JF⊨Det[F];
W⊨1/2 λ (JF-1)^2+μ (1/2 (Tr[Ce]-3)-Log[JF]);
Rg⊨Jed SMSD[W,pe];
Kg⊨SMSD[Rg,pe];
```

using differentiation of W with respect to the nodal displacements $\mathbf{p}_e$. This is done by the command 'SMSD.' Multiplication with the determinant of the Jacobian of the isoparametric transformation is necessary to obtain the correct volume integral in physical space of the element. Here, a different way is used to obtain the residual. The virtual work of the stresses in (48) can be written with the variation $\boldsymbol{\eta} = \delta\mathbf{u}$ by using the chain rule as

$$\mathbf{P} \cdot \text{Grad}\, \delta\mathbf{u} = \frac{\partial W}{\partial \mathbf{F}} \cdot \frac{\partial \mathbf{F}}{\partial \mathbf{u}}\, \delta\mathbf{u} = \frac{\partial W}{\partial \mathbf{u}} \cdot \delta\mathbf{u}. \tag{77}$$

Hence, the virtual work, see for the classical formulation (48), can now be redefined, leading to the residual is

$$\int_V \frac{\partial W}{\partial \mathbf{u}} \cdot \delta\mathbf{u}\, dV - \int_V \rho_0\, (\bar{\mathbf{b}} - \dot{\mathbf{v}}) \cdot \delta\mathbf{u}\, dV - \int_{\partial V_\sigma} \bar{\mathbf{t}} \cdot \delta\mathbf{u}\, dA = 0. \tag{78}$$

The tangent matrix $\mathbf{K}_T$ then simply follows from the derivative of the residual $\mathbf{R}$ with respect to the nodel displacements $\mathbf{p}_e$.

The only operation that is left is the export of the symbolically generated residual and tangent matrix to a subroutine that can be called as user element in one of the finite element codes mentioned above. This is done as depicted in Fig. 5. The last statement closes the loop over the Gauss points, see Fig. 3.

Fig. 5 Code generation

```
SMSExport[wgp Rg,p$$,"AddIn"→True];
SMSExport[wgp Table[Kg⟦i,j⟧,{i,1,np},{j,i,np}],
  Table[s$$[i,j],{i,1,np},{j,i,np}],"AddIn"→True];
SMSEndDo[];
```

Formulation based on the weak form. Often it is not possible to develop finite elements starting from a strain energy function, e.g., in case of inelastic materials, and thus, the more general approach is to use the weak form. In that case, one can start from the weak form (50)

$$\int_V \mathbf{S}(\mathbf{u}) \cdot \frac{1}{2} \delta \mathbf{C} \, dV - \int_V \rho_0 \, \bar{\mathbf{b}} \cdot \delta \mathbf{u} \, dV - \int_{\partial V_\sigma} \bar{\mathbf{t}} \cdot \delta \mathbf{u} \, dA = 0. \tag{79}$$

where the second Piola–Kirchhoff stress tensor $\mathbf{S}(\mathbf{u})$ has to be expressed using a constitutive equation, $\mathbf{C}(\mathbf{u})$ is again the right Cauchy-Green tensor. The last two integrals are related to body and surface forces.

Now the isogeometric formulation with the same ansatz function as before is employed, and hence, Eqs. (73)–(76) are still valid. The only new equation that is needed is the stress response related to the strain energy defined in (72). This is obtained from

$$\mathbf{S} = 2 \frac{\partial W}{\partial \mathbf{C}} = \lambda J (J - 1) \, \mathbf{C}^{-1} + \mu \, (\mathbf{1} - \mathbf{C}^{-1}\,). \tag{80}$$

Now the code included in Fig. 4 can be exchanged by the code depicted in Fig. 6. Here in the second line the Cauchy-Green tensor and its inverse are defined. Then in the next line the response function for the second Piola–Kirchhoff stress $\mathbf{S}$, see (80), is provided. After that follows a loop over all unknowns. Here, the variation of $\mathbf{C}$ is computed in the fifth line by differentiation of the right Cauchy-Green tensor with respect to the displacement field. This is justified by

$$\delta \mathbf{C}_e(\mathbf{u}) = \delta \mathbf{C}_e(\mathbf{p}_e) = \frac{\partial \mathbf{C}_e}{\partial \mathbf{p}_e} \, \delta \mathbf{p}_e. \tag{81}$$

Note that $\delta \mathbf{p}_e$ is not needed in the formulation since it will be put in front of the residual vector, see, e.g., Wriggers (2008). The residual follows in the sixth line its form stems from the first integral in (79). Here, the last two integrals, defining the loading are omitted to have shorter code.

Fig. 6 Code generation using weak form

```
F⊨IdentityMatrix[3]+H;JF⊨Det[F];
Ct⊨Fᵀ.F;Cei⊨SMSInverse[Ct];
S⊨λ JF (JF-1) Cei+μ (IdentityMatrix[3]-Cei);
SMSDo[m,1,np];
  δCt⊨SMSD[Ct,pe,m];
  Rgm⊨Jed 1/2 Tr[δCt.S];
  SMSExport[wgp Rgm,p$$[m],"AddIn"→True];
  SMSDo[n,m,np];
    Kgmn⊨SMSD[Rgm,pe,n];
    SMSExport[wgp Kgmn,s$$[m,n],"AddIn"→True];
  SMSEndDo[];
SMSEndDo[];
```

Fig. 7 Code generation using pseudo potential

```
Ct⊨Fᵀ.F;Cei⊨SMSInverse[Ct];
S⊢SMSFreeze[λ JF (JF-1) Cei+μ (IdentityMatrix[3]-Cei)];
WP⊨1/2 Tr[Ct.Transpose[S]];
SMSDo[m,1,np];
  Rgm⊨Jed SMSD[WP,pe,m,"Constant"→S];
  SMSExport[wgp Rgm,p$$[m],"AddIn"→True];
```

After that the standard procedure to export the code and to obtain the tangent stiffnesss matrix by differentiation of the residual with respect to the displacement field ends the formulation of the element based on the weak form.

Another possibility when using *AceGen* is to compute the residual in the following form

$$\mathbf{R} = \frac{J_e}{2}\,\mathbf{S}_e(\mathbf{p}_e)\cdot\frac{\partial \mathbf{C}_e(\mathbf{p}_e)}{\partial \mathbf{p}_e} = \left.\frac{\partial W^P(\mathbf{p}_e)}{\partial \mathbf{p}_e}\right|_{\mathbf{S}_e = const.} \tag{82}$$

with the pseudo potential of the specific strain energy $W = \int_V \bar{W}\,dV$

$$\bar{W}^P(\mathbf{p}_e) = \frac{1}{2}\,\mathbf{S}_e(\mathbf{p}_e)\cdot\mathbf{C}_e(\mathbf{p}_e) = \frac{1}{2}\,\mathrm{tr}[\mathbf{S}_e(\mathbf{p}_e)^T\mathbf{C}_e(\mathbf{p}_e)]. \tag{83}$$

Thus, the part that computes the residual in Fig. 6 can be exchanged with Fig. 7.

It is obvious that all formulation have to produce exactly the same results as the elements derived directly from the same strain energy. Here, the interesting question is: which of these formulations produces the more efficient code? Actually, it can be shown that the time to generate the code for the finite element is the shortest when the strain energy (72) is used. Using the weak form or the pseudo potential is 10–20 % slower. This also holds for the time to compute residual vector and tangent matrix since it takes about 70 % more time to evaluate the residual vector and tangent matrix.

Another observation is that the development of such element, once the basic structure of the toolbox is set, is very fast. The type of element, that is described in Fig. 4 only needs about 10–15 s to generate. Compared to the time one needs to write the code, and before, to derive residual, and tangent including debugging in all phases is a lot slower. Thus, it pays off to invest in understanding modern development software that can do derivations automatically like *AceGen*.

2.4 Mixed Elements for Incompressibility

Pure displacement elements tend to lock and thus will generate a too stiff response. In case of incompressible materials this phenomenon stems from the volumetric constraint $J = \det \mathbf{F} = 1$, and hence, is called volume locking. Mixed finite element methods can be introduced to avoid volume locking, see, e.g., Zienkiewicz and Taylor (1989) and Brezzi and Fortin (1991).

Finite elements that are derived form mixed methods have to fulfill additional mathematical conditions that guarantee the stability of the element formulation. For incompressibility this condition is known as BB-condition, named after its inventors Babuska and Brezzi, see e.g. Brezzi and Fortin (1991). A numerical method to investigate fulfillment of the BB-condition can be found in Chapelle and Bathe (1993). For general nonlinear applications there exists no direct formulation of the BB-condition, however, discussions in this direction can be found in Auricchio et al. (2013).

Different ways can be followed to construct mixed elements. Here, the specific strain energy is used to shorten notation.

Lagrangian multipliers The method of Lagrangian multipliers provides a classical way to enforce constraints. Here, as an example the incompressibility constraint $G(J) = J - 1 = 0$ (with $J = \det \mathbf{F}$) is considered which is directly introduced

$$\bar{W}(\mathbf{u}, p) = \bar{W}(\mathbf{C}(\mathbf{u})) + p\, G(J(\mathbf{u})). \tag{84}$$

The strain energy function $\bar{W}(\mathbf{C}(\mathbf{u}))$ is given by any of the functions defined in Sect. 1.3. In the finite element formulation occur now additional unknowns associated with the Lagrange multiplier p in addition to the displacements. In the case of incompressibility the Lagrange multiplier can be viewed as the pressure. Early finite element formulations that use this approach can be found in Oden and Key (1970) and Duffet and Reddy (1983).

Perturbed Lagrangian formulation This formulation is based on the following specific strain energy function

$$\bar{W}(\mathbf{u}, p) = \bar{W}(\mathbf{C}(\mathbf{u})) + p\, G(J(\mathbf{u})) - \frac{1}{2\,\epsilon}\, p^2. \tag{85}$$

The constraint condition is again given as in (84). $\epsilon > 0$ is a perturbation parameter. Choosing now continuous ansatz function for displacements and pressure. Due to the third term in (85) the pressures can be removed from the system. When discontinuous ansatz functions are used for the pressure p then they can be eliminated at element level and lead to a formulation that is equivalent to a *penalty* formulation.

Note that the solution now depends on the perturbation or penalty parameter. For small values of ϵ the influence of the constraint condition disappears. For large values of ϵ the constraint is fulfilled more and more exactly but the condition number of the equation system that appears in a Newton typer solution will be very large. Many papers regarding various variants of (85) have been published. For the large strain case of rubber elasticity see, e.g., Häggblad and Sundberg (1983) and Sussman and Bathe (1987).

Hu–Washizu **functional**. In this functional the incompressibility constraint is formulated via a constitutive equation for the pressure, see (62). In short the specific strain energy

$$\bar{W}(\mathbf{u}, p, \theta) = \bar{W}(\widehat{\mathbf{C}}(\mathbf{u})) + K\,[\,G(\theta)]^2 + p\,(J(\mathbf{u}) - \theta) \tag{86}$$

is formulated. Here ansatz functions can be selected for the displacement $\mathbf{u}$, the pressure p and the volumetric strain θ. $G(\theta)$ defines the constitutive equation for the pressure term, here K is the modulus of compression. The formulation of $W(\widehat{\mathbf{C}})$ is provided by (39). A finite element based on this approach was firstly presented in Simo et al. (1985).

2.5 *Mixed Element for Finite Deformations*

The program *AceGen*, see Korelc (2011), is used now to generate finite elements for the cases discussed above.

Mixed element T2-P1 with Lagrangian multipliers In order to fulfill the equation for the volume constraint $J - 1$ exactly a pure mixed element is considered. It is based on a tetrahedral element with quadratic ansatz functions for the displacement field $\mathbf{u}$ and linear ansatz functions for the pressure p. Since the most efficient formulation in *AceGen* is generated using the strain energy function. We can start from

$$W(\mathbf{u}, p) = \int\limits_V \left\{ \frac{\mu}{2}\,[\mathrm{tr}\mathbf{C}(\mathbf{u}) - 3] - \mu\,\log J(\mathbf{u}) + p\,[\,J(\mathbf{u}) - 1\,)\,] \right\} dV \tag{87}$$

where the last term is related to the constraint. The mixed strain energy is closely related to the strain energy (72), only the first term in (72) is replaced by the third term in the strain energy in (87).

The mixed element has now to be generated using different ansatz functions for displacements and pressure. This can be achieved within *AceGen* in a way that is described in Fig. 8. The SMSTemplate defines the topology, here 'O2' means that a quadratic tetrahedra element is generated. It has 10 nodes and for the first 4 nodes 4 degrees of freedom (displacement and pressure variables) and the last 6 nodes 3 degrees of freedom (displacement variables). This is also denoted by the SMSNodeID where 'Mix -L' stands for pressure and displacements and 'D' for displacements. Coordinates of the nodes are defined as usual. However, the unknown nodal values are contained in a table from which one has to extract the displacement and pressure unknowns. This is done in line 10 of Fig. 8.

Additionally a body force is defined in 12 and 13 where q is the load vector and 'mult' the load factor.

After this definition of the general element quantities the isoparametric ansatz and the physical problem can be described. In Fig. 9 the first three lines define the loop

```
SMSInitialize["T2-P1-3d", "Environment" → "AceFEM", "Mode" → "Optimal"];
SMSTemplate[
 "SMSTopology" → "O2", "SMSNoNodes" → 10, "SMSDOFGlobal" → {4, 4, 4, 4, 3, 3, 3, 3, 3, 3},
 "SMSNodeID" → {"MIX -L", "MIX -L", "MIX -L", "MIX -L", "D", "D", "D", "D", "D", "D"},
 "SMSDomainDataNames" -> {"μ", "Qx", "Qy", "Qz"},
 "SMSDefaultData" -> {42, 0, 0, 0}, "SMSSymmetricTangent" → True]
SMSStandardModule["Tangent and residual"];
nen = SMSNoNodes; ndim = SMSNoDimensions; np = SMSNoDOFGlobal;
𝕏IO ⊢ Table[SMSReal[nd$$[i, "X", j]], {i, nen}, {j, ndim}];
𝕡eIO ⊢ SMSReal[Table[nd$$[i, "at", j], {i, nen}, {j, SMSDOFGlobal⟦i⟧}]];
𝕡e = Flatten[𝕡eIO];
𝕦IO = 𝕡eIO⟦{1, 2, 3, 4, 5, 6, 7, 8, 9, 10}, {1, 2, 3}⟧; pIO = 𝕡eIO⟦{1, 2, 3, 4}, 4⟧;
{μ, Qx, Qy, Qz} ⊨ SMSReal[Table[es$$["Data", i], {i, 4}]];
q = {Qx, Qy, Qz};
mult ⊨ SMSReal[rdata$$["Multiplier"]];
```

Fig. 8 First part of *AceGen* code for a mixed element

```
SMSDo[Ig, 1, SMSInteger[es$$["id", "NoIntPoints"]]];
Ξ = {ξ, η, ζ} ⊢ Table[SMSReal[es$$["IntPoints", i, Ig]], {i, 3}];
wgp ⊢ SMSReal[es$$["IntPoints", 4, Ig]];
κ = 1 - ξ - η - ζ;
Nh ⊨ {(2 ξ - 1) ξ, (2 η - 1) η, (2 ζ - 1) ζ, (2 κ - 1) κ, 4 ξ η, 4 η ζ, 4 ζ ξ, 4 ξ κ, 4 η κ, 4 ζ κ};
Np ⊨ {ξ, η, ζ, κ};
𝕏 ⊢ SMSFreeze[Nh.𝕏IO]; 𝕦 ⊨ Nh.𝕦IO; p = Np.pIO;
𝕁e ⊨ SMSD[𝕏, Ξ]; Jed ⊨ Det[𝕁e];
ℍ ⊨ SMSD[𝕦, 𝕏, "Dependency" → {Ξ, 𝕏, SMSInverse[𝕁e]}];
𝔽 ⊨ IdentityMatrix[3] + ℍ;
Ct ⊨ 𝔽ᵀ.𝔽; JF ⊨ Det[𝔽];
W ⊨ μ 1/2 (Tr[Ct] - 3) - μ Log[JF] + p (JF - 1) - mult q.𝕦;
```

Fig. 9 Second part of *AceGen* code for a mixed element

over the Gauss points, the isoparametric coordinates and the weighting factors related to the Gauss points within the numerical integration procedure. The fourth to fifth line defines the quadratic shape functions 'Nh' for the displacement interpolation

$$
\begin{aligned}
&N_1 = (2\xi - 1)\xi,\ N_2 = (2\eta - 1)\eta,\ N_3 = (2\zeta - 1)\zeta,\ N_4 = (2\kappa - 1)\kappa,\\
&N_5 = 4\xi\eta,\ N_6 = 4\eta\zeta,\ N_7 = 4\zeta\xi,\ N_8 = 4\xi\kappa,\ N_9 = 4\eta\kappa,\ N_{10} = 4\zeta\kappa,
\end{aligned}
$$

with $\kappa = 1 - \xi - \eta - \zeta$. Furthermore, in the sixth line, the linear shape functions 'Np' are defined

$$
Np_1 = \xi, \quad Np_2 = \eta, \quad Np_3 = \zeta, \quad Np_4 = \kappa
$$

that will be used to interpolate the pressure variables within the element. The ansatz for u and p is defined in the seventh line of Fig. 9. The next line describes the derivation of the Jacobi matrix that defines the isoparametric map. After that the displacement, $\mathbf{H}$, and the deformation gradient $\mathbf{F}$, see (8), are computed as well as the right Cauchy-Green tensor $\mathbf{C}$ and the determinant J of $\mathbf{F}$. For more details, see also the description in Fig. 4. These kinematic variables are needed to define the specific strain energy $\bar{W}$ in the last line of Fig. 9.

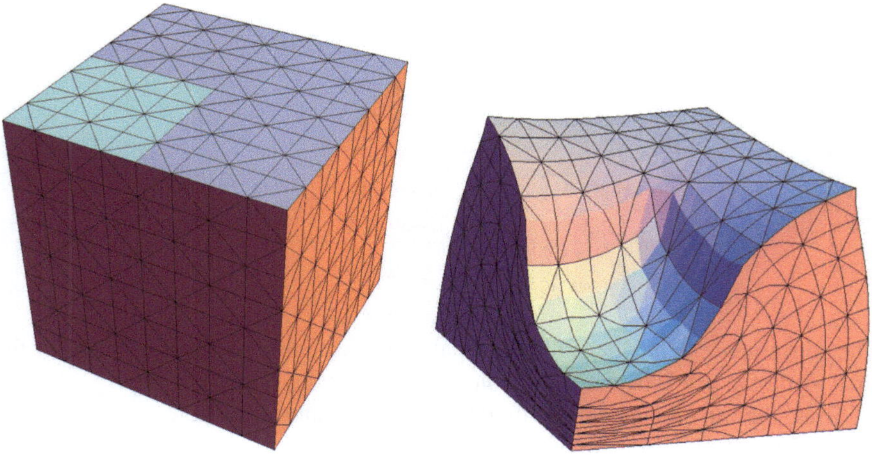

Fig. 10 Undeformed and severely deformed configuration of a block (a quarter of the mesh)

The residual and tangent matrix follows then by the same procedure that is provided in Fig. 4. No difference has been made with respect to displacement and pressure variables, since these are combined in the vector $\mathbf{p}_e$ and the command SMSD differentiates the strain energy function with respect to all variables and thus will produce the full residual vector that has the length of the 34 unknowns. Thus, the tangent matrix has the size 34×34 where zero elements appear on the diagonal related to the pressure degree of freedom. Thus special equation solvers have to be used to solve engineering problems in a reliable way.

As an example to highlight the performance of the mixed element a block which is partly loaded by a constant surface load is considered. Figure 10 depicts a quarter of a mesh and deformed configuration of a block of rubber-like material. The block is discretized using the mixed T2-P1 element descibed above. The block has dimensions of $100 \times 100 \times 50$.[6] It is on rollers at the bottom. Furthermore symmetry conditions are applied to be able to compute only one quarter of the block. Furthermore, the displacements at the top of the block are fixed in X and Y direction. The block is, as shown in Fig. 10, discretized by $8 \times 8 \times 8$ T2-P1 elements. This leads to a total of 2592 elements and 12521 unknowns. Material data need only be provided for the shear modulus it is selected as $\mu = 1.61148$. A surface load of $q = 9$ is applied at the part that is colored in turquois in three load steps.

This loading leads to a severely deformed configuration. Newton's method is applied to solve the problem. It should be noted that the solution within each load step converges quadratically. The element depicts a very robust behavior since only three load steps, and within each load step only six iterations were needed to converge.

[6] All data are provided as dimensionless constants, it is assumed that the dimensions match real physical data.

Table 1 Convergence of the mid-displacement using the T2-P1 element

Mesh division	d.o.f	u_z	Total iterations
$2 \times 2 \times 2$	263	−30,803	19
$4 \times 4 \times 4$	1733	−38,344	19
$8 \times 8 \times 8$	12521	−39,271	18
$16 \times 16 \times 16$	95057	−39,349	18
$24 \times 24 \times 16$	315193	−39,372	18

The convergence behavior of the maximal displacement in vertical direction for the final load of $q = 9$ is depicted in Table 1 for a variety of discretizations.

Thus, the mixed element is efficient and robust and applicable for finite strain cases with severe deformations.

Mixed element Q1-P0 based on Hu–Washizu's principle In this section, we derive a large deformation finite element which is based on the Hu–Washizu variational formulation. This element, developed in Simo et al. (1985), is implemented in many existing finite element codes and uses linear shape functions for the deformation field related to the deviatoric kinematical variables. The shape functions for the displacements are the same as in (74). The mixed/enhanced variables, pressure p, and volumetric strain θ, are approximated by shape functions that are assumed to be constant.

The formulation starts from the special Hu–Washizu principle (62). Here, the specific strain energy $\bar{W}$ is given as

$$\bar{W}(\mathbf{u}, p, \theta) = \frac{\mu}{2}\left(J^{-\frac{2}{3}}\,\mathrm{tr}\mathbf{C} - 3\right) + p\left[J - \frac{J_e}{J_{e0}}\,\theta\right] + \frac{K}{2}\left[\frac{J_e}{J_{e0}}\,\theta - 1\right]^2 . \tag{88}$$

Here $Je = \det \mathbf{J}_e$ is the determinant of Jacobi matrix of the isoparametric transformation and $J_{e0} = \det \mathbf{J}_{e0}$ is the same determinant evaluated at the element center. This special form of $\bar{W}$ was selected in order to obtain the same element formulation as provided in Simo et al. (1985) in which the integrals for the mixed parts are evaluated differently using one point integration.

In this formulation the two additional unknowns p and θ can be eliminated at element level by using the Schur complement. This is automatically done in *AceGen* when variables are defined by the "SMSNoDOFCondense" command in the template, see Fig. 11. The associated variables are defined in the 10th line of the input and are combined with the displacement variables to the general unknown vector in line 11. All the rest is standard input where "H1" in the second line defines the topology of the 8 noded hexahedral element.

The second part of the *AceGen* input for the Q1-P0 element is provided in Fig. 12. The shape function for the element are defined in lines 4 and 5. Then the standard procedure as before leads to the displacement interpolation in the element. New is the computation of $\det \mathbf{J}_{e0}$ in line 9 and the definition of the mixed variables p and

```
SMSInitialize["Q1P0", "Environment" → "AceFEM", "Mode" → "Optimal"];
SMSTemplate["SMSTopology" → "H1"
  , "SMSNoDOFCondense" → 2
   , "SMSDomainDataNames" -> {"κ", "μ", "Qx", "Qy", "Qz"}
  , "SMSDefaultData" -> {501, 1.61148, 0, 0, 0}];
SMSStandardModule["Tangent and residual"];
nen = SMSNoNodes; ndim = SMSNoDimensions; np = SMSNoDOFGlobal; nhe = SMSNoDOFCondense;
XIO ⊢ Table[SMSReal[nd$$[i, "X", j]], {i, nen}, {j, ndim}];
peIO ⊢ SMSReal[Table[nd$$[i, "at", j], {i, nen}, {j, SMSDOFGlobal[[i]]}]];
heIO ⊢ SMSReal[Array[ed$$["ht", #] &, nhe]];
pe = Flatten[{peIO, heIO}];
{κ, μ, Qx, Qy, Qz} ⊨ SMSReal[Array[es$$["Data", #1] &, 5]];
q = {Qx, Qy, Qz};
mult ⊨ SMSReal[rdata$$["Multiplier"]];
```

Fig. 11 First part of *AceGen* code for the Q1-P0 element based on the Hu–Washizu principle

```
SMSDo[Ig, 1, SMSInteger[es$$["id", "NoIntPoints"]]];
Ξ = {ξ, η, ζ} ⊢ Table[SMSReal[es$$["IntPoints", i, Ig]], {i, 3}];
wgp ⊢ SMSReal[es$$["IntPoints", 4, Ig]];
{ξi, ηi, ζi} = {{-1, 1, 1, -1, -1, 1, 1, -1}, {-1, -1, 1, 1, -1, -1, 1, 1}, {-1, -1, -1, -1, 1, 1, 1, 1}};
Nh ⊨ MapThread[1 / 8 (1 + ξ #1) (1 + η #2) (1 + ζ #3) &, {ξi, ηi, ζi}];
X ⊢ SMSFreeze[Nh.XIO]; u ⊨ Nh.peIO;
Je ⊨ SMSD[X, Ξ]; Jed ⊨ Det[Je];
H ⊨ SMSD[u, X, "Dependency" → {Ξ, X, SMSInverse[Je]}];
Je0 ⊨ SMSReplaceAll[Je, {ξ → 0, η → 0, ζ → 0}]; J0 ⊨ Det[Je0];
θ ⊨ heIO[[1]] + 1; p ⊨ heIO[[2]];
F ⊨ IdentityMatrix[3] + H;
JF ⊨ Det[F]; Ct ⊨ F^T.F;
W ⊨ p (JF - θ J0 / Jed) + μ / 2 (JF ^ (-2 / 3) Tr[Ct] - 3) + κ / 2 (θ J0 / Jed - 1) ^2 - mult q.u;
SMSDo[m, 1, np + nhe];
    Rgm ⊨ Jed SMSD[W, pe, m];
    SMSExport[wgp Rgm, p$$[m], "AddIn" → True];
    SMSDo[n, m, np + nhe];
        Kgmn ⊨ SMSD[Rgm, pe, n];
        SMSExport[wgp Kgmn, s$$[m, n], "AddIn" → True];
    SMSEndDo[];
SMSEndDo[];
SMSEndDo[];
SMSWrite[];
```

Fig. 12 Second part of *AceGen* code for the Q1-P0 element based on the Hu–Washizu principle

θ in line 10. Note that the constant 1 is added to θ in order to obtain a zero energy state in the initial configuration. After that all is standard. One only has to take care that the loops now run over all displacements and the mixed variables, see line 14 and 17.

The same example of the block under surface load, as in the section before, is computed to depict the efficiency and robustness of this element formulation. The mesh and deformed mesh of a discretization with $32 \times 32 \times 32$ elements for an applied load of $q = 9$ is shown in Fig. 13. This again depicts the finite deformation state under the given load.

The convergence behavior of the maximal displacement when using the Q1-P0 element in vertical direction for the final load of $q = 9$ is depicted in Table 2 for a variety of discretizations. It can be seen that contrary to the T2-P1 element this Q1-P0 formulation converges in a different way. Small numbers of elements yield a solution that is to weak. However, both formulations converge to the same value.

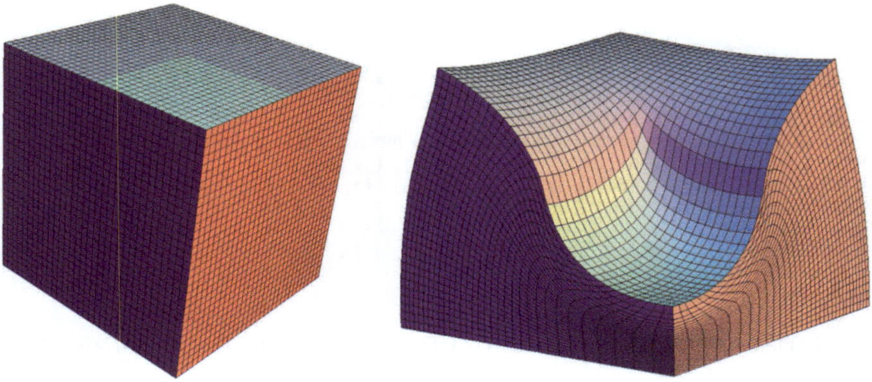

Fig. 13 Undeformed and severely deformed configuration of a block (a quarter of the mesh)

Table 2 Convergence of the mid-displacement using the Q1-P0 element

Mesh division	d.o.f	u_z	Total iterations
$2 \times 2 \times 4$	42	−45,085	18
$4 \times 4 \times 4$	260	−41,324	20
$8 \times 8 \times 8$	1800	−40,562	20
$16 \times 16 \times 16$	13328	−39,675	20
$32 \times 32 \times 32$	102432	−39,463	20
$48 \times 48 \times 32$	341040	−39,425	20

Mixed element for anisotropic material The last example is concerned with a three-dimensional finite element for anisotropic material behavior at finite strains. In this example the formulation accounts for transversely isotropic material behavior by using a mixed approach. It is assumed that the material is not extendable in the given direction **a**.

The isotropic strain energy function W^{iso} which describes the behavior of the isotropic part of the material is given by

$$W^{iso}(\mathbf{u}) = \frac{\mu}{2}(\,\mathrm{tr}\,\mathbf{C} - 3 - 2\,\log J\,) + \frac{\lambda}{4}(J^2 - 1 - 2\log J\,) \tag{89}$$

where μ and λ are the Lamé constants. The enforcement of the constraint that ensures that the material does not extend in the direction of **a** leads to the following condition

$$\mathbf{a} \cdot \mathbf{E}\,\mathbf{a} = 0 \tag{90}$$

where **E** is the Green-Lagrangian strain tensor, see (9). Since it is simpler to work with the right Cauchy-Green tensor **C** this constraint can be written as

$$\mathbf{a} \cdot \mathbf{E}\,\mathbf{a} = \mathbf{a} \cdot (\mathbf{C} - \mathbf{1})\mathbf{a} = \mathbf{a} \cdot \mathbf{C}\,\mathbf{a} - 1 \quad \text{for } \|\mathbf{a}\| = 1. \tag{91}$$

Furthermore we can write

$$\mathbf{a} \cdot \mathbf{C}\,\mathbf{a} = \mathbf{C} \cdot \mathbf{M} = \text{tr}[\,\mathbf{CM}\,] \quad \text{with } \mathbf{M} = \mathbf{a} \otimes \mathbf{a}. \tag{92}$$

Thus, the Lagrange multiplier term related to the constraint of a material that is not extendable in the direction **a** yields

$$W^{ti}(\mathbf{u}, p) = p\,(\text{tr}[\,\mathbf{CM}\,] - 1) \tag{93}$$

which then leads to the final form of the strain energy

$$W(\mathbf{u}, p) = W^{iso}(\mathbf{u}) + W^{ti}(\mathbf{u}, p). \tag{94}$$

A tetraeder element with quadratic interpolation for the displacement field **u** and linear interpolation for the mixed variable p is selected.[7] Thus, the same shape functions as for the first element, T2-P1, in Sect. 2.5 can be used which leads to an *AceGen* input that is equivalent for the first part with the code depicted in Fig. 8. The only difference is that the vector **a** defining the direction of the anisotropy has to be defined as element input as $\mathbf{a} = \{\, a_x, a_y, a_z \,\}$.

The second part of the code is shown in Fig. 14. Again the shape functions for the displacement and the mixed variable are defined as well as the kinematical relations. New in this formulation is the computation of the structure tensor **M**, see (92), and the tensor product **CM**. After that the strain energy can be formulated. It is given in the last line of the input. The first two terms are related to the isotropic part W^{iso} while the last term defines the Lagrange multiplier term, W^{ti}, including the constraint (91).

An example that will show a clear anisotropic response is the Cook's membrane problem of a tapered cantilever, clamped at the left end. The structure is loaded at the right end by a constant vertical load, as depicted in left part of Fig. 15 The selected date for the Lame constants are $\mu = 500$ and $\lambda = 1000$. The direction of anisotropy is given by $\mathbf{a} = \frac{1}{\sqrt{3}}\{\, 1, 1, 1 \,\}$.

Different mesh densities where used to compute the solution. The deformed mesh in Fig. 15 was computed with a mesh of $16 \times 16 \times 4$ elements which lead to a total number of 26885 degree of freedoms. The deformation clearly depicts the twist in the deformed shape due to the anisotopic constraint at large deformations. The solution was computed with several load steps. In total 10 load steps were applied for all discretizations reported in Table 3. The convergence behavior was robust, six iterations per load step were needed for all discretizations to obtain convergence. In this solution, procedure Newton type convergence was observed.

Table 3 shows the convergence behavior of the end displacement at the top node $(X = 48, Y = 60, Z = 0)$ for the Z-displacement u_{zu} and the Y-displacement u_y. Furthermore the displacement u_{zb} at the node $(X = 48, Y = 44, Z = 0)$ is reported to depict the torsion of the cross section.

[7] Here the variable p is the stress component related to the constraint, e.g., the stress in direction of **a**. It has to be scaled in order to yield the correct stress.

```
SMSDo[Ig, 1, SMSInteger[es$$["id", "NoIntPoints"]]];
Ξ = {ξ, η, ζ} ⊢ Table[SMSReal[es$$["IntPoints", i, Ig]], {i, 3}];
wgp ⊢ SMSReal[es$$["IntPoints", 4, Ig]];
κ = 1 - ξ - η - ζ;
Nh ⊨ {(2 ξ - 1) ξ, (2 η - 1) η, (2 ζ - 1) ζ,
    (2 κ - 1) κ, 4 ξ η, 4 η ζ, 4 ζ ξ, 4 ξ κ, 4 η κ, 4 ζ κ};
Np ⊨ {ξ, η, ζ, κ};
𝕏 ⊢ SMSFreeze[Nh.𝕏IO];
𝕦 ⊨ Nh.𝕦IO; p = Np.pIO;
Je ⊨ SMSD[𝕏, Ξ]; Jed ⊨ Det[Je];
ℍ ⊨ SMSD[𝕦, 𝕏, "Dependency" → {Ξ, 𝕏, SMSInverse[Je]}];
𝔽 ⊨ IdentityMatrix[3] + ℍ;
Ct ⊨ 𝔽ᵀ.𝔽; JF ⊨ Det[𝔽];
𝕒 ⊨ {ax, ay, az};
M ⊨ Outer[Times, 𝕒 , 𝕒];
CM ⊨ Ct.M;
W ⊨ μ 1/2 (Tr[Ct] - 3 - 2 Log[JF]) + λ / 4 (JF^2 - 1 - 2 Log[JF]) + (Tr[CM] - 1) p;
```

Fig. 14 Second part of *AceGen* code for the mixed element with transversely anisotropic material

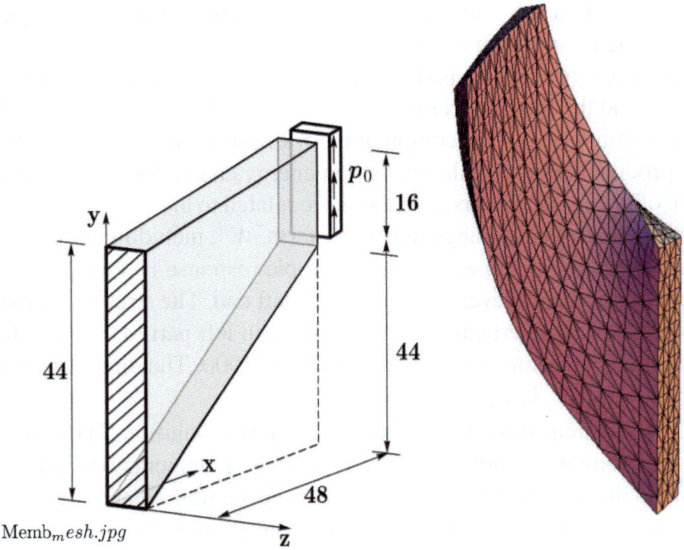

Fig. 15 Undeformed and severely deformed configuration of a block (a quarter of the mesh)

Note that the convergence is not completely monotonic. This results from the fact that the mesh density in Z-direction was kept constant. This choice would not make a difference in the isotropic case, however, due to anisotropic material behavior the cantilever deforms in all directions, and thus, the discretization also in Z-direction plays a role.

Table 3 Convergence of the end displacements using the T2-P1 element for an anisotropic material

Mesh division	d.o.f	u_y	u_{zu}	u_{zb}
$2 \times 2 \times 4$	537	19,859	−1,1639	−7,6436
$4 \times 4 \times 4$	1877	21,684	−0,3660	−8,5128
$8 \times 8 \times 4$	6981	21,966	−0,2167	−8,8043
$16 \times 16 \times 4$	26885	22,009	−0,2303	−8,8332
$32 \times 32 \times 4$	105477	22,021	−0,2326	−8,8186

3 Conclusions

Finite element development has to be based on rigorous continuum mechanics. This is necessary in order to treat finite strain cases correctly and to be able to investigate the best formulations for efficient and robust numerical solution schemes. With a general automated approach, like the methodology provided in *AceGen* it is possible to explore many different but theoretically equivalent continuum formulations in order to find the most efficient discretization scheme.

This contribution was focussed on both aspects. Due to limited space not all continuum formulations were discussed in the light of the associated finite element formulations. However several—even non standard aspects—like the introduction of pseudo potential formulations were discussed in the light of efficient and simple formulations for the development of finite elements.

This contribution has focused on elastic materials, but the same ideas can also be applied to inelastic material response. An in depth discussion of the related theoretical and numerical framework would have been too extensive and thus could not be covered. This is left for future discussions.

References

Altenbach, J., & Altenbach, H. (1994). *Einführung in die Kontinuumsmechanik*. Stuttgart: Teubner-Verlag.

Arnold, D. N., Brezzi, F., & Douglas, J. (1984). Peers: A new mixed finite element for plane elasticity. *Japan Journal of Applied Mathematics*, *1*, 347–367.

Auricchio, F., da Velga, L. B., Lovadina, C., Reali, A., Taylor, R. L., & Wriggers, P. (2013). Approximation of incompressible large deformation elastic problems: some unresolved issues. *Computational Mechanics*, *52*, 1153–1167.

Becker, E., & Bürger, W. (1975). *Kontinuumsmechanik*. Stuttgart: B.G. Teubner.

Belytschko, T., Ong, J. S. J., Liu, W. K., & Kennedy, J. M. (1984). Hourglass control in linear and nonlinear problems. *Computer Methods in Applied Mechanics and Engineering*, *43*, 251–276.

Braess, D. (1992). *Finite elemente*. Berlin: Springer.

Brezzi, F., & Fortin, M. (1991). *Mixed and hybrid finite element methods*. Berlin: Springer.

Chadwick, P. (1999). *Continuum mechanics, Concise theory and problems*. Mineola: Dover Publications.

Chapelle, D., & Bathe, K. J. (1993). The inf-sup test. *Computers and Structures*, *47*, 537–545.
Ciarlet, P. G. (1988). *Mathematical elasticity I: Three-dimensional elasticity*. Amsterdam: North-Holland.
Cottrell, J. A., Hughes, T. J. R., & Bazilevs, Y. (2009). *Isogeometric analysis: Toward integration of CAD and FEA*. New York: Wiley.
Duffet, G., & Reddy, B. D. (1983). The analysis of incompressible hyperelastic bodies by the finite element method. *Computer Methods in Applied Mechanics and Engineering*, *41*, 105–120.
Eringen, A. (1967). *Mechanics of Continua*. New York: Wiley.
Flory, P. (1961). Thermodynamic relations for high elastic materials. *Transactions of the Faraday Society*, *57*, 829–838.
Fraeijs de Veubeke, B. M. (1975). Stress function approach. In *World Congress on the Finite Element Method in Structural Mechanics* (pp. 1–51). Bournmouth.
Häggblad, B., & Sundberg, J. A. (1983). Large strain solutions of rubber components. *Computers and Structures*, *17*, 835–843.
Holzapfel, G. A. (2000). *Nonlinear solid mechanics*. Chichester: Wiley.
Hughes, T. R. J. (1987). *The finite element method*. Englewood Cliffs: Prentice Hall.
Korelc, J. (1997). Automatic generation of finite-element code by simultaneous optimization of expressions. *Theoretical Computer Science*, *187*, 231–248.
Korelc, J. (2002). Multi-language and multi-environment generation of nonlinear finite element codes. *Engineering with Computers*, *18*, 312–327.
Korelc, J. (2011). *AceGen and AceFEM user manual.Technical report*. University of Ljubljana. http://www.fgg.uni-lj.si/symech/.
Malvern, L. E. (1969). *Introduction to the mechanics of a continuous medium*. Englewood Cliffs: Prentice-Hall.
Marsden, J. E., & Hughes, T. J. R. (1983). *Mathematical foundations of elasticity*. Englewood Cliffs: Prentice-Hall.
Mathematica. (2011). http://www.wolfram.com.
Oden, J. T., & Key, J. E. (1970). Numerical analysis of finite axisymmetrical deformations of incompressible elastic solids of revolution. *International Journal of Solids and Structures*, *6*, 497–518.
Ogden, R. W. (1984). *Non-linear elastic deformations*. Chichester: Ellis Horwood and Wiley.
Reese, S. (2005). On a physically stabilized one point finite element formulation for three-dimensional finite elasto-plasticity. *Computer Methods in Applied Mechanics and Engineering*, *194*, 4685–4715.
Reese, S., & Wriggers, P. (2000). A new stabilization concept for finite elements in large deformation problems. *International Journal for Numerical Methods in Engineering*, *48*, 79–110.
Simo, J. C., & Armero, F. (1992). Geometrically non-linear enhanced strain mixed methods and the method of incompatible modes. *International Journal for Numerical Methods in Engineering*, *33*, 1413–1449.
Simo, J. C., Armero, F., & Taylor, R. L. (1993). Improved versions of assumed enhanced strain tri-linear elements for 3D finite deformation problems. *Computer Methods in Applied Mechanics and Engineering*, *110*, 359–386.
Simo, J. C., & Rifai, M. S. (1990). A class of assumed strain methods and the method of incompatible modes. *International Journal for Numerical Methods in Engineering*, *29*, 1595–1638.
Simo, J. C., Taylor, R. L., & Pister, K. S. (1985). Variational and projection methods for the volume constraint in finite deformation elasto-plasticity. *Computer Methods in Applied Mechanics and Engineering*, *51*, 177–208.
Stenberg, R. (1988). A family of mixed finite elements for elasticity problems. *Numerische Mathematik*, *48*, 513–538.
Sussman, T., & Bathe, K. J. (1987). A finite element formulation for nonlinear incompressible elastic and inelastic analysis. *Computers and Structures*, *26*, 357–409.
Taylor, R. L. (2011). *FEAP: A finite element analysis program*. Technical report. http://www.ce.berkeley.edu/feap.

Truesdell, C., & Noll, W. (1965). The nonlinear field theories of mechanics. In S. Flügge (Ed.), *Handbuch der Physik III/3*. Berlin: Springer.

Truesdell, C., & Toupin, R. (1960). The classical field theorie. *Handbuch der Physik III/1*. Berlin: Springer.

Washizu, K. (1975). *Variational methods in elasticity and plasticity* (2nd ed.). Oxford: Pergamon Press.

Wriggers, P. (2008). *Nonlinear finite elements*. Berlin: Springer.

Zienkiewicz, O. C., & Taylor, R. L. (1989). *The finite element method* (4th ed., Vol. 1). London: McGraw Hill.

Zienkiewicz, O. C., & Taylor, R. L. (2000). *The finite element method* (5th ed., Vol. 2). Oxford: Butterworth-Heinemann.

Zienkiewicz, O. C., Taylor, R. L., & Too, J. M. (1971). Reduced integration technique in general analysis of plates and shells. *International Journal for Numerical Methods in Engineering*, *3*, 275–290.

Three-Field Mixed Finite Element Methods in Elasticity

Batmanathan Dayanand Reddy

Abstract This chapter comprises a unified account of three-field mixed formulations for problems in elasticity. Various well-known formulations such as mixed enhanced strains and enhanced assumed strains are shown to be special cases of the general formulation. Conditions for locking-free convergence of finite element approximations are established for the linear problem. The linearized incremental problem arising from an application of Newton's method is analyzed along similar lines, and conditions for locking-free behaviour and uniform convergence established. Numerical examples illustrate the performance of the mixed formulations.

1 Introduction

Mixed formulations of boundary value problems are popular for a variety of reasons. For example, in the context of continuum mechanics, these occur in a natural way when formulations include as the primary unknown variables not only the displacement but also the stress and perhaps the strain. In the context of finite element approximations, mixed approaches constitute an effective means of circumventing the problems associated with volumetric or shear locking: standard displacement-based approaches with the use of low-order elements generally lead to poor approximations for near-incompressible or incompressible behaviour, and for very thin bodies such as beams, plates and shells.

The objective of this chapter is to present a unified and general account of three-field mixed formulations for elastic problems, together with the conditions under which finite element approximations of the general approach and various special cases are convergent. Particular well-known formulations are shown to be special cases of the general theory. The first part of the development will take place in the

B.D. Reddy (✉)
Centre for Research in Computational and Applied Mechanics, University of Cape Town, Rondebosch, South Africa
e-mail: daya.reddy@uct.ac.za

J. Schröder and P. Wriggers (eds.), *Advanced Finite Element Technologies*,
CISM International Centre for Mechanical Sciences 566,
DOI 10.1007/978-3-319-31925-4_3

context of small strains and linear elasticity and draws on the works Djoko et al. (2006), Lamichhane et al. (2006).

There has also been extensive activity related to the development of stable, locking-free methods for nonlinear elasticity. Prominent among these methods have been enhanced strains: the extension to nonlinear problems was carried out in Simo and Armero (1992), Simo et al. (1993), with further extensions to avoid numerical instabilities particularly for problems of extreme compression (see for example Mueller-Hoeppe et al. 2009, Wriggers and Reese 1996). Mixed approaches have been the subject of the studies in Chavan et al. (2007), Kasper and Taylor (2000b). More recently, the issue of numerical stability for the enhanced strain method has been approached from a different direction, viz., that of adding stabilization or penalty terms which would also serve to avoid locking in the incompressible limit (Ten Eyck and Lew 2010).

The above works deal with the construction of numerically stable schemes. The issue of physical instabilities relate to the need to identify bifurcation points, from which two or more solutions are possible, or limit points. The computational challenge is then one of the constructing solutions for the primary and secondary paths: see for example Duffett and Reddy (1986). Two-dimensional formulations have been studied by Auricchio et al. (2005), Auricchio et al. (2013) to determine stability ranges and in the latter work to examine their behaviour in the region of bifurcation and limit points. Further details and discussion may be found, for example, in Wriggers (2008).

The rest of the chapter is organized as follows. Section 2 introduces the linear problem and reviews the phenomenon of locking as well as the method of enhanced assumed strains as one particular remedy. Section 3 is devoted to the formulation of the three-field problem in general terms, while finite element approximations are discussed in Sect. 4. The methods of mixed enhanced strains and enhanced assumed strains are shown to be special cases of the general formulation. Conditions for well-posedness and locking-free behaviour of solutions are presented. Section 5 is devoted to an extension of the treatment in the earlier sections to the linearized problem associated with nonlinear elasticity, where analogous conditions for well-posedness are presented.

2 Governing Equations

Consider an elastic body with domain $\Omega \subset \mathbb{R}^d$ $(d = 2, 3)$ and boundary Γ with unit outward normal $\boldsymbol{n}$. The boundary is assumed to comprise nonoverlapping parts Γ_u and Γ_t, where $\Gamma_u \neq \emptyset$. The governing equations of the equilibrium problem for a linear isotropic elastic body are as follows:

Equilibrium:	$-\operatorname{div}\boldsymbol{\sigma} = \boldsymbol{f},$	(1a)
Elasticity relation:	$\boldsymbol{\sigma} = \mathbb{C}\boldsymbol{\varepsilon},$	(1b)
Strain-displacement relation:	$\boldsymbol{\varepsilon}(\boldsymbol{u}) = \nabla_s \boldsymbol{u} := \frac{1}{2}(\nabla \boldsymbol{u} + [\nabla \boldsymbol{u}]^T),$	(1c)
Boundary conditions:	$\boldsymbol{u} = \boldsymbol{0}$ on Γ_u, $\quad \boldsymbol{\sigma n} = \bar{\boldsymbol{t}}$ on Γ_t.	(1d)

Here, $\boldsymbol{u}$ is the displacement vector, $\boldsymbol{\sigma}$ the Cauchy stress, $\boldsymbol{\varepsilon}$ the linearized strain tensor, $\boldsymbol{f}$ the body force, and $\bar{\boldsymbol{t}}$ the prescribed traction. We use the notation $\nabla_s \boldsymbol{u}$ rather than $\boldsymbol{\varepsilon}(\boldsymbol{u})$ for the expression of the strain in terms of the displacement, to avoid confusion with the strain as an independent variable $\boldsymbol{\varepsilon}$. The elasticity tensor $\mathbb{C}$ is given by

$$\mathbb{C}\boldsymbol{\varepsilon} = \lambda(\operatorname{tr}\boldsymbol{\varepsilon})\boldsymbol{I} + 2\mu\boldsymbol{\varepsilon} \tag{2}$$

in which λ and μ are the Lamé constants, given in terms of Young's modulus E and Poisson's ratio ν by $\lambda = E\nu/(1+\nu)(1-2\nu)$ and $\mu = E/2(1+\nu)$.

The weak formulation We define the space $V = \{\boldsymbol{v} \mid v_i \in H^1(\Omega),\ \boldsymbol{v} = \boldsymbol{0} \text{ on } \Gamma_u\}$ of admissible displacements. Then, the standard weak formulation for the displacement problem is as follows: find $\boldsymbol{u} \in V$ that satisfies

$$\int_\Omega \mathbb{C}\nabla_s \boldsymbol{u} : \nabla_s \boldsymbol{v}\, dx = \int_\Omega \boldsymbol{f} \cdot \boldsymbol{v}\, dx + \int_{\Gamma_t} \bar{\boldsymbol{t}} \cdot \boldsymbol{v}\, dx \quad \text{for all } \boldsymbol{v} \in V. \tag{3}$$

The weak formulation may be written more compactly by defining the bilinear form $a : V \times V \to \mathbb{R}$ and linear functional $\ell : V \to \mathbb{R}$ by

$$a(\boldsymbol{u}, \boldsymbol{v}) = \int_\Omega \mathbb{C}\nabla_s \boldsymbol{u} : \nabla_s \boldsymbol{v}\, dx, \quad \ell(\boldsymbol{v}) = \int_\Omega \boldsymbol{f} \cdot \boldsymbol{v}\, dx + \int_{\Gamma_t} \bar{\boldsymbol{t}} \cdot \boldsymbol{v}\, dx; \tag{4}$$

then the problem becomes one of finding $\boldsymbol{u} \in V$ that satisfies

$$a(\boldsymbol{u}, \boldsymbol{v}) = \ell(\boldsymbol{v}) \quad \text{for all } \boldsymbol{v} \in V. \tag{5}$$

The weak formulation is equivalent to the principle of minimum potential energy

$$\boldsymbol{u} = \operatorname{argmin}_{v \in V} J(\boldsymbol{v}) := \frac{1}{2}\int_\Omega \nabla_s \boldsymbol{v} : \mathbb{C}\nabla_s \boldsymbol{v}\, dx - \int_\Omega \boldsymbol{f} \cdot \boldsymbol{v}\, dx - \int_{\Gamma_t} \bar{\boldsymbol{t}} \cdot \boldsymbol{v}\, dx.$$

Finite element approximations We will be interested primarily in low-order approximations on quadrilaterals and hexahedra, and with this in mind introduce the finite-dimensional subspace V^h of continuous, piecewise-bilinear (trilinear in three dimensions) functions, which will be denoted by Q_1 henceforth. We denote by $\mathcal{T}_h$ a regular quadrilateral (resp., hexahedral) triangulation of Ω, which we assume for convenience to be polygonal (resp., polyhedral). An arbitrary element $K \in \mathcal{T}_h$ is generated by an isoparametric map from reference element $\hat{K} = (-1, 1)^d$. The diameter of K is denoted by h_K and the mesh size of the triangulation defined by $h = \max_K h_K$.

The finite element approximation of problem (5) is then as follows: find $\boldsymbol{u}_h \in V^h$ that satisfies

$$a(\boldsymbol{u}_h, \boldsymbol{v}_h) = \ell(\boldsymbol{v}_h) \quad \text{for all } \boldsymbol{v}_h \in V^h. \tag{6}$$

Locking The discrete problem has a unique solution $\boldsymbol{u}_h \in V^h$. A standard derivation of the estimate of the error shows, however, that

$$\|\boldsymbol{u} - \boldsymbol{u}_h\|_V \leq C(\lambda)h;$$

that is, the constant depends on λ which becomes unbounded in the incompressible limit. From a practical point of view one finds that the discrete problem leads to a poor approximation in the incompressible limit: this is the phenomenon of locking, which is illustrated in the simple example in Fig. 1: the bottom curve shows the result using the Q_1 element. A variety of approaches have been developed to circumvent the problem of locking while retaining the use of low-order elements. One such approach is the method of enhanced assumed strains (Simo and Rifai 1990). In this formulation, we introduce as additional unknowns the stress $\boldsymbol{\sigma}_h$ and an enhanced strain $\widetilde{\boldsymbol{\varepsilon}}_h$, and replace (1c) by

$$\boldsymbol{\varepsilon}_h = \nabla_s \boldsymbol{u}_h + \widetilde{\boldsymbol{\varepsilon}}_h. \tag{7}$$

We define discrete spaces S^h and $\widetilde{E}^h$ of stresses and enhanced strains. We then have set of three equations in weak form: that is,

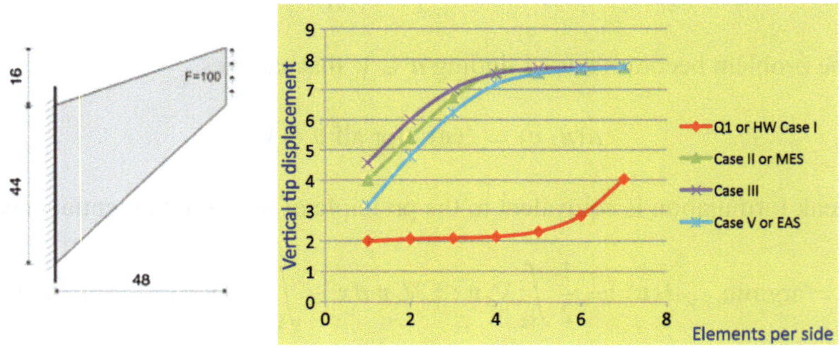

Fig. 1 The Cook problem: the *bottom curve* shows the locking behaviour associated with the Q_1 element; the other *three curves* show the results for mixed enhanced strains, enhanced assumed strains, and a third stable formulation.

$$\text{Equilibrium:} \qquad \int_\Omega \boldsymbol{\sigma}_h : \nabla_s \boldsymbol{v}_h \, dx = \ell(\boldsymbol{v}_h), \tag{8a}$$

$$\text{Elasticity relation:} \qquad \int_\Omega [\boldsymbol{\sigma}_h - \mathbb{C}\boldsymbol{\varepsilon}_h] : \widetilde{\boldsymbol{e}}_h \, dx = 0, \tag{8b}$$

$$\text{Strain-displacement relation:} \qquad \int_\Omega [\boldsymbol{\varepsilon}_h - (\nabla_s \boldsymbol{u}_h + \widetilde{\boldsymbol{\varepsilon}}_h)] : \boldsymbol{\tau}_h \, dx = 0, \tag{8c}$$

for $\boldsymbol{v}_h \in V^h$, $\widetilde{\boldsymbol{e}}_h \in \widetilde{E}^h$, and $\boldsymbol{\tau}_h \in S^h$. Now, we impose the condition that the stresses and enhanced strains be mutually L^2-orthogonal: $S^h \perp \widetilde{E}^h$, or $(\boldsymbol{\tau}_h, \widetilde{\boldsymbol{e}}_h)_{L^2} = 0$. The terms involving products of stresses and enhanced strains in (8) fall out, and we may combine the remaining equations into a single equation:

$$\int_\Omega \mathbb{C}(\nabla_s \boldsymbol{u}_h + \widetilde{\boldsymbol{\varepsilon}}_h) : (\nabla_s \boldsymbol{v}_h + \widetilde{\boldsymbol{e}}_h) \, dx = \ell(\boldsymbol{v}_h), \quad \boldsymbol{v}_h \in V^h, \ \widetilde{\boldsymbol{e}}_h \in \widetilde{E}^h. \tag{9}$$

This is the method of enhanced assumed strains (EAS). It will be seen later that this formulation may be recovered as a special case in a general three-field formulation.

3 A General Three-Field Formulation

We formulate the general problem in terms of three variables: the displacement $\boldsymbol{u}$, the strain $\boldsymbol{\varepsilon}$, and the stress $\boldsymbol{\sigma}$. We define spaces of stresses S and strains E by $S = E = \{\boldsymbol{e} \mid e_{ji} = e_{ij}, \ e_{ij} \in L^2(\Omega)\}$; the space of displacements V has been previously defined. We also set $W = V \times E$. Next, take the inner product, respectively, of (1a), (1b) and (1c) with arbitrary members $\boldsymbol{v} \in V$, $\boldsymbol{\tau} \in S$ and $\boldsymbol{e} \in E$, and carry out integration by parts on the equilibrium equation only, as before. By combining appropriately this gives the set of weak equations

$$\int_\Omega \mathbb{C}\boldsymbol{\varepsilon} : \boldsymbol{e} \, dx + \int_\Omega [\nabla_s \boldsymbol{v} - \boldsymbol{e}] : \boldsymbol{\sigma} \, dx = \int_\Omega \boldsymbol{f} \cdot \boldsymbol{v} \, dx, \tag{10a}$$

$$\int_\Omega [\nabla_s \boldsymbol{u} - \boldsymbol{\varepsilon}] : \boldsymbol{\tau} \, dx = 0. \tag{10b}$$

This set of equations may be packaged in more compact form by defining the bilinear forms

$$\overline{a} : W \times W \to \mathbb{R}, \quad \overline{a}((\boldsymbol{u}, \boldsymbol{\varepsilon}); (\boldsymbol{v}, \boldsymbol{e})) = \int_\Omega \mathbb{C}\boldsymbol{\varepsilon} : \boldsymbol{e} \, dx, \tag{11a}$$

$$\overline{b} : W \times S \to \mathbb{R}, \quad \overline{b}((\boldsymbol{u}, \boldsymbol{\varepsilon}), \boldsymbol{\tau}) = \int_\Omega (\nabla_s \boldsymbol{u} - \boldsymbol{\varepsilon}) : \boldsymbol{\tau} \, dx. \tag{11b}$$

Then the weak formulation (10) may be written as follows: find $(\boldsymbol{u}, \boldsymbol{\varepsilon}) \in V \times E$ and $\boldsymbol{\sigma} \in S$ that satisfy

$$\overline{a}((\boldsymbol{u},\boldsymbol{\varepsilon});(\boldsymbol{v},\boldsymbol{e}))+\overline{b}((\boldsymbol{v},\boldsymbol{e}),\boldsymbol{\sigma})=\ell(\boldsymbol{v})\quad\text{for all }(\boldsymbol{v},\boldsymbol{e})\in V\times E, \tag{12a}$$

$$\overline{b}((\boldsymbol{u},\boldsymbol{\varepsilon}),\boldsymbol{\tau})=0\quad\text{for all }\boldsymbol{\tau}\in S. \tag{12b}$$

This is in the standard form for mixed variational problems. Furthermore, it can be shown that the problem (12) is equivalent to the minimax problem

$$\min_{(\boldsymbol{v},\boldsymbol{e})\in W}\max_{\boldsymbol{\tau}\in S} H((\boldsymbol{v},\boldsymbol{e}),\boldsymbol{\tau})=\int_\Omega\left[\tfrac{1}{2}\boldsymbol{e}:\mathbb{C}\boldsymbol{e}+(\boldsymbol{e}-\nabla_s\boldsymbol{v}):\boldsymbol{\tau}-\boldsymbol{f}\cdot\boldsymbol{v}\right]dx. \tag{13}$$

This is the de Veubeke–Hu–Washizu formulation (Fraeijs de Veubeke 1951; Hu 1955; Washizu 1955). Unfortunately a direct approach to the mixed formulation does not allow us to demonstrate λ-independence of finite element approximations. To be able to carry out such an analysis, we modify the three-field formulation in a consistent way as follows.

Returning to the functional H defined in (13), we assume that the test functions $\boldsymbol{e}\in E$ and $\boldsymbol{\tau}\in S$ satisfy the volumetric part of the elasticity relation, that is, $\operatorname{tr}\boldsymbol{\tau}=\kappa\operatorname{tr}\boldsymbol{e}$ where $\kappa=3\lambda+2\mu$ is ($3\times$) the bulk modulus. Exploiting this constraint, we arrive at the modified formulation: find $\boldsymbol{u}\in V$, $\boldsymbol{\varepsilon}\in E$ and $\boldsymbol{\sigma}\in S$ that satisfy

$$\int_\Omega 2\mu\boldsymbol{\varepsilon}:\boldsymbol{e}\,dx+\int_\Omega\left[\nabla_s\boldsymbol{v}-\boldsymbol{e}+\frac{\lambda}{\kappa}(\operatorname{tr}\boldsymbol{e})\boldsymbol{I}\right]:\boldsymbol{\sigma}\,dx=\int_\Omega\boldsymbol{f}\cdot\boldsymbol{v}\,dx, \tag{14a}$$

$$-\int_\Omega\frac{\lambda}{2\kappa^2}(\operatorname{tr}\boldsymbol{\sigma})(\operatorname{tr}\boldsymbol{\tau})+\int_\Omega\left[\nabla_s\boldsymbol{u}-\boldsymbol{\varepsilon}+\frac{\lambda}{\kappa}(\operatorname{tr}\boldsymbol{\varepsilon})\boldsymbol{I}\right]:\boldsymbol{\tau}\,dx=0, \tag{14b}$$

for all $\boldsymbol{v}\in V$, $\boldsymbol{e}\in E$ and $\boldsymbol{\tau}\in S$. The problem can be cast in the usual format for mixed formulations by defining the bilinear forms

$$a:W\times W\to\mathbb{R},\quad a((\boldsymbol{u},\boldsymbol{\varepsilon});(\boldsymbol{v},\boldsymbol{e}))=\int_\Omega 2\mu\boldsymbol{\varepsilon}:\boldsymbol{e}\,dx, \tag{15a}$$

$$b:W\times S\to\mathbb{R},\quad b((\boldsymbol{u},\boldsymbol{\varepsilon}),\boldsymbol{\tau})=\int_\Omega\left(\nabla_s\boldsymbol{u}-\boldsymbol{\varepsilon}+\frac{\lambda}{\kappa}(\operatorname{tr}\boldsymbol{e})\boldsymbol{I}\right):\boldsymbol{\tau}\,dx, \tag{15b}$$

$$c:S\times S\to\mathbb{R},\quad c(\boldsymbol{\sigma},\boldsymbol{\tau})=\int_\Omega(\operatorname{tr}\boldsymbol{\sigma})(\operatorname{tr}\boldsymbol{\tau})\,dx. \tag{15c}$$

Then the problem (14) may be written *equivalently* in the form

$$a((\boldsymbol{u},\boldsymbol{\varepsilon});(\boldsymbol{v},\boldsymbol{e}))+b((\boldsymbol{v},\boldsymbol{e}),\boldsymbol{\sigma})=\ell(\boldsymbol{v})\quad\text{for all }(\boldsymbol{v},\boldsymbol{e})\in W, \tag{16a}$$

$$-\frac{\lambda}{\kappa^2}c(\boldsymbol{\sigma},\boldsymbol{\tau})+b((\boldsymbol{u},\boldsymbol{\varepsilon}),\boldsymbol{\tau})=0\quad\text{for all }\boldsymbol{\tau}\in S. \tag{16b}$$

This problem can be shown to have a unique solution with a constant that is *independent of* λ.

4 Finite Element Formulations of the Mixed Problems

We define finite element subspaces $V^h \subset V$, $E^h \subset E$ and $S^h \subset S$. From (12) and (16), the discrete versions of the standard and modified three-field formulations are then as follows: find $(\boldsymbol{u}_h, \boldsymbol{\varepsilon}_h) \in V^h \times E^h := W^h$ and $\boldsymbol{\sigma}_h \in S^h$ that satisfy, for $(\boldsymbol{v}_h, \boldsymbol{e}_h) \in W^h$ and $\boldsymbol{\tau}_h \in S^h$,

$$\overline{a}((\boldsymbol{u}_h, \boldsymbol{\varepsilon}_h); (\boldsymbol{v}_h, \boldsymbol{e}_h)) + \overline{b}((\boldsymbol{v}_h, \boldsymbol{e}_h), \boldsymbol{\sigma}_h) = \ell(\boldsymbol{v}_h), \tag{17a}$$

$$\overline{b}((\boldsymbol{u}_h, \boldsymbol{\varepsilon}_h), \boldsymbol{\tau}_h) = 0; \tag{17b}$$

or, for the modified problem,

$$a((\boldsymbol{u}_h, \boldsymbol{\varepsilon}_h); (\boldsymbol{v}_h, \boldsymbol{e}_h)) + b((\boldsymbol{v}_h, \boldsymbol{e}_h), \boldsymbol{\sigma}_h) = \ell(\boldsymbol{v}_h), \tag{18a}$$

$$-\frac{\lambda}{\kappa^2} c(\boldsymbol{\sigma}_h, \boldsymbol{\tau}_h) + b((\boldsymbol{u}_h, \boldsymbol{\varepsilon}_h), \boldsymbol{\tau}_h) = 0. \tag{18b}$$

Conditions for equivalence of the formulations The standard and modified three-field formulations are of course equivalent in the continuous case. In the discrete case, however, the equivalence is not automatic and will depend on the choice of finite element spaces. It has been shown by Lamichhane et al. (2006) that problems (17) and (18) are *equivalent provided that*

$$S^h \subseteq E^h \quad \text{and} \quad (\operatorname{tr} \boldsymbol{\varepsilon}_h)\boldsymbol{I} \in E^h. \tag{19}$$

If all components of the strain $\boldsymbol{\varepsilon}_h$ are spanned by complete polynomials then condition $(19)_2$ holds. However, take for example the case in which $\varepsilon_{11} = a + bx$ and $\varepsilon_{22} = c + dy$ and $\varepsilon_{12} = 0$. Then $(\operatorname{tr} \boldsymbol{\varepsilon}_h) = \lambda(a + c + bx + dy) + 2\mu(a + bx)$ and $(\operatorname{tr} \boldsymbol{\varepsilon}_h)$ is spanned by complete polynomials of degree 1, unlike ε_{11}.

From the general formulations (17) and (18) various well-known mixed formulations can be extracted as special cases. For example, if the spaces S^h and E^h are chosen such that conditions (19) are satisfied then, first, we know that the standard and modified formulations are equivalent. Furthermore, we can eliminate $\boldsymbol{\varepsilon}_h$ to obtain the *Hellinger-Reissner* formulation

$$\int_\Omega [\mathbb{C}^{-1}\boldsymbol{\sigma}_h - \nabla_s \boldsymbol{u}_h] : \boldsymbol{\tau}_h = 0, \tag{20a}$$

$$\int_\Omega \boldsymbol{\sigma}_h : \nabla_s \boldsymbol{v}_h \, dx = \int_\Omega \boldsymbol{f} \cdot \boldsymbol{v}_h \, dx. \tag{20b}$$

The method of mixed enhanced strains due to Kasper and Taylor (2000a) may be obtained from the general three-field formlulation as follows. We choose E^h and S^h such that $E^h = S^h \oplus \widetilde{E}^h$ with $\widetilde{E}^h \perp S^h$, where $\perp$ denotes L^2-orthogonality. Thus, the strain comprises a component in the space of stresses plus an enhanced strain orthogonal to the stresses. In Kasper and Taylor (2000a), the following choices have been made on the reference element:

$$\sigma_{h,11} = a_0 + a_1 y, \quad \sigma_{h,22} = b_0 + b_1 x, \quad \sigma_{h,12} = c, \tag{21a}$$

$$\widetilde{\varepsilon}_{h,11} = d_0 + d_1 x, \quad \widetilde{\varepsilon}_{h,22} = e_0 + e_1 y, \quad \widetilde{\varepsilon}_{h,12} = 0. \tag{21b}$$

Thus, all components of E^h comprise complete polynomials of degree 1 so that the conditions for the modified and standard formulations to be equivalent are satisfied.

Furthermore, the problem may be written as a displacement formulation by eliminating the stress and strain from the mixed formulation. First, recall the definition of the L^2-orthogonal projection P onto the space of stresses:

$$\int_\Omega \left(P\varepsilon_h - \varepsilon_h\right) : \boldsymbol{\tau}_h \, dx = 0 \quad \text{for all } \boldsymbol{\tau}_h \in S^h. \tag{22}$$

In practical terms, we write $\boldsymbol{\varepsilon}_h = \boldsymbol{B}\boldsymbol{\alpha}$ and $\boldsymbol{\tau}_h = \boldsymbol{G}\boldsymbol{\beta}$, with $\boldsymbol{\alpha}$ and $\boldsymbol{\beta}$ the local degrees of freedom of $\boldsymbol{\varepsilon}_h$ and $\boldsymbol{\tau}_h$, respectively, and $\boldsymbol{B}$ and $\boldsymbol{G}$ the respective matrices of basis functions. Then, $P\boldsymbol{\varepsilon}_h = \boldsymbol{G}\boldsymbol{\gamma}$ for some $\boldsymbol{\gamma}$ which is found from (22). This process, applied to the weak formulation, allows us to eliminate the stress and to write the problem in the symmetric form

$$\int_\Omega \mathbb{C}[P\nabla_s \boldsymbol{u}_h + \widetilde{\varepsilon}_h] : (P\nabla_s \boldsymbol{v}_h + \widetilde{e}_h) \, dx = \int_\Omega \boldsymbol{f} \cdot \boldsymbol{v} \, dx. \tag{23}$$

Furthermore, as described in Kasper and Taylor (2000a), the enhanced strain may be condensed out at element level. The details are omitted.

The method of enhanced assumed strains (EAS) follows from the choice $E^h = \nabla_s(V^h) + \widetilde{E}^h$ with $\widetilde{E}^h \perp S^h$: that is, we write the strain in the form $\varepsilon_h = \nabla_s \boldsymbol{u}_h + \widetilde{\varepsilon}_h$ with the enhanced strains being orthogonal to the space of stresses. Then, (10b) is identically satisfied and from (10a) we recover the formulation (9).

Figure 1 shows results obtained using inter alia enhanced assumed strains and mixed enhanced strains. Further results may be found in Djoko et al. (2006).

4.1 Well-Posedness

The discrete mixed problem, or its modified counterpart, assuming that the conditions for equivalence are met, has a unique solution provided that the following two conditions are satisfied (Lamichhane et al. 2006):

(a) Ellipticity or coerciveness: there exists a constant $c_0 > 0$ such that $\|P\nabla_s \boldsymbol{v}_h\|_S \geq c_0 \|\nabla_s \boldsymbol{v}_h\|$ for all $\boldsymbol{v}_h \in V^h$;
(b) Inf-sup condition: given $\boldsymbol{\tau}_h \in S^h$ construct a corresponding discrete pressure p_h such that $p_h \boldsymbol{I}$ belongs to S^h. Then the pair $(\boldsymbol{v}_h, p_h)$ must be stable in relation to the incompressibility condition div $\boldsymbol{v} = 0$; that is,

$$\inf_{q_h}\sup_{v_h} \frac{\int_\Omega p_h \operatorname{div} \boldsymbol{v}_h \, dx}{\|\boldsymbol{v}_h\|_V \|q_h\|_{L^2}} \geq \beta_h > 0. \tag{24}$$

Furthermore, the solution converges uniformly with respect to λ: that is, there exists a constant C independent of h and of λ such that

$$\|\boldsymbol{u} - \boldsymbol{u}_h\|_V + \|\boldsymbol{\varepsilon} - \boldsymbol{\varepsilon}_h\|_E + \|\boldsymbol{\sigma} - \boldsymbol{\sigma}_h\|_S \leq Ch.$$

The ellipticity condition is readily checked for a range of choices of discrete spaces S^h, and shown to be satisfied. The discrete inf-sup condition requires more effort though. First, we have to define the discrete pressure. To see the need for this, note that if we decompose $\boldsymbol{\sigma}_h \in S^h$ into its spherical and deviatoric components $\operatorname{sph} \boldsymbol{\sigma}_h = \frac{1}{d}(\operatorname{tr} \boldsymbol{\sigma}_h)\boldsymbol{I}$ and $\operatorname{dev}\boldsymbol{\sigma}_h = \boldsymbol{\sigma}_h - \operatorname{sph} \boldsymbol{\sigma}_h$, then it may be that $\operatorname{sph} \boldsymbol{\sigma}_h$ and $\operatorname{dev} \boldsymbol{\sigma}_h$ do not belong to S^h! The way around this dilemma is to define the discrete deviatoric and spherical parts of $\boldsymbol{\sigma}_h$ by

$$\operatorname{dev}_h\boldsymbol{\sigma}_h = P[\operatorname{dev} \boldsymbol{\sigma}_h], \quad \operatorname{sph}_h \boldsymbol{\sigma}_h = \boldsymbol{\sigma}_h - \operatorname{dev}_h\boldsymbol{\sigma}_h = p_h \boldsymbol{I}. \tag{25}$$

These belong to S^h.

Consider the case in which the basis for S^h on the reference element is given by (21a). The spherical part is given by $\operatorname{sph} \boldsymbol{\sigma}_h = \frac{1}{2}[(a_0 + b_0) + a_1 y + b_1 x]\boldsymbol{I} \notin S^h$; likewise, the deviatoric part also does not belong to S^h. Using the definition $(25)_1$, on the other hand, we find that $\operatorname{sph}_h \boldsymbol{\sigma}_h = \frac{1}{2}(a_0 + b_0)\boldsymbol{I}$ which belongs to S^h. It follows that $\operatorname{dev}_h \boldsymbol{\sigma}_h \in S^h$ as well.

It can be shown (Lamichhane et al. 2006) that for the well-known methods including mixed enhanced strains and enhanced assumed strains, the discrete pressures p_h correspond to piecewise-constants; so (24) amounts to an inf-sup condition for the $Q_1 - P_0$ pair. The properties of this pair are well known: in particular, it does not satisfy a uniform inf-sup condition and it is necessary to extract or filter out a checkerboard mode. This does not, however, affect the displacement, which is uniformly convergent.

5 Extension to Problems of Nonlinear Elasticity

In this section, we draw on the analysis presented earlier to formulate and analyze the incremental or linearized problem that has to be solved, for example, as an iteration in a Newton scheme for problems in nonlinear elasticity. The focus of this section is close in spirit to that of Ten Eyck and Lew (2010), in that we design a numerically stable finite element scheme assuming that there are no physical instabilities. The presentation in this section is based on Chama and Reddy (2015).

Consider the motion of a body $\Omega_0 \subset \mathbb{R}^d$ from its initial configuration at time t_0 to its current, deformed configuration at time t. The motion of the body is described by the map

$$x = \varphi(X, t) = X + u(X, t), \tag{26}$$

where X is the position of a material particle in Ω_0, x its current position in $\Omega = \varphi(\Omega_0, t)$, and $u(X, t)$ is the displacement. The deformation of the body is characterized by the deformation gradient

$$F = \frac{\partial \varphi(X, t)}{\partial X} = \nabla \varphi(X, t) = I + \nabla u(X, t), \tag{27}$$

where ∇ is the gradient operator in the reference configuration. In what follows, the determinant of F is denoted by J, with $J > 0$. The left and right Cauchy–Green tensors are defined, respectively, by

$$B = FF^T \quad \text{and} \quad C = F^T F. \tag{28}$$

The equilibrium equation in the reference configuration is given by

$$\text{Div}\, P + f = 0 \tag{29}$$

in which P is the first Piola–Kirchhoff stress, Div is the divergence operator in the reference configuration, and f is the body force per unit reference volume.

We confine attention to isotropic hyperelastic materials, which are defined by specification of a strain energy function $\Psi = \Psi(F)$ or $\Psi = \Psi(C)$. Here and henceforth, we denote functions and their values by the same symbol. The constitutive equation for hyperelastic materials is then given by

$$P = \frac{\partial \Psi}{\partial F}. \tag{30}$$

For example, the compressible neo-Hookean material has the strain energy function

$$\Psi(C) = \frac{\mu}{2}\left(\text{tr}\, C - 3\right) - \mu \ln J + \frac{\lambda}{2}\,(\ln J)^2, \tag{31}$$

where tr$A$ denotes the trace of a tensor A and μ and λ are the Lamé moduli.

For convenience, we assume the homogeneous Dirichlet boundary condition $u = 0$ on the boundary Γ_0 of Ω_0.

The application of Newton's method to the nonlinear system (27), (29) and (30) results in the following linearized problem: given u_{n-1}, find (u_n, F_n, P_n) that satisfy

$$-\text{Div}\, \Delta P = R_n, \tag{32a}$$

$$\Delta F = \nabla \Delta u, \tag{32b}$$

$$\Delta P = \mathbb{A} \Delta F \tag{32c}$$

together with the boundary condition $\Delta \boldsymbol{u} = \boldsymbol{0}$. Here, $\Delta(\,) := (\,)_n - (\,)_{n-1}$ denotes the difference between iterates $n-1$ and n, $\boldsymbol{R}_n$ is the residual, given by $\boldsymbol{R}_n = \boldsymbol{f}_n + \operatorname{Div} \boldsymbol{P}_{n-1}$, and

$$\mathbb{A} = \left. \frac{\partial^2 \Psi}{\partial \boldsymbol{F} \partial \boldsymbol{F}} \right|_{n-1} \tag{33}$$

is the first elasticity tensor, which has the major symmetry $\mathbb{A}_{ijkl} = \mathbb{A}_{klij}$, and whose only nonzero entries are $\mathbb{A}_{iiii}$, $\mathbb{A}_{iijj}$, $\mathbb{A}_{ijij}$, $\mathbb{A}_{ijji}$, with $i \neq j$ and no summation on repeated indices (see Chadwick and Ogden 1971).

We will be interested in mixed variational formulations of the problem (32), and with this in mind define the spaces of increments of displacements $\mathcal{V}$, deformation gradients $\mathcal{D}$ and stresses $\mathcal{S}$ by

$$\mathcal{V} = [H_0^1(\Omega)]^d, \quad \mathcal{D} = \mathcal{S} = \{\boldsymbol{Q} : \Omega \to \mathbb{R}^{d \times d} \mid Q_{ij} \in L^2(\Omega) \quad i, j = 1, \ldots, d\}.$$

It is assumed in what follows that the elasticity tensor $\mathbb{A}$ is $\mathcal{V}-$elliptic (see also Ten Eyck and Lew 2010): that is, there exists $\kappa > 0$ such that

$$\int_{\Omega_0} \nabla \boldsymbol{u} : \mathbb{A} \nabla \boldsymbol{u} \, \mathrm{d}V \geq \kappa \|\boldsymbol{u}\|_{\mathcal{V}}^2, \quad \forall \boldsymbol{u} \in \mathcal{V}. \tag{34}$$

In this way, we preclude physical instabilities or bifurcation (see Auricchio et al. 2013 for consideration of these phenomena).

Mixed three-field formulation of the linearized problem By an abuse of notation, for convenience, we henceforth denote the increments by superposed dots: that is, we write $\dot{(\,)} := \Delta(\,)$. Then, denoting the L^2-inner product by $(\cdot, \cdot)_0$, the three-field mixed formulation of the linearized system is as follows: given $\boldsymbol{u}_{n-1}$, find $(\dot{\boldsymbol{u}}, \dot{\boldsymbol{F}}, \dot{\boldsymbol{P}}) \in \mathcal{V} \times \mathcal{D} \times \mathcal{S}$ that satisfy

$$(\mathbb{A}\dot{\boldsymbol{F}} - \dot{\boldsymbol{P}}, \boldsymbol{G})_0 = 0 \qquad \forall \boldsymbol{G} \in \mathcal{D}, \tag{35a}$$

$$(\dot{\boldsymbol{F}} - \nabla \dot{\boldsymbol{u}}, \boldsymbol{Q})_0 = 0 \qquad \forall \boldsymbol{Q} \in \mathcal{S}, \tag{35b}$$

$$(\dot{\boldsymbol{P}}, \nabla \boldsymbol{v})_0 = (\boldsymbol{R}_n, \boldsymbol{G})_0 \qquad \forall \boldsymbol{v} \in \mathcal{V}. \tag{35c}$$

Define the bilinear forms $a(\cdot, \cdot)$ and $b(\cdot, \cdot)$ and the linear functional $\ell(\cdot)$ by

$$\begin{aligned} a : (\mathcal{V} \times \mathcal{D}) \times (\mathcal{V} \times \mathcal{D}) \to \mathbb{R}, &\quad a\big((\dot{\boldsymbol{F}}, \dot{\boldsymbol{u}}), (\boldsymbol{G}, \boldsymbol{v})\big) = (\mathbb{A}\dot{\boldsymbol{F}}, \boldsymbol{G})_0, \\ b : (\mathcal{V} \times \mathcal{D}) \times \mathcal{S} \to \mathbb{R}, &\quad b\big((\boldsymbol{G}, \boldsymbol{v}), \dot{\boldsymbol{P}}\big) = \big(\nabla \boldsymbol{v} - \boldsymbol{G}, \dot{\boldsymbol{P}}\big)_0, \\ \ell : \mathcal{V} \to \mathbb{R}, &\quad \ell(\boldsymbol{v}) = (\boldsymbol{R}_n, \boldsymbol{v})_0. \end{aligned}$$

Then the problem (35) is as follows: given $\boldsymbol{u}_{n-1}$, $\boldsymbol{F}_{n-1}$, $\boldsymbol{P}_{n-1}$ and $\boldsymbol{f}_n$, find $(\dot{\boldsymbol{u}}, \dot{\boldsymbol{F}}, \dot{\boldsymbol{P}}) \in \mathcal{V} \times \mathcal{D} \times \mathcal{S}$ such that

$$a\big((\dot{F}\,,\,\dot{u}),(G\,,\,v)\big)+b\big((G\,,\,v),\dot{P}\big)=\ell(v)\quad\forall\,(v\,,\,G)\in\mathcal{V}\times\mathcal{D}\,,\tag{37a}$$

$$b\big((\dot{F}\,,\,\dot{u}),Q\big)=0\qquad\forall\,Q\in\mathcal{S}.\tag{37b}$$

This can be shown to be equivalent to the saddlepoint problem of finding $\dot{u},\ \dot{F},\ \dot{P}\in\mathcal{V},\ \mathcal{D},\ \mathcal{S}$ that satisfy

$$(\dot{u},\ \dot{F},\ \dot{P})=\operatorname{argmin}\int_\Omega\left[\frac{1}{2}\dot{F}:\mathbb{A}\dot{F}+(\operatorname{Grad}\dot{u}-\dot{F}):\dot{P}-R_n\cdot\dot{u}\right]dx.$$

Modified weak formulation Assume that the elasticity tensor $\mathbb{A}$ has the form

$$\mathbb{A}=\mathbb{D}+\lambda\mathbb{E},\tag{38}$$

in which λ is the Lamé parameter, and $\mathbb{D}$ and $\mathbb{E}$ are independent of λ and inherit the symmetry properties of $\mathbb{A}$. For the neo-Hookean material (31), the nonzero components are, with respect to the principal basis, and bearing in mind the symmetries,

$$D_{111}=D_{222}=\frac{\mu}{J}(a_i^2+1),\quad D_{1212}=\frac{\mu}{J}a_1^2,\quad D_{2121}=\frac{\mu}{J}a_2^2,$$

$$E_{1111}=E_{2222}=\frac{1}{J}(1-\ln J),\quad E_{1221}=-\ln J.$$

We construct a modified form of (37) by introducing the bilinear forms

$$a_0\big((\dot{F}\,,\,\dot{u})\,,\,(G\,,\,v)\big)=\big(\mathbb{D}\dot{F}\,,\,G\big)_0,\tag{39a}$$

$$b_1\big((G\,,\,v)\,,\,\dot{P}\big)=\big(\nabla v,\dot{P}\big)_0-\big(\mathbb{D}\mathbb{A}^{-1}\dot{P}\,,\,G\big)_0,\tag{39b}$$

$$b_2\big((G\,,\,v)\,,\,\dot{P}\big)=b\big((G\,,\,v),\dot{P}\big)=(\nabla v-G,\dot{P})_0.\tag{39c}$$

Then the modified problem is: find $(\dot{u},\dot{F},\dot{P})\in\mathcal{V}\times\mathcal{D}\times\mathcal{S}$ such that

$$a_0\big((\dot{F}\,,\,\dot{u})\,,\,(G\,,\,v)\big)+b_1\big((G\,,\,v)\,,\,\dot{P}\big)=\ell(v)\quad\forall\,(v\,,\,G)\in\mathcal{V}\times\mathcal{D},\tag{40a}$$

$$b_2\big((\dot{F}\,,\,\dot{u})\,,\,Q\big)=0\qquad\forall\,Q\in\mathcal{S},\tag{40b}$$

or equivalently

$$\big(\mathbb{D}(\mathbb{A}^{-1}\dot{P}-\dot{F})\,,\,G\big)_0=0\qquad\forall G\in\mathcal{D},\tag{41a}$$

$$\big(\dot{F}-\nabla\dot{u}\,,\,Q\big)_0=0\qquad\forall Q\in\mathcal{S},\tag{41b}$$

$$\big(\nabla v\,,\,\dot{P}\big)_0=\ell(v)\qquad\forall v\in\mathcal{V}.\tag{41c}$$

The equivalence between the original and modified continuous formulations (37) and (40) is readily established.

Finite element approximations We consider finite element approximations based on shape-regular triangulations of a polygonal or polyhedral domain Ω_0, with mesh size h. A triangulation is denoted by $\mathcal{T}_h$ with K an arbitrary member of $\mathcal{T}_h$. We define $\mathbb{A}_h$ to be the piecewise-constant function equal to the mean value of $\mathbb{A}$ on K that is,

$$\mathbb{A}_h|_K = \frac{1}{\text{Vol}(K)} \int_K \mathbb{A}|_K \, \mathrm{d}x,$$

where $\text{Vol}(K)$ is the volume or area of K. The discrete moduli $\mathbb{D}_h$ and $\mathbb{E}_h$ are defined similarly. We introduce next the discrete counterparts of the standard and modified three-field problems (37) and (40).

The discrete standard three-field formulation Let $\mathcal{V}_h \subset \mathcal{V}, \mathcal{S}_h \subset \mathcal{S}$ and $\mathcal{D}_h \subset \mathcal{D}$ be finite-dimensional spaces for the increments of displacement, first Piola–Kirchhoff stress and deformation gradient. Then from (35), we define the discrete form of the standard formulation as the problem of finding $(\dot{\boldsymbol{u}}_h\,,\ \dot{\boldsymbol{F}}_h\,,\ \dot{\boldsymbol{P}}_h) \in \mathcal{V}_h \times \mathcal{D}_h \times \mathcal{S}_h$ such that

$$\left(\mathbb{A}_h \dot{\boldsymbol{F}}_h - \dot{\boldsymbol{P}}_h\,,\ \boldsymbol{G}_h\right)_0 = 0 \quad \forall \boldsymbol{G}_h \in \mathcal{D}_h, \tag{42a}$$

$$\left(\dot{\boldsymbol{F}}_h - \nabla \dot{\boldsymbol{u}}_h\,,\ \boldsymbol{Q}_h\right)_0 = 0 \quad \forall \boldsymbol{Q}_h \in \mathcal{S}_h, \tag{42b}$$

$$\left(\dot{\boldsymbol{P}}_h, \nabla \boldsymbol{v}_h\right)_0 = \ell(\boldsymbol{v}_h) \quad \forall \boldsymbol{v}_h \in \mathcal{V}_h. \tag{42c}$$

The discrete modified formulation The modified formulation is given by the set of equations

$$\left(\mathbb{D}_h(\mathbb{A}_h^{-1}\dot{\boldsymbol{P}}_h - \dot{\boldsymbol{F}}_h)\,,\ \boldsymbol{G}_h\right)_0 = 0 \qquad \forall \boldsymbol{G}_h \in \mathcal{D}_h, \tag{43a}$$

$$\left(\dot{\boldsymbol{F}}_h - \nabla \dot{\boldsymbol{u}}_h\,,\ \boldsymbol{Q}_h\right)_0 = 0 \qquad \forall \boldsymbol{Q}_h \in \mathcal{S}_h, \tag{43b}$$

$$\left(\nabla \boldsymbol{v}_h\,,\ \dot{\boldsymbol{P}}_h\right)_0 = \ell(\boldsymbol{v}_h) \qquad \forall \boldsymbol{v}_h \in \mathcal{V}_h. \tag{43c}$$

Note that only the first equations in these two formulations differ. We have the following results on equivalence of the discrete and modified formulations, and on well-posedness and convergence of solutions to the discrete problem (Chama and Reddy 2015).

Lemma 5.1 *Assume that the spaces $\mathcal{S}_h$ and $\mathcal{D}_h$ are such that*

$$\mathcal{S}_h \subset \mathcal{D}_h\,, \quad \mathbb{A}_h\mathcal{D}_h \subset \mathcal{D}_h \quad \textit{and} \quad \mathbb{D}_h^{-1}\mathbb{A}_h\mathcal{D}_h \subset \mathcal{D}_h. \tag{44}$$

Then the discrete formulations (42) and (43) are equivalent.

Theorem 5.2 *The discrete mixed problem (17) or its modified counterpart (18), assuming that the conditions for equivalence are met, has a unique solution provided that the following two conditions are satisfied:*

(*a*) *Ellipticity or coerciveness: There exists a constant* $c_0 > 0$ *such that* $\|P\nabla \boldsymbol{v}_h\|_{\mathcal{S}} \geq c_0\|\nabla \boldsymbol{v}_h\|_{\mathcal{V}}$ *for all* $\boldsymbol{v}_h \in \mathcal{V}^h$*, where* P *denotes the* L^2*-orthogonal projection onto* $\mathcal{S}^h$*;*
(*b*) *Inf-sup condition: there exists a mesh-independent constant* $\beta_h > 0$ *such that*

$$\inf_{\mathcal{Q}_h \in \mathcal{S}^h} \sup_{v_h \in \mathcal{V}^h} \frac{\int_\Omega \nabla \boldsymbol{v}_h : \boldsymbol{Q}_h \, dx}{\|\boldsymbol{v}_h\|_{\mathcal{V}} \|\boldsymbol{Q}_h\|_{\mathcal{S}}} \geq \beta_h. \tag{45}$$

Furthermore, the error satisfies the uniform estimate

$$\|\boldsymbol{u} - \boldsymbol{u}_h\|_{\mathcal{V}} + \|\varepsilon - \varepsilon_h\|_{\mathcal{E}} + \|\boldsymbol{\sigma} - \boldsymbol{\sigma}_h\|_{\mathcal{S}} \leq Ch.$$

5.1 An Application: Mixed Enhanced Strains

We choose as an example the set of discrete spaces corresponding to the mixed enhanced formulation for nonlinear problems (Kasper and Taylor 2000b). It is shown in Chama and Reddy (2015) that this choice of elements satisfies the conditions in Theorem 5.2 for well-posedness.

As an example, consider a problem of a body with reference domain $\Omega_0 = (0,\ 0.25) \times (0,\ 0.5) \times (0,\ 1)$. A uniform traction $\bar{t} = 1.0 \times 10^{-2} N/m$ is applied on the surface $z = 1$ in the x direction, and a homogeneous Dirichlet boundary condition on the face $z = 0$. The remaining sides are traction-free. Figure 2 shows the deformed configuration of the domain with the use of the mixed and Q_1 elements, respectively. The locking behaviour in the latter case is evident.

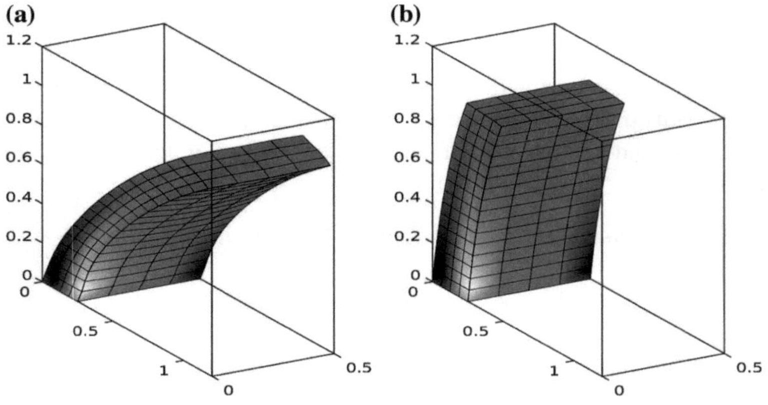

Fig. 2 Deformed shape of a *square cylindrical* domain subjected to shear: **a** using the three-field formulation; **b** using the standard Q_1 element

Acknowledgments The support of the South African Department of Science and Technology and National Research Foundation through the South African Research Chair in Computational Mechanics is gratefully acknowledged.

References

Auricchio, F., Beirão da Veiga, L., Lovadina, C., & Reali, A. (2005). A stability study of some mixed finite elements for large deformation elasticity problems. *Computer Methods in Applied Mechanics and Engineering*, *194*, 1075–1092.

Auricchio, F., Beirão da Veiga, L., Lovadina, C., Reali, A., Taylor, R. L., & Wriggers, P. (2013). Approximation of incompressible large deformation elastic problems: some unresolved issues. *Computational Mechanics*, *52*, 1153–1167.

Chadwick, P., & Ogden, R. W. (1971). On the definition of elastic moduli. *Archive for Rational Mechanics and Analysis*, *44*, 41–53.

Chama, A. & Reddy, B. D. (2015). Three-field mixed finite element approximations of problems in nonlinear elastcity. *Preprint*.

Chavan, K. S., Lamichhane, B. P., & Wohlmuth, B. I. (2007). Locking-free finite element methods for linear and nonlinear elasticity in 2D and 3D. *Computer Methods in Applied Mechanics and Engineering*, *196*, 4075–4086.

Djoko, J. K., Lamichhane, B. P., Reddy, B. D., & Wohlmuth, B. I. (2006). Conditions for equivalence between the Hu-Washizu and related formulations, and computational behavior in the incompressible limit. *Computer Methods in Applied Mechanics and Engineering*, *195*, 4161–4178.

Duffett, G. A., & Reddy, B. D. (1986). The solution of multi-parameter systems of equations with application to problems in nonlinear elasticity. *Computer Methods in Applied Mechanics and Engineering*, *59*(2), 179–213.

Fraeijs de Veubeke, B. M. *Diffusion des Inconnues Hyperstatiques dans les Voilures à Longeron Couplés, Bull. Serv. Technique de l'Aeronautique, Imprimérie Marcel Hayez*. Bulletin du Service Technique de l'Aéronautique 24. Hayez, 1951.

Hu, H. (1955). On some variational principles in the theory of elasticity and the theory of plasticity. *Scientia Sinica*, *4*, 33–54.

Kasper, E. P., & Taylor, R. L. (2000a). A mixed-enhanced strain method: Part I: Geometrically linear problems. *Computers and Structures*, *75*, 237–250.

Kasper, E. P., & Taylor, R. L. (2000b). A mixed-enhanced strain method: Part II: Geometrically nonlinear problems. *Computers and Structures*, *75*, 251–260.

Lamichhane, B. P., Reddy, B. D., & Wohlmuth, B. I. (2006). Convergence in the incompressible limit of finite element approximations based on the Hu-Washizu formulation. *Numerische Mathematik*, *104*, 151–175.

Mueller-Hoeppe, D. S., Loehnert, S., & Wriggers, P. (2009). A finite deformation brick element with inhomogeneous mode enhancement. *International Journal for Numerical Methods in Engineering*, *78*, 1164–1187.

Simo, J. C., & Armero, F. (1992). Geometrically non-linear enhanced strain mixed methods and the method of incompatible modes. *International Journal for Numerical Methods in Engineering*, *33*, 1413–1449.

Simo, J. C., & Rifai, M. S. (1990). A class of mixed assumed strain methods and the method of incompatible modes. *International Journal for Numerical Methods in Engineering*, *29*, 1595–1638.

Simo, J. C., Armero, F., & Taylor, R. L. (1993). Improved versions of assumed enhanced strain tri-linear elements for 3d finite deformation problems. *International Journal for Numerical Methods in Engineering*, *110*, 359–386.

Ten Eyck, A., & Lew, A. (2010). An adaptive stabilization strategy for enhanced strain methods in non-linear elasticity. *International Journal for Numerical Methods in Engineering*, *81*, 1387–1416.
Washizu, K. (1995). *On the variational principles of elasticity and plasticity*. Report 25–18, M.I.T. Aeroelastic and Structures Research Laboratory.
Wriggers, P. (1998). *Nonlinear finite element methods*. Berlin: Springer.
Wriggers, P., & Reese, S. (1996). A note on enhanced strain methods for large deformations. *Computer Methods in Applied Mechanics and Enginering*, *135*(3–4), 201–209.

Stress-Based Finite Element Methods in Linear and Nonlinear Solid Mechanics

Benjamin Müller and Gerhard Starke

Abstract A comparison of stress-based finite element methods is given for the prototype problem of linear elasticity and then extended to finite-strain hyperelasticity. Of particular interest is the accuracy of traction forces in reasonable Sobolev norms with an emphasis on uniform approximation behavior in the incompressible limit. The mixed formulation of Hellinger–Reissner type leading to a saddle-point problem as well as a first-order system least-squares approach are investigated and the strong connections between these two methods are studied. In addition, we also discuss stress reconstruction techniques based on displacement approximations by nonconforming finite elements.

1 Introduction

The accurate resolution of stresses associated with numerical simulations in solid mechanics is of paramount importance in many applications. Large stress components may cause plastic behavior or even damage and therefore need to be approximated well. In particular, if surface traction forces are of interest, finite element approximations in spaces which allow the safe evaluation of boundary traces need to be used. For standard displacement-based or (in the incompressible case) displacement-pressure approaches, the associated stresses are only contained in L^2 which means that the (normal component of the) boundary traces are not defined. Variational principles which involve stresses in $H(\text{div})$-like saddle-point formulations of Hellinger–Reissner-type or first-order system least-squares approaches overcome this problem directly. Another option is to reconstruct stresses in $H(\text{div})$ from sufficiently accurate L^2 approximations in analogy to the flux reconstruction procedures described, e.g., in Luce and Wohlmuth (2004), Nicaise et al. (2008),

B. Müller · G. Starke (✉)
Fakultät für Mathematik, Universität Duisburg-Essen, 45127 Essen, Germany
e-mail: gerhard.starke@uni-due.de

B. Müller
e-mail: benjamin.mueller@uni-due.de

J. Schröder and P. Wriggers (eds.), *Advanced Finite Element Technologies*,
CISM International Centre for Mechanical Sciences 566,
DOI 10.1007/978-3-319-31925-4_4

Braess et al. (2009), Cai and Zhang (2012), Ern and Vohralík (2015). For an equilibration approach to stress reconstruction in two-dimensional linear elasticity see also Parés et al. (2006). Particularly attractive in the context of incompressible elasticity are the quadratic nonconforming elements introduced by Fortin and Soulie (1983) and, in three space dimensions, Fortin (1985). Flux and stress reconstruction procedures working in an element-wise way were studied for these elements by Kim (2012).

The history of mixed finite element methods of saddle-point type for the approximation of stresses in $H(\text{div})$ in linear elasticity models goes back for at least 30 years with early contributions by Arnold et al. (1984a, b) and Stenberg (1988) among others, see Boffi et al. (2013, Chap. 9), for more details. Later, this approach also received much attention in the engineering community, see e.g., Klaas et al. (1995). For the class of first-order system least-squares methods, the state of the art is presented in Bochev and Gunzburger (2009) with a focus of fluid rather than solid mechanics. The $H(\text{div})$-based stress formulation which will be our starting point in this contribution was studied in Cai and Starke (2004) for the linear elasticity case and extended to hyperelastic material models in Müller et al. (2014).

The investigation of hyperelastic models in Sect. 5 will be presented in detail for the specific example of a neo-Hookean material law. For background on the analytical and numerical treatment of hyperelasticity, we refer to Ciarlet (1988) and LeTallec (1994). Concerning a priori finite element error estimates associated with such models, see Carstensen and Dolzmann (2004). Our focus in Sect. 5 of this contribution will again be on approaches which remain robust in the incompressible limit. Similar to the linear elasticity case this may be achieved either by adding an auxiliary pressure variable (cf. Auricchio et al. 2013) or by inverting the stress-strain relation, cf. Wriggers (2008), Sect. 10.3.

The elasticity problems under our consideration are based on an open, bounded, and connected domain $\Omega \subset \mathbb{R}^d$ $(d = 2, 3)$ with Lipschitz-continuous boundary, which constitutes the reference configuration of the undeformed state. The boundary is divided into two disjoint subsets Γ_D and Γ_N, for simplicity, both assumed to be nonempty. On Γ_D, homogeneous displacement boundary conditions $\mathbf{u} = \mathbf{0}$ are imposed, while surface traction forces $\boldsymbol{\sigma} \cdot \mathbf{n} = \mathbf{g}$ are prescribed on Γ_N. The linear elasticity model may then be written as the first-order system

$$\begin{aligned} \operatorname{div} \boldsymbol{\sigma} + \mathbf{f} &= \mathbf{0} \\ \boldsymbol{\sigma} - \mathcal{C}\varepsilon(\mathbf{u}) &= \mathbf{0} \end{aligned} \tag{1}$$

in Ω subject to the above boundary conditions with $\varepsilon(\mathbf{u}) = (\nabla\mathbf{u} + (\nabla\mathbf{u})^T)/2$ and

$$\mathcal{C}\varepsilon = 2\mu\varepsilon + \lambda(\operatorname{tr} \varepsilon)\mathbf{I}. \tag{2}$$

The system (1) may be derived from minimizing the energy associated with the deformed system given by

$$\int_\Omega \psi(\varepsilon(\mathbf{v}))\, dx - \int_\Omega \mathbf{f} \cdot \mathbf{v}\, dx - \int_{\Gamma_N} \mathbf{g} \cdot \mathbf{v}\, ds, \tag{3}$$

where the stored energy function is given by

$$\psi(\varepsilon) = \mu|\varepsilon|^2 + \frac{\lambda}{2}(\mathrm{tr}\,\varepsilon)^2. \tag{4}$$

The necessary conditions for a stationary point of (3) are then equivalent to (1).

While we can always scale the units such that μ is on the order of 1, an important issue is the behavior of the formulations in the incompressible limit $\lambda \to \infty$. It is already apparent from (2) that a naive numerical approach to the above minimization problem will cause problems for incompressible or nearly incompressible materials. One possible remedy consists in replacing $\lambda(\mathrm{tr}\,\varepsilon)$ by a new variable p which has the physical interpretation of a pressure. Another option is to use the inverse $\mathcal{C}^{-1}$ instead of $\mathcal{C}$ in the variational formulation. A straightforward calculation shows that

$$\mathcal{C}^{-1}\boldsymbol{\sigma} = \frac{1}{2\mu}\left(\boldsymbol{\sigma} - \frac{\lambda}{2\mu + d\lambda}\mathrm{tr}(\boldsymbol{\sigma})\mathbf{I}\right) \overset{\lambda\to\infty}{\to} \frac{1}{2\mu}\left(\boldsymbol{\sigma} - \frac{1}{d}\mathrm{tr}(\boldsymbol{\sigma})\mathbf{I}\right) = \frac{1}{2\mu}\mathbf{dev}\,\boldsymbol{\sigma},$$

i.e., the operator $\mathcal{C}^{-1}$ remains well-defined in the incompressible limit, where it constitutes the orthogonal projection onto the trace-free matrices **dev**. Since $\mathcal{C}^{-1}$ itself is not invertible any more in the incompressible limit, we also write $\mathcal{A}$ instead of $\mathcal{C}^{-1}$ in order to avoid misunderstandings. The first-order system (1) turns into

$$\begin{aligned} \mathrm{div}\,\boldsymbol{\sigma} + \mathbf{f} &= \mathbf{0} \\ \mathcal{A}\boldsymbol{\sigma} - \varepsilon(\mathbf{u}) &= \mathbf{0}. \end{aligned} \tag{5}$$

Of course, any variational approach based on (5) needs to use the stress $\boldsymbol{\sigma}$ as an independent variable in the formulation. Such approaches will be presented in the next sections.

We will make use of norms and inner products associated with different spaces throughout this paper. Since $L^2(\Omega)$ (and its vector and matrix variants $L^2(\Omega)^d$ and $L^2(\Omega)^{d\times d}$, respectively) occurs most often, we abbreviate the associated norm simply by $\|\cdot\|$ and the corresponding inner product by $(\cdot\,,\,\cdot)$. Since we assume $\Gamma_D \neq \emptyset$ (more precisely, a subset of $\partial\Omega$ of positive measure), Korn's inequality is valid in the form

$$\|\nabla \mathbf{v}\| \le C_K \|\varepsilon(\mathbf{v})\| \text{ for all } \mathbf{v} \in H^1_{\Gamma_D}(\Omega)^d. \tag{6}$$

Our general regularity assumption is that $\Omega \subset \mathbb{R}^d$, $\Gamma_N \subset \partial\Omega$ and $\Gamma_D \subset \partial\Omega$ are such that, for any $\mathbf{f} \in L^2(\Omega)^d$ the solution of (5) satisfies $(\boldsymbol{\sigma}, \mathbf{u}) \in H^\alpha(\Omega)^{d\times d} \times H^{1+\alpha}(\Omega)^d$ such that

$$\|\boldsymbol{\sigma}\|_{H^\alpha(\Omega)} + \|\mathbf{u}\|_{H^{1+\alpha}(\Omega)} \le C_R \|\mathbf{f}\| \tag{7}$$

holds for some constant $C_R > 0$ and some $\alpha > 0$.

2 Stress-Based Mixed Formulation Based on the Hellinger–Reissner Principle

This section is focussed on the approximation of stresses in the Sobolev space $H(\text{div}, \Omega)^d$. The subspaces

$$\begin{aligned} H_{\Gamma_N}(\text{div}, \Omega)^d &= \{\boldsymbol{\tau} \in H(\text{div}, \Omega)^d : \boldsymbol{\tau} \cdot \mathbf{n} = \mathbf{0} \text{ on } \Gamma_N\}, \\ H^0_{\Gamma_N}(\text{div}, \Omega)^d &= \{\boldsymbol{\tau} \in H_{\Gamma_N}(\text{div}, \Omega)^d : \text{div}\, \boldsymbol{\tau} = \mathbf{0}\} \end{aligned}$$

will also be used. From the stress–strain relation $\boldsymbol{\varepsilon}(\mathbf{u}) = \mathcal{C}^{-1}\boldsymbol{\sigma}$, integration by parts leads to

$$(\mathcal{C}^{-1}\boldsymbol{\sigma}, \boldsymbol{\tau}) + (\mathbf{u}, \text{div}\, \boldsymbol{\tau}) + (\boldsymbol{\gamma}, \mathbf{as}\, \boldsymbol{\tau}) = 0, \tag{8}$$

for all $\boldsymbol{\tau} \in H_{\Gamma_N}(\text{div}, \Omega)^d$, where $\mathbf{as}\, \boldsymbol{\tau} = (\boldsymbol{\tau} - \boldsymbol{\tau}^T)/2$ denotes the asymmetric part and $\boldsymbol{\gamma}$ is a new variable introduced for $\mathbf{as}\, \nabla\mathbf{u}$. Together with the two equations

$$\begin{aligned} (\text{div}\, \boldsymbol{\sigma} + \mathbf{f}, \mathbf{v}) &= 0 \text{ for all } \mathbf{v} \in L^2(\Omega)^d, \\ (\mathbf{as}\, \boldsymbol{\sigma}, \boldsymbol{\theta}) &= 0 \text{ for all } \boldsymbol{\theta} \in L^2(\Omega)^{d\times d,\text{as}}, \end{aligned} \tag{9}$$

where $L^2(\Omega)^{d\times d,\text{as}}$ denotes the subspace of $L^2(\Omega)^{d\times d}$ with vanishing symmetric part, the mixed variational formulation of Hellinger–Reissner consists in finding $(\boldsymbol{\sigma}, \mathbf{u}, \boldsymbol{\gamma}) \in (\boldsymbol{\sigma}^N + H_{\Gamma_N}(\text{div}, \Omega)^d) \times L^2(\Omega)^d \times L^2(\Omega)^{d\times d,\text{as}}$ such that (8) and (9) hold. An alternative way of deriving this mixed variational formulation consists in viewing it as the KKT conditions for the minimization of the energy $(\mathcal{C}^{-1}\boldsymbol{\sigma}, \boldsymbol{\sigma})/2$ subject to the constraints (9). In this context, $\mathbf{u}$ and $\boldsymbol{\gamma}$ are Lagrange parameters for the momentum balance and symmetry conditions, respectively, in (9). For the well-posedness of the system (8) and (9), the following result is of crucial importance.

Theorem 2.1 *Assume that $\Gamma_N \subseteq \partial\Omega$ consists of a finite number of connected components each of which has positive $(d-1)$-dimensional measure. Then,*

$$\|\boldsymbol{\tau}\| \lesssim \|\mathbf{dev}\, \boldsymbol{\tau}\| \tag{10}$$

holds for all $\boldsymbol{\tau} \in H^0_{\Gamma_N}(\text{div}, \Omega)^d$.

Theorem 2.1 follows from the more general result of Theorem 3.1 in Sect. 3. A direct and simple proof for the two-dimensional case is given in the following.

Proof (*for $d = 2$*) Without loss of generality assume that $\Gamma_N \subsetneq \partial\Omega$ and that Γ_N is connected with $|\Gamma_N| > 0$ (just replace Γ_N by one of its connected components, cut something off if $\Gamma_N = \partial\Omega$).

Using (Girault and Raviart 1986, Theorem I.3.1), we can write $\boldsymbol{\tau} = \mathbf{curl}\, \boldsymbol{\phi}$ with $\boldsymbol{\phi} \in H^1(\Omega)^2$. The boundary conditions $(\mathbf{curl}\, \boldsymbol{\phi}) \cdot \mathbf{n} = \mathbf{0}$ imply $\boldsymbol{\phi}$ to be constant on

Γ_N, which we may choose to be zero, i.e., $\phi \in H^1_{\Gamma_N}(\Omega)^2$. Therefore, ϕ satisfies Korn's inequality (6) and we obtain

$$
\begin{aligned}
\|\boldsymbol{\tau}\| = \|\mathbf{curl}\,\phi\| = \|\nabla\phi\| &\lesssim \|\varepsilon(\phi)\| \\
&= \|\begin{pmatrix} \partial_1\phi_1 & \frac{1}{2}(\partial_2\phi_1 + \partial_1\phi_2) \\ \frac{1}{2}(\partial_2\phi_1 + \partial_1\phi_2) & \partial_2\phi_2 \end{pmatrix}\| \\
&= \|\begin{pmatrix} \frac{1}{2}(\partial_2\phi_1 + \partial_1\phi_2) & -\partial_1\phi_1 \\ \partial_2\phi_2 & -\frac{1}{2}(\partial_2\phi_1 + \partial_1\phi_2) \end{pmatrix}\| \\
&= \|\mathbf{dev}\begin{pmatrix} \partial_2\phi_1 & -\partial_1\phi_1 \\ \partial_2\phi_2 & -\partial_1\phi_2 \end{pmatrix}\| = \|\mathbf{dev}\,\mathbf{curl}\,\phi\| = \|\mathbf{dev}\,\boldsymbol{\tau}\|.
\end{aligned}
$$

□

Theorem 2.1 implies

$$
(\mathcal{C}^{-1}\boldsymbol{\tau}, \boldsymbol{\tau}) \geq \frac{1}{2\mu}\|\mathbf{dev}\,\boldsymbol{\tau}\|^2 \gtrsim \|\boldsymbol{\tau}\|^2 \text{ for all } \boldsymbol{\tau} \in H^0_{\Gamma_N}(\mathrm{div}, \Omega)^d. \tag{11}
$$

Since $H^0_{\Gamma_N}(\mathrm{div}, \Omega)^d$ contains the null space of the constraints (9), the required coercivity condition is satisfied. As a second ingredient to the well-posedness, the inf-sup condition has to be established for (9), see (Boffi et al. 2013, Proposition 9.3.2).

For the discretization of (8) and (9), finite element spaces $\mathbf{\Pi}_h \subset H_{\Gamma_N}(\mathrm{div}, \Omega)^d$, $\mathbf{Z}_h \subset L^2(\Omega)^d$ and $\mathbf{\Theta}_h \subset L^2(\Omega)^{d\times d,\mathrm{as}}$ are inserted into (8) and (9) leading to a mixed finite element approximation $(\boldsymbol{\sigma}_h^{HR}, \mathbf{z}_h^{HR}, \gamma_h^{HR})$. To this end, various finite element combinations which satisfy the discrete inf-sup condition have been proposed, starting with the famous PEERS element by Arnold et al. (1984a). For a systematic treatment of this topic see Boffi et al. (2013, Chap. 9). It is interesting to note that for $k \geq 1$, the triple of finite element spaces

$$
(\mathbf{\Pi}_h, \mathbf{Z}_h, \mathbf{\Theta}_h) = RT_k(\mathcal{T}_h)^d \times DP_k(\mathcal{T}_h)^d \times P_k(\mathcal{T}_h)^{d\times d,\mathrm{as}}
$$

is inf-sup stable (see Boffi et al. 2009, 2013, Example 9.4.1). An important property of this approach is that the momentum balance is best possible, i.e.,

$$
\|\mathrm{div}\,\boldsymbol{\sigma}_h^{HR} + \mathbf{f}\| = \|\mathbf{f} - \pi_h\mathbf{f}\| = \inf_{\mathbf{z}_h \in \mathbf{Z}_h} \|\mathbf{f} - \mathbf{z}_h\|.
$$

From the ellipticity and the inf-sup conditions, optimal order accuracy also follows for the stress approximation with respect to the $L^2(\Omega)$-norm, i.e.,

$$
\|\boldsymbol{\sigma} - \boldsymbol{\sigma}_h^{HR}\| \lesssim \inf_{\boldsymbol{\tau}_h \in \mathbf{\Pi}_h} \|\boldsymbol{\sigma} - \boldsymbol{\tau}_h\| \lesssim h^\alpha \|\boldsymbol{\sigma}\|_{H^\alpha(\Omega)}. \tag{12}
$$

3 Stress-Displacement First-Order System Least Squares

In this section, we consider the first-order system least-squares approach based on

$$\mathcal{R}(\boldsymbol{\sigma}, \mathbf{u}) := \begin{pmatrix} \operatorname{div} \boldsymbol{\sigma} + \mathbf{f} \\ \mathcal{A}\boldsymbol{\sigma} - \boldsymbol{\varepsilon}(\mathbf{u}) \end{pmatrix}, \tag{13}$$

i.e., the minimization of

$$\mathcal{F}(\boldsymbol{\tau}, \mathbf{v}) := \|\mathcal{R}(\boldsymbol{\tau}, \mathbf{v})\|^2 = \|\operatorname{div} \boldsymbol{\tau} + \mathbf{f}\|^2 + \|\mathcal{A}\boldsymbol{\tau} - \boldsymbol{\varepsilon}(\mathbf{v})\|^2 \tag{14}$$

among all $\boldsymbol{\tau} \in \boldsymbol{\sigma}^N + H_{\Gamma_N}(\operatorname{div}, \Omega)^3$ and $\mathbf{v} \in H^1_{\Gamma_D}(\Omega)^3$. The minimizer $(\boldsymbol{\sigma}, \mathbf{u})$ of (14) satisfies

$$\begin{aligned} (\operatorname{div} \boldsymbol{\sigma}, \operatorname{div} \boldsymbol{\tau}) + (\mathcal{A}\boldsymbol{\sigma} - \boldsymbol{\varepsilon}(\mathbf{u}), \mathcal{A}\boldsymbol{\tau}) &= -(\mathbf{f}, \operatorname{div} \boldsymbol{\tau}), \\ -(\mathcal{A}\boldsymbol{\sigma} - \boldsymbol{\varepsilon}(\mathbf{u}), \boldsymbol{\varepsilon}(\mathbf{v})) &= 0 \end{aligned} \tag{15}$$

for all $(\boldsymbol{\tau}, \mathbf{v}) \in H_{\Gamma_N}(\operatorname{div}, \Omega)^3 \times H^1_{\Gamma_D}(\Omega)^3$, which constitutes a linear variational problem. The well-posedness of (15) follows from the coercivity and continuity of the bilinear form

$$\mathcal{B}((\boldsymbol{\sigma}, \mathbf{u}); (\boldsymbol{\tau}, \mathbf{v})) := (\operatorname{div} \boldsymbol{\sigma}, \operatorname{div} \boldsymbol{\tau}) + (\mathcal{A}\boldsymbol{\sigma} - \boldsymbol{\varepsilon}(\mathbf{u}), \mathcal{A}\boldsymbol{\tau} - \boldsymbol{\varepsilon}(\mathbf{v})) \tag{16}$$

with respect to $H_{\Gamma_N}(\operatorname{div}, \Omega)^3 \times H^1_{\Gamma_D}(\Omega)^3$ uniformly in the incompressible limit. This property was shown under our assumptions on Ω, Γ_N and Γ_D in Cai and Starke (2004). A consequence of its validity in the incompressible limit is the following result.

Theorem 3.1 *Assume that $\Gamma_N \subseteq \partial\Omega$ consists of a finite number of connected components each of which has positive $(d-1)$-dimensional measure. Then,*

$$\|\boldsymbol{\tau}\| \lesssim \|\mathbf{dev}\, \boldsymbol{\tau}\| + \|\operatorname{div} \boldsymbol{\tau}\| \tag{17}$$

holds for all $\boldsymbol{\tau} \in H_{\Gamma_N}(\Omega)^d$.

The result of Theorem 3.1 was proved in Arnold et al. (1984b) for the case $\Gamma_N = \emptyset$ under the additional constraint $(\operatorname{tr} \boldsymbol{\tau}, 1) = 0$ (see also (Boffi et al. 2013, Proposition 9.1.1)) and in the general two-dimensional case in Carstensen and Dolzmann (1998).

The discrete first-order system least-squares approximation is obtained by minimizing (14) among all $\boldsymbol{\tau}_h = \boldsymbol{\sigma}^N + \boldsymbol{\Pi}_h$ and $\mathbf{v}_h \in \mathbf{V}_h$, where $\boldsymbol{\Pi}_h \subset H_{\Gamma_N}(\operatorname{div}, \Omega)^3$ and $\mathbf{V}_h \subset H^1_{\Gamma_D}(\Omega)^3$ are suitable finite element spaces. The approximate solution $\boldsymbol{\sigma}_h^{LS} \in \boldsymbol{\sigma}^N + \boldsymbol{\Pi}_h$, $\mathbf{u}_h^{LS} \in \mathbf{V}_h$ is determined by

$$\begin{aligned} (\operatorname{div} \boldsymbol{\sigma}_h^{LS}, \operatorname{div} \boldsymbol{\tau}_h) + (\mathcal{A}\boldsymbol{\sigma}_h^{LS} - \boldsymbol{\varepsilon}(\mathbf{u}_h^{LS}), \mathcal{A}\boldsymbol{\tau}_h) &= -(\mathbf{f}, \operatorname{div} \boldsymbol{\tau}_h), \\ -(\mathcal{A}\boldsymbol{\sigma}_h^{LS} - \boldsymbol{\varepsilon}(\mathbf{u}_h^{LS}), \boldsymbol{\varepsilon}(\mathbf{v}_h)) &= 0 \end{aligned} \tag{18}$$

for all $(\boldsymbol{\tau}_h, \mathbf{v}_h) \in \boldsymbol{\Pi}_h \times \mathbf{V}_h$. Due to the coercivity and continuity of the underlying bilinear form we obtain a quasi-optimal approximation, i.e.,

$$\begin{aligned} \|\boldsymbol{\sigma} - \boldsymbol{\sigma}_h^{LS}\|_{\text{div},\Omega} &\lesssim \inf_{\boldsymbol{\tau}_h \in \boldsymbol{\Pi}_h} \|\boldsymbol{\sigma} - \boldsymbol{\tau}_h\|_{\text{div},\Omega}, \\ \|\mathbf{u} - \mathbf{u}_h^{LS}\|_{1,\Omega} &\lesssim \inf_{\mathbf{v}_h \in \mathbf{V}_h} \|\mathbf{u} - \mathbf{v}_h\|_{1,\Omega}. \end{aligned} \tag{19}$$

In particular, using, for some $l \geq 1$, Raviart–Thomas spaces of degree $l - 1$ for $\boldsymbol{\Pi}_h$ combined with standard conforming finite elements of degree l for $\mathbf{V}_h$, one gets

$$\begin{aligned} \|\boldsymbol{\sigma} - \boldsymbol{\sigma}_h^{LS}\|_{\text{div},\Omega} &\lesssim h^l \left(|\boldsymbol{\sigma}|_{l,\Omega} + |\text{div}\,\boldsymbol{\sigma}|_{l,\Omega}\right), \\ \|\mathbf{u} - \mathbf{u}_h^{LS}\|_{1,\Omega} &\lesssim h^l |\mathbf{u}|_{l+1,\Omega}, \end{aligned}$$

if $\boldsymbol{\sigma} \in H^l(\Omega)^{3\times3}$ with div $\boldsymbol{\sigma}(= -\mathbf{f}) \in H^l(\Omega)^3$ and $\mathbf{u} \in H^{l+1}(\Omega)^3$ is satisfied. It may also be worth noting that within the first-order least-squares approach, piecewise linear conforming displacement approximations are of optimal order uniformly in the incompressible limit. Of course, this requires the simultaneous computation of stress approximations in the lowest-order Raviart–Thomas spaces which may be considered too costly if these quantities are not of particular interest. If the solution is less regular, then the optimal approximation order may be retained with adaptively refined triangulations based on using the local evaluation of the functional as an a posteriori error estimator, cf. Cai et al. (2005). For domains with curved boundaries in association with the higher-order case $l > 1$, parametric finite element spaces would be needed in order to retain the optimal approximation order. This would involve the parametric Raviart–Thomas spaces studied in Bertrand et al. (2014) for $\boldsymbol{\Pi}_h$ in combination with standard isoparametric elements, cf. Brenner and Scott (2008, Sect. 10.4).

It is important to keep in mind that the two terms in the functional defined by (14) need to be scaled appropriately in order to get reasonable approximations. This is due to the fact that the constants involved in the above estimates must not become exceedingly large. The two main ingredients which influence these constants are the Lamé parameter μ in the material law (2) and C_K in Korn's inequality (6). If both are on the order of one, then the scaling in (14) is adequate. This can be achieved by the choice of suitable units for measuring forces and lengths. Our computational experience suggests that it is generally less harmful to weight the momentum balance term too strong than too weak with respect to the above rules.

In contrast to the mixed approximation $\boldsymbol{\sigma}_h^{HR}$, our least-squares approximation $\boldsymbol{\sigma}_h^{LS}$ does not satisfy the momentum balance exactly if $\mathbf{f} \in \text{div}\,\Pi_h$. We will now show that, in fact, the momentum balance term in the functional (14) converges faster than the overall functional. The proof is inspired by the techniques used in Brandts et al. (2006) for the investigation of the relations between saddle point and least-squares formulations for the first-order system formulation of the Poisson equation.

Theorem 3.2 *Under our regularity assumptions, the momentum balance accuracy associated with the first-order system least-squares approximation satisfies*

$$\|\text{div}\,\boldsymbol{\sigma}_h^{LS} + \mathbf{f}\| \lesssim h^\alpha \left(\|\boldsymbol{\sigma} - \boldsymbol{\sigma}_h^{LS}\| + \|\varepsilon(\mathbf{u}) - \varepsilon(\mathbf{u}_h^{LS})\|\right) + \inf_{\mathbf{z}_h \in \mathbf{Z}_h} \|\mathbf{f} - \mathbf{z}_h\|. \quad (20)$$

Proof With $\mathbf{f}_h = \pi_h \mathbf{f} \in \mathbf{Z}_h$, the triangle inequality leads to

$$\|\text{div}\,\boldsymbol{\sigma}_h^{LS} + \mathbf{f}\| \le \|\text{div}\,\boldsymbol{\sigma}_h^{LS} + \mathbf{f}_h\| + \|\mathbf{f} - \mathbf{f}_h\| = \|\text{div}\,\boldsymbol{\sigma}_h^{LS} + \mathbf{f}_h\| + \|\mathbf{f} - \pi_h \mathbf{f}\|. \quad (21)$$

The first term on the right-hand side in (21) can be written as

$$\begin{aligned} \|\text{div}\,\boldsymbol{\sigma}_h^{LS} + \mathbf{f}_h\| &= \sup_{\mathbf{z}_h \in \mathbf{Z}_h} \frac{(\text{div}\,\boldsymbol{\sigma}_h^{LS} + \mathbf{f}_h, \mathbf{z}_h)}{\|\mathbf{z}_h\|} \\ &= \sup_{\mathbf{z}_h \in \mathbf{Z}_h} \frac{(\text{div}\,\boldsymbol{\sigma}_h^{LS} + \mathbf{f}, \mathbf{z}_h)}{\|\mathbf{z}_h\|} = \sup_{\mathbf{z}_h \in \mathbf{Z}_h} \frac{(\text{div}\,(\boldsymbol{\sigma}_h^{LS} - \boldsymbol{\sigma}), \mathbf{z}_h)}{\|\mathbf{z}_h\|}. \end{aligned} \quad (22)$$

For any $\mathbf{z}_h \in \mathbf{Z}_h$, the following auxiliary boundary value problem may be defined: Find $\Xi \in H_{\Gamma_N}(\text{div}, \Omega)^3$ and $\boldsymbol{\eta} \in H^1_{\Gamma_D}(\Omega)^3$ such that

$$\begin{aligned} \text{div}\,\Xi &= \mathbf{z}_h, \\ \mathcal{A}\Xi - \varepsilon(\boldsymbol{\eta}) &= \mathbf{0} \end{aligned} \quad (23)$$

holds. Let $\Xi_h^{HR} \in \boldsymbol{\Pi}_h$ be the mixed finite element approximation of Hellinger–Reissner type to (23) and let $\boldsymbol{\eta}_h \in \mathbf{V}_h$ be any approximation to $\boldsymbol{\eta}$, then

$$\begin{aligned} &(\text{div}\,(\boldsymbol{\sigma} - \boldsymbol{\sigma}_h^{LS}), \mathbf{z}_h) \\ &\quad = (\text{div}\,(\boldsymbol{\sigma} - \boldsymbol{\sigma}_h^{LS}), \text{div}\,\Xi) \\ &\quad = (\text{div}\,(\boldsymbol{\sigma} - \boldsymbol{\sigma}_h^{LS}), \text{div}\,\Xi) + (\mathcal{A}(\boldsymbol{\sigma} - \boldsymbol{\sigma}_h^{LS}) - \varepsilon(\mathbf{u} - \mathbf{u}_h^{LS}), \mathcal{A}\Xi - \varepsilon(\boldsymbol{\eta})) \\ &\quad = (\text{div}\,(\boldsymbol{\sigma} - \boldsymbol{\sigma}_h^{LS}), \text{div}\,(\Xi - \Xi_h^{HR})) \\ &\qquad + (\mathcal{A}(\boldsymbol{\sigma} - \boldsymbol{\sigma}_h^{LS}) - \varepsilon(\mathbf{u} - \mathbf{u}_h^{LS}), \mathcal{A}(\Xi - \Xi_h^{HR}) - \varepsilon(\boldsymbol{\eta} - \boldsymbol{\eta}_h)) \\ &\quad = (\mathcal{A}(\boldsymbol{\sigma} - \boldsymbol{\sigma}_h^{LS}) - \varepsilon(\mathbf{u} - \mathbf{u}_h^{LS}), \mathcal{A}(\Xi - \Xi_h^{HR}) - \varepsilon(\boldsymbol{\eta} - \boldsymbol{\eta}_h)) \end{aligned}$$

holds due to (15), (18) and the fact that $\text{div}\,\Xi_h^{HR} = \text{div}\,\Xi = \mathbf{z}_h$ is satisfied. Combining this with (22) leads to

$$\begin{aligned} \|\text{div}\,\boldsymbol{\sigma}_h^{LS} + \mathbf{f}_h\| &\le \|\mathcal{A}(\boldsymbol{\sigma} - \boldsymbol{\sigma}_h^{LS}) - \varepsilon(\mathbf{u} - \mathbf{u}_h^{LS})\| \\ &\qquad \sup_{\Xi} \frac{\|\mathcal{A}(\Xi - \Xi_h^{HR}) - \varepsilon(\boldsymbol{\eta} - \boldsymbol{\eta}_h)\|}{\|\text{div}\,\Xi\|} \\ &\lesssim \left(\|\boldsymbol{\sigma} - \boldsymbol{\sigma}_h^{LS}\| + \|\varepsilon(\mathbf{u} - \mathbf{u}_h^{LS})\|\right) \\ &\qquad \sup_{\Xi} \frac{\|\Xi - \Xi_h^{HR}\| + \|\varepsilon(\boldsymbol{\eta} - \boldsymbol{\eta}_h)\|}{\|\text{div}\,\Xi\|} \\ &\lesssim h^\alpha \left(\|\boldsymbol{\sigma} - \boldsymbol{\sigma}_h^{LS}\| + \|\varepsilon(\mathbf{u} - \mathbf{u}_h^{LS})\|\right) \end{aligned} \quad (24)$$

due to (12) and our general regularity assumption from Sect. 1. □

Theorem 3.2 states that the error associated with momentum balance converges of higher order. In particular, if $\mathbf{f} \in H^\alpha(\Omega)^d$ is assumed for the right-hand side, then

$$\|\mathrm{div}\,(\boldsymbol{\sigma} - \boldsymbol{\sigma}_h^{LS})\| \lesssim h^\alpha \left(\|\boldsymbol{\sigma} - \boldsymbol{\sigma}_h^{LS}\| + \|\varepsilon(\mathbf{u}) - \varepsilon(\mathbf{u}_h^{LS})\| + \|\mathbf{f}\|\right) \tag{25}$$

holds. One implication of (25) is concerned with the approximation of boundary traces. For the approximation of the resultant traction forces,

$$\langle(\boldsymbol{\sigma} - \boldsymbol{\sigma}_h^{LS}) \cdot \mathbf{n}, \mathbf{e}\rangle_{L^2(\partial\Omega)} = (\mathrm{div}\,(\boldsymbol{\sigma} - \boldsymbol{\sigma}_h^{LS}), \mathbf{e}) \lesssim \|\mathrm{div}\,(\boldsymbol{\sigma} - \boldsymbol{\sigma}_h^{LS})\|\,\|\mathbf{e}\| \tag{26}$$

holds for any constant displacement field $\mathbf{e} \in \mathbb{R}^d$. A further implication, which is seen best directly in (24), is that the second term in the least-squares functional (14) dominates if $(\boldsymbol{\sigma}_h^{LS}, \mathbf{u}_h^{LS})$ is inserted. This property can be of use, in particular, in the study of the functional as an a posteriori error estimator.

4 Stress Reconstruction for Displacement-Pressure Approaches

The most commonly used approach to compute finite element approximations for the linear elasticity model is based on minimizing the energy in (3) among all $\mathbf{v} \in H^1_{\Gamma_D}(\Omega)^d$. The solution $\mathbf{u} \in H^1_{\Gamma_D}(\Omega)^d$ satisfies

$$\int_\Omega (2\mu\, \varepsilon(\mathbf{u}) : \varepsilon(\mathbf{v}) + \lambda\, (\mathrm{div}\, \mathbf{u})\, (\mathrm{div}\, \mathbf{v}))\; dx = \int_\Omega \mathbf{f} \cdot \mathbf{v}\, dx + \int_{\Gamma_N} \mathbf{g} \cdot \mathbf{v}\, ds$$

or, in short notation,

$$2\mu\, (\varepsilon(\mathbf{u}), \varepsilon(\mathbf{v})) + \lambda\, (\mathrm{div}\, \mathbf{u}, \mathrm{div}\, \mathbf{v}) = (\mathbf{f}, \mathbf{v}) + \langle \mathbf{g}, \mathbf{v}\rangle_{L^2(\Gamma_N)} \tag{27}$$

for all $\mathbf{v} \in H^1_{\Gamma_D}(\Omega)^d$. Obviously, this formulation becomes problematic as the Lamé parameter λ tends to ∞ which is the case for incompressible materials. One possible remedy is to introduce a new pressure-like variable $p = \lambda\, \mathrm{div}\, \mathbf{u}$ which leads to the saddle-point problem

$$\begin{aligned} 2\mu\, (\varepsilon(\mathbf{u}), \varepsilon(\mathbf{v})) + (p, \mathrm{div}\, \mathbf{v}) &= (\mathbf{f}, \mathbf{v}) + \langle \mathbf{g}, \mathbf{v}\rangle_{L^2(\Gamma_N)} \\ (\mathrm{div}\, \mathbf{u}, q) - \frac{1}{\lambda}(p, q) &= 0 \end{aligned} \tag{28}$$

for all $\mathbf{v} \in H^1_{\Gamma_D}(\Omega)^d$ and $q \in L^2(\Omega)$. This saddle-point problem is a regular perturbation of the Stokes problem modeling incompressible fluid flow (which coincides with the limiting case $\lambda = \infty$) and as such can be treated with any inf-sup stable finite element pair $(\mathbf{V}_h, Q_h)$ for the Stokes equations, cf. Boffi et al. (2013, Sect. 4.3). The resulting finite-dimensional saddle-point problem is then to find $\mathbf{u}_h \in \mathbf{V}_h$ and

$p_h \in Q_h$ such that

$$\begin{aligned} 2\mu\,(\varepsilon(\mathbf{u}_h), \varepsilon(\mathbf{v}_h)) + (p_h, \operatorname{div} \mathbf{v}_h) &= (\mathbf{f}, \mathbf{v}_h) + \langle \mathbf{g}, \mathbf{v}_h \rangle_{L^2(\Gamma_N)} \\ (\operatorname{div} \mathbf{u}_h, q_h) - \frac{1}{\lambda}(p_h, q_h) &= 0 \end{aligned} \tag{29}$$

holds for all $\mathbf{v}_h \in \mathbf{V}_h$ and $q_h \in Q_h$. Since we are interested in the approximation quality of the stresses $\boldsymbol{\sigma}(\mathbf{u}, p) = 2\mu\varepsilon(\mathbf{u}) + p\,\mathbf{I}$ computed from approximations to $\mathbf{u}$ and p, combinations seem favorable, where the error $\|\varepsilon(\mathbf{u} - \mathbf{u}_h)\|$ converges at the same order as $\|p - p_h\|$. Such a combination is given, for example, by the Taylor–Hood elements (continuously quadratic for $\mathbf{V}_h$ with continuously linear for Q_h) and their higher-order generalizations, cf. Boffi et al. (2013, Sect. 8.8). Another possibility is the use of the quadratic nonconforming elements introduced in Fortin and Soulie (1983) for $\mathbf{V}_h$ combined with discontinuous piecewise linears. Since these elements have some favorable properties with respect to the associated derived stresses $\boldsymbol{\sigma}(\mathbf{u}_h, p_h)$, we will investigate them more closely.

The Quadratic Nonconforming Elements by Fortin–Soulie

With respect to a triangulation $\mathcal{T}_h$ of Ω with the corresponding set of sides (edges for $d = 2$, faces for $d = 3$) denoted by $\mathcal{S}_h$, the quadratic nonconforming finite element space is defined by

$$\begin{aligned} \mathbf{V}_h^{FS} = \{\mathbf{v}_h \in L^2(\Omega)^d : \ & \mathbf{v}_h|_T \in P_2(T)^d \text{ for all } T \in \mathcal{T}_h, \\ & \langle [\![\mathbf{v}_h]\!]_S, \mathbf{z} \rangle_{L^2(S)} = 0 \text{ for all } \mathbf{z} \in P_1(S)^d,\ S \in \mathcal{S}_h \cap \Omega, \\ & \langle \mathbf{v}_h, \mathbf{z} \rangle_{L^2(S)} = 0 \text{ for all } \mathbf{z} \in P_1(S)^d,\ S \in \mathcal{S}_h \cap \Gamma_D\}, \end{aligned} \tag{30}$$

where $[\![\cdot]\!]_S$ denotes the jump across the side S. It is necessary to go to quadratic nonconforming elements since the linear nonconforming elements by Crouzeix–Raviart do not satisfy the discrete Korn's inequality, in general, if $\Gamma_N \neq \emptyset$. That such an inequality, which reads

$$\sum_{T \in \mathcal{T}_h} \|\nabla \mathbf{v}_h\|_{L^2(T)}^2 \lesssim \sum_{T \in \mathcal{T}_h} \|\varepsilon(\mathbf{v}_h)\|_{L^2(T)}^2 \text{ for all } \mathbf{v}_h \in \mathbf{V}_h, \tag{31}$$

holds for $\mathbf{V}_h = \mathbf{V}_h^{FS}$ (under our assumption that $\Gamma_D \neq \emptyset$) is a consequence of (Brenner 2003, Theorem 3.1). The validity of (31) is required for the well-posedness of the variational formulation which now consists in finding $\mathbf{u}_h \in \mathbf{V}_h^{FS}$ and $p_h \in Q_h^{FS} := \{q_h \in L^2(\Omega) : q_h|_T \in P_1(T)\}$ such that

$$\begin{aligned} 2\mu \sum_{T \in \mathcal{T}_h} (\varepsilon(\mathbf{u}_h), \varepsilon(\mathbf{v}_h))_{L^2(T)} + \sum_{T \in \mathcal{T}_h} (p_h, \operatorname{div} \mathbf{v}_h)_{L^2(T)} &= (\mathbf{f}, \mathbf{v}_h) + \langle \mathbf{g}, \mathbf{v}_h \rangle_{0,\Gamma_N} \\ \sum_{T \in \mathcal{T}_h} (\operatorname{div} \mathbf{u}_h, q_h)_{L^2(T)} - \frac{1}{\lambda}(p_h, q_h) &= 0 \end{aligned} \tag{32}$$

is valid for all $\mathbf{v}_h \in \mathbf{V}_h^{FS}$, $q_h \in Q_h^{FS}$. As was already described in the original papers by Fortin and Soulie (1983) and Fortin (1985), the quadratic nonconforming space can be written as $\mathbf{V}_h^{FS} = \mathbf{V}_h^{TH} + \mathbf{B}_h^{NC}$, where $\mathbf{V}_h^{TH}$ is (component-wise) the standard space of conforming quadratic elements and $\mathbf{B}_h^{NC}$ denotes (again component-wise) a suitable space of nonconforming bubble functions. In the two-dimensional case, this nonconforming bubble space is given by

$$\begin{aligned}\mathbf{B}_h^{NC,2} = \{\mathbf{b}_h \in L^2(\Omega)^2 : \mathbf{b}_h|_T \in P_2(T)^2 \text{ for all } T \in \mathcal{T}_h,\\ \langle \mathbf{v}_h, \mathbf{z}\rangle_{L^2(S)} = 0 \text{ for all } \mathbf{z} \in P_1(S)^2,\ S \in \mathcal{S}_h\},\end{aligned}$$

i.e., there is exactly one nonconforming bubble function in $\mathbf{B}_h^{NC,2}$ per triangle. We denote the corresponding one-dimensional space by $\mathbf{B}_h^{NC,2}(T)$. In the three-dimensional case, the nonconforming bubble space is given by

$$\begin{aligned}\mathbf{B}_h^{NC,3} = &\{\mathbf{b}_h \in L^2(\Omega)^3 : \mathbf{b}_h|_T \in P_2(T)^3 \text{ for all } T \in \mathcal{T}_h,\\ &\langle \mathbf{v}_h, \mathbf{z}\rangle_{L^2(S)} = 0 \text{ for all } \mathbf{z} \in P_1(S)^3,\ S \in \mathcal{S}_h\}\\ &+ \{\mathbf{b}_h \in L^2(\Omega)^3 : \mathbf{b}_h|_T \in P_2(T)^3 \text{ for all } T \in \mathcal{T}_h,\ \mathbf{v}_h|_S \in \mathbf{B}_h^{NC,2}(S)\\ &\text{and } \langle [\![\mathbf{v}_h]\!]_S, \mathbf{z}\rangle_{L^2(S)} = 0 \text{ for all } \mathbf{z} \in P_1(S)^3,\ S \in \mathcal{S}_h\}.\end{aligned}$$

The first part of $\mathbf{B}_h^{NC,3}$ consists of exactly one nonconforming bubble function per tetrahedra, again denoted by $\mathbf{B}_h^{NC,3}(T)$. The second part is made up of two-dimensional nonconforming bubble functions $B_h^{NC,2}(S)$ for each face $S \in \mathcal{S}_h$ extended suitably into the two neighboring tetrahedra. It should be kept in mind that the representation $\mathbf{V}_h^{FS} = \mathbf{V}_h^{TH} + \mathbf{B}_h^{NC}$ is not a direct sum. Globally constant functions can be expressed in two different ways in these subspaces, in general. Moreover, in the three-dimensional case, the representation of conforming piecewise linear functions is not unique.

The following result was also already contained in the original papers by Fortin and Soulie (1983) and Fortin (1985) including the proof given below.

Proposition 4.1 *Assume that* $\mathbf{f} \in L^2(\Omega)^d$ *is piecewise constant with respect to the triangulation* $\mathcal{T}_h$ *of* $\Omega \subset \mathbb{R}^d$, $d = 2$ *or* 3 *and that* $\mathbf{g} \in L^2(\Gamma_N)^d$ *is piecewise linear with respect to the subset of sides* $\mathcal{S}_{h,N} = \mathcal{S}_h \cap \Gamma_N$ *associated with the Neumann boundary. If we denote by* $\mathcal{S}_{h,i} = \mathcal{S}_h \cap \Omega$ *the subset of sides interior to the domain, then the (piecewise linear) stresses* $\tilde{\boldsymbol{\sigma}}_h(\mathbf{u}_h, p_h) = 2\mu\varepsilon(\mathbf{u}_h) + p_h\,\mathbf{I}$ *computed from the solution* $(\mathbf{u}_h, p_h) \in \mathbf{V}_h^{FS} \times Q_h^{FS}$ *of (29) satisfy*

$$\mathbf{f} + \operatorname{div} \tilde{\boldsymbol{\sigma}}_h(\mathbf{u}_h, p_h) = 0 \textit{ piecewise for all } T \in \mathcal{T}_h, \tag{33}$$

$$\begin{aligned}\langle \mathbf{g} - \tilde{\boldsymbol{\sigma}}_h(\mathbf{u}_h, p_h)\cdot\mathbf{n}, \mathbf{e}_i\rangle_{L^2(S)} &= 0 \textit{ for all } S \in \mathcal{S}_{h,N},\\ \langle [\![\tilde{\boldsymbol{\sigma}}_h(\mathbf{u}_h, p_h)\cdot\mathbf{n}]\!]_S, \mathbf{e}_i\rangle_{L^2(S)} &= 0 \textit{ for all } S \in \mathcal{S}_{h,i},\end{aligned} \tag{34}$$

where $\mathbf{e}_i \in \mathbb{R}^d$ *denotes the* i*-th unit vector.*

Proof Inserting a nonconforming bubble function $\mathbf{b}_T \in \mathbf{B}_h^{NC}$ with support restricted to T as test function into (32) leads to

$$\begin{aligned} 0 &= \langle \mathbf{g}, \mathbf{b}_T \rangle_{0,\Gamma_N \cap \partial T} + (\mathbf{f}, \mathbf{b}_T)_{L^2(T)} - (\widetilde{\boldsymbol{\sigma}}_h(\mathbf{u}_h, p_h), \varepsilon(\mathbf{b}_T))_{L^2(T)} \\ &= \langle \mathbf{g}, \mathbf{b}_T \rangle_{0,\Gamma_N \cap \partial T} - \langle \widetilde{\boldsymbol{\sigma}}_h(\mathbf{u}_h, p_h) \cdot \mathbf{n}, \mathbf{b}_T \rangle_{0,\partial T} + (\mathbf{f} + \operatorname{div} \widetilde{\boldsymbol{\sigma}}_h(\mathbf{u}_h, p_h), \mathbf{b}_T)_{L^2(T)} \\ &= (\mathbf{f} + \operatorname{div} \widetilde{\boldsymbol{\sigma}}_h(\mathbf{u}_h, p_h), \mathbf{b}_T)_{L^2(T)}, \end{aligned}$$

where the fact was used that $\langle \mathbf{s}, \mathbf{b}_T \rangle_{L^2(S)} = 0$ for all $\mathbf{s} \in P_1(S)^d$, $S \subset \partial T$.

The term $\mathbf{f} + \operatorname{div} \widetilde{\boldsymbol{\sigma}}_h(\mathbf{u}_h, p_h)$, constant on T, must therefore vanish.

For all test functions $\mathbf{v}_h \in \mathbf{V}_h^{FS}$, we therefore get from (32) that

$$\begin{aligned} 0 &= \langle \mathbf{g}, \mathbf{v}_h \rangle_{L^2(\Gamma_N)} + (\mathbf{f}, \mathbf{v}_h) - \sum_{T \in \mathcal{T}_h} (\widetilde{\boldsymbol{\sigma}}_h(\mathbf{u}_h, p_h), \varepsilon(\mathbf{v}_h))_{L^2(T)} \\ &= \sum_{S \in \mathcal{S}_{h,N}} \langle \mathbf{g} - \widetilde{\boldsymbol{\sigma}}_h(\mathbf{u}_h, p_h) \cdot \mathbf{n}, \mathbf{v}_h \rangle_{L^2(S)} - \sum_{S \in \mathcal{S}_{h,i}} \langle [\![\widetilde{\boldsymbol{\sigma}}_h(\mathbf{u}_h, p_h) \cdot \mathbf{n}]\!]_S, \mathbf{v}_h \rangle_{L^2(S)} \end{aligned}$$

holds. We pick one of the sides $S \in \mathcal{S}_h$ and choose the test function $\mathbf{v}_h \in \mathbf{V}_h^{FS}$ in such a way that in the sum above only the term associated with this particular side does not vanish. In two dimensions, this is achieved using a conforming piecewise quadratic function that vanishes on all edges besides S. In three dimensions, the nonconforming bubble function corresponding to the face S has the desired properties (note that $[\![\widetilde{\boldsymbol{\sigma}}_h(\mathbf{u}_h, p_h) \cdot \mathbf{n}]\!]_S$ is of degree 1 on all faces). The symmetry properties of the chosen test functions with respect to S finally implies (34). □

The properties of $\widetilde{\boldsymbol{\sigma}}_h(\mathbf{u}_h, p_h)$ proven in Proposition 4.1 can be used to get an efficient stress reconstruction $\boldsymbol{\sigma}_h^R \in H(\operatorname{div}, \Omega)^d$ by Raviart–Thomas elements of next-to-lowest order $\mathbf{\Pi}_h^1$. We will now explain how such a construction can be done in an element-wise fashion. In the two-dimensional case this is equivalent to the technique described in Kim (2012). The stress reconstruction $\boldsymbol{\sigma}_h^R \in \mathbf{\Pi}_h^1$ is determined on each element $T \in \mathcal{T}_h$ by the following conditions:

$$\begin{aligned} \boldsymbol{\sigma}_h^R\big|_T \cdot \mathbf{n} &= \{\{\widetilde{\boldsymbol{\sigma}}_h(\mathbf{u}_h, p_h)\}\}_S \cdot \mathbf{n} \text{ for all } S \subset \partial T, \\ \operatorname{div} \boldsymbol{\sigma}_h^R\big|_T &= \pi_h^1 \mathbf{f}\big|_T, \end{aligned} \tag{35}$$

where $\{\{\cdot\}\}_S$ stands for the average value on S between the two adjacent elements (set $\{\{\widetilde{\boldsymbol{\sigma}}_h(\mathbf{u}_h, p_h)\}\}_S \cdot \mathbf{n} = \mathbf{g}$ on all sides $S \subset \Gamma_N$) and π_h^1 denotes the $L^2(\Omega)$ projection onto the piecewise linear (possibly discontinuous) functions on $\mathcal{T}_h$. The first line in (35) coincides with the standard interpolation conditions on the sides $S \subset \partial T$ for next-to-lowest-order Raviart–Thomas elements, cf. Boffi et al. (2013, Example 2.5.3). It remains to be shown that, in the situation encountered here, the remaining d interpolation conditions in (Boffi et al. 2013, Example 2.5.3) are equivalent to the second line in (35). To this end, note that

$$
\begin{aligned}
(\operatorname{div} \boldsymbol{\sigma}_h^R, \mathbf{e}_i)_{L^2(T)} &= \langle \boldsymbol{\sigma}_h^R \cdot \mathbf{n}, \mathbf{e}_i \rangle_{L^2(\partial T)} = \langle \widetilde{\boldsymbol{\sigma}}_h(\mathbf{u}_h, p_h) \cdot \mathbf{n}, \mathbf{e}_i \rangle_{L^2(\partial T)} \\
&= (\operatorname{div} \widetilde{\boldsymbol{\sigma}}_h(\mathbf{u}_h, p_h), \mathbf{e}_i)_{L^2(T)} = (\pi_h^0 \mathbf{f}, \mathbf{e}_i)_{L^2(T)}
\end{aligned}
$$

holds, where (34) is used in the first line and (33) in the second line. This means that $\pi_h^0(\operatorname{div} \boldsymbol{\sigma}_h^R)\big|_T = \pi_h^0 \mathbf{f}\big|_T$ and the second condition of (35) consists of only d linear equations at most which may be used to satisfy the remaining interpolation conditions.

The construction is rather simple and consists of the following two steps:

(i) Compute, on each element T, an affine function $\boldsymbol{\sigma}_h^{R,0}\Big|_T \in P_1(T)^d$ which satisfies the first set of conditions in (35). These are $d(d+1)$ conditions for $d(d+1)$ coefficients and amounts to the assignment of the appropriate degrees of freedom depending on the finite element basis used. This results in an approximation $\boldsymbol{\sigma}_h^{R,0} \in H(\operatorname{div}, \Omega)^d$ with $\boldsymbol{\sigma}_h^{R,0} \cdot \mathbf{n} = \mathbf{g}$ on Γ_N and piecewise constant $\operatorname{div} \boldsymbol{\sigma}_h^{R,0}$ (which may also be interpreted as approximation in the BDM_1 space, cf. Kim 2012).

(ii) Update for better divergence approximation (if $\mathbf{f}$ is not constant on T) by adjusting the coefficients associated with the interior degrees of freedom.

The above reconstruction results in a stress approximation with similar properties as for the Hellinger–Reissner formulation using the Boffi–Brezzi–Fortin elements studied at the end of Sect. 2. In particular, the momentum balance error is minimized and optimal order approximation of the stress is achieved with respect to the $L^2(\Omega)$ norm.

Computational Comparison for Incompressible Linear Elasticity

We close the part of this contribution associated with linear elasticity by some two-dimensional computational results in order to provide some insight on the actual behavior of the methods introduced above.

Example 1 The underlying domain is a quadrilateral with vertices at (0, 0), (0.48, 0.44), (0.48, 0.6) and (0, 0.44), commonly known as Cook's membrane. It is fixed ($\mathbf{u} = \mathbf{0}$) at the left edge of the boundary ($x_1 = 0$) while a uniform traction force pointing upwards ($\boldsymbol{\sigma} \cdot \mathbf{n} = (0, 1)$) is applied at the right edge ($x_1 = 0.48$). At the remaining part of the boundary it is kept in equilibrium ($\boldsymbol{\sigma} \cdot \mathbf{n} = \mathbf{0}$). All the computations are done for the incompressible limit $\lambda = \infty$ while μ is set to 1. Figure 1 shows the initial triangulation with 44 elements. The results on a sequence of uniform refinements starting from this initial triangulation are compared for different methods.

Table 1 shows the resultant traction force

$$\int_{\Gamma_D} \mathbf{n} \cdot (\boldsymbol{\sigma} \cdot \mathbf{n}) \, ds$$

in normal direction acting on the fixed left boundary part calculated from different finite element approximations of displacement-pressure type. Due to the divergence

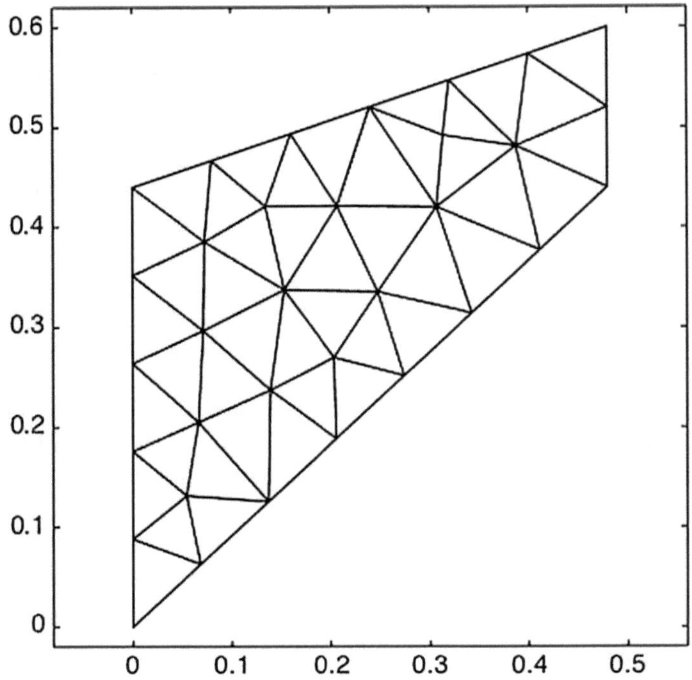

Fig. 1 Initial triangulation for Cook's membrane

Table 1 Resultant normal traction for displacement-pressure methods

l	$\|\mathcal{T}_h\|$	Taylor–Hood	P2/P0	Fortin–Soulie
0	44	3.8467×10^{-2}	3.3047×10^{-2}	-4.0246×10^{-16}
1	176	2.8816×10^{-2}	2.4458×10^{-2}	2.3592×10^{-16}
2	704	2.0867×10^{-2}	1.7671×10^{-2}	2.3384×10^{-15}
3	2816	1.4829×10^{-2}	1.2547×10^{-2}	2.5180×10^{-14}
4	11264	1.0383×10^{-2}	8.7743×10^{-3}	-1.6708×10^{-14}
5	45056	7.1988×10^{-3}	6.0760×10^{-3}	-2.0714×10^{-13}
6	180224	4.9631×10^{-3}	4.1841×10^{-3}	-3.4016×10^{-13}

theorem the exact value is 0. Obviously, the evaluation of the Taylor–Hood and P2/P0 (piecewise constant pressure) approximations on the boundary does not reproduce this resultant traction force exactly while this is the case for the Fortin–Soulie approximations in accordance with Proposition 4.1. Another interpretation of these results is that the piecewise linear stress approximations in $L^2(\Omega)^{d \times d}$ are not suitable for their evaluation on the boundary, in general. For the nonconforming Fortin–Soulie approximations, the trace on the Dirichlet boundary coincides with those associated with the recovered stress in $H(\mathrm{div}, \Omega)^d$ and does therefore produce a good approximation of the boundary tractions.

Figure 2 shows the quality of the normal traction approximation for different displacement-pressure elements (after six uniform refinements) in the neighborhood of the singularity at the left upper vertex of Cook's membrane. The shaded graph is associated with the Taylor–Hood element pair, the dotted lines are for P2/P0 and the solid straight lines for the Fortin–Soulie elements. The black curve represents the correct traction force distribution and was computed on an adaptively refined triangulation by the least-squares approach of Sect. 3. Away from the singularity all approximations are quite accurate, while severe differences are visible in the quality how well the singular behavior is resolved. The Fortin–Soulie element does perform much better and therefore justifies its larger number of degrees of freedom.

Table 2 lists the same quantities as Table 1 but this time compares the different methods from Sects. 2, 3 and 4. Due to the exact momentum conservation, the approximations with the Boffi–Brezzi–Fortin elements based on the Hellinger–Reissner principle produce the resultant traction forces perfectly (up to roundoff errors). The first-order system least-squares approach does not compute the resultant traction force exactly but to quite acceptable accuracy while the numbers associated with the

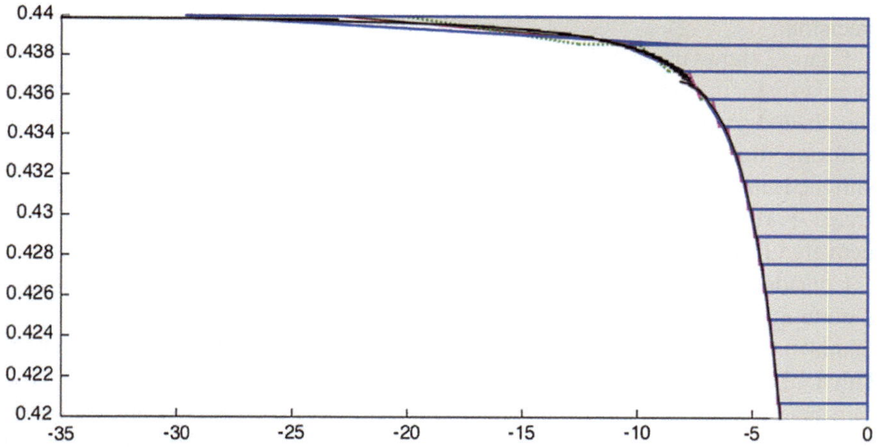

Fig. 2 Approximation of normal traction near singularity

Table 2 Resultant normal traction for stress-based methods

l	$\|\mathcal{T}_h\|$	HR (BBF)	LS (RT1/P2)	Recov. from FS
0	44	-1.6676×10^{-13}	1.9608×10^{-4}	7.6328×10^{-16}
1	176	-1.0358×10^{-12}	9.2856×10^{-5}	-9.7145×10^{-17}
2	704	1.5558×10^{-11}	4.3819×10^{-5}	-4.8260×10^{-15}
3	2816	3.2884×10^{-10}	2.0679×10^{-5}	-1.5449×10^{-14}
4	11264	1.8848×10^{-10}	9.7529×10^{-6}	-6.0172×10^{-14}
5	45056	8.9723×10^{-9}	4.6026×10^{-6}	-2.2891×10^{-14}

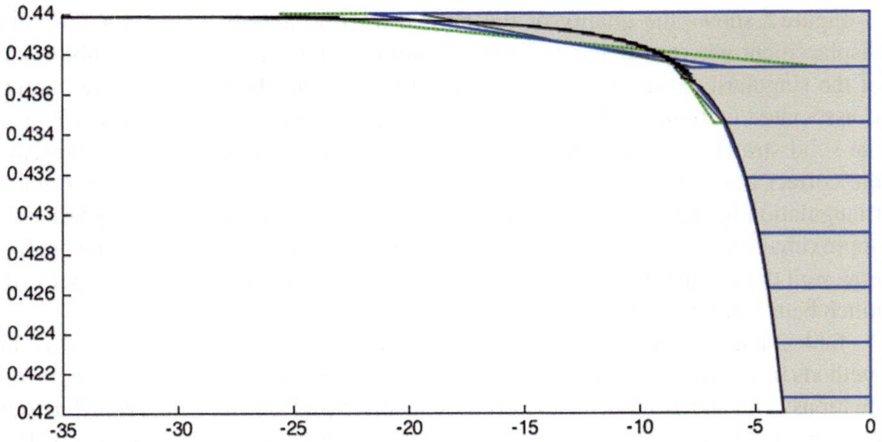

Fig. 3 Approximation of normal traction near singularity

recovered stresses from the Fortin–Soulie elements are again correct up to working precision.

Figure 3 shows the distribution of the normal traction at the left boundary near the singularity for the three different approaches of Table 2 after five uniform refinements. The shaded graph belongs to the first-order system least-squares approach which performs slightly worse than the two alternatives. The dotted lines are associated with the Hellinger–Reissner principle using the finite element combination of Boffi–Brezzi–Fortin and seem to resolve the singularity slightly better than the stresses recovered from the Fortin–Soulie elements (solid straight lines).

Considering linear elasticity computations, the stress reconstruction approach is quite attractive since the global system that needs to be solved involves fewer unknowns and the reconstructed stresses are of a similar accuracy as those obtained with a mixed method of saddle-point or least-squares type. The situation may, however, be different for more complicated models where the stress is involved more directly. This is the case, for instance, in the context of inelastic behavior caused by stress components exceeding a certain limit where the direct treatment of stresses in the variational formulation is advantageous (cf. Reddy 1992 for a mixed approach of saddle-point type, Starke 2007; Schwarz et al. 2009 for a least-squares type approach). A comparison of the different approaches for the nonlinear problems arising in association with hyperelastic material models will be given in Sect. 5.

5 Extension to Finite-Strain Hyperelasticity

In the previous sections, the linear elasticity model was considered which is derived under the assumption of small strains. Now we switch to the more general case of finite strains with hyperelastic material models. More details on the validity of these models and their mathematical aspects can be found, e.g., in (Ciarlet 1988, Chap. 4). Based on the deformation gradient given by $\boldsymbol{F} = \boldsymbol{F}(\mathbf{u}) := \mathbf{I} + \nabla\mathbf{u}$, the left and right Cauchy–Green strain tensors are defined as $\mathbf{B} = \mathbf{B}(\mathbf{u}) := \boldsymbol{F}(\mathbf{u})\boldsymbol{F}(\mathbf{u})^T$ and $\boldsymbol{C} = \boldsymbol{C}(\mathbf{u}) := \boldsymbol{F}(\mathbf{u})^T\boldsymbol{F}(\mathbf{u})$, respectively. This nonlinear dependence of strains to displacements constitutes the geometrically nonlinear nature of this model. In addition, there is also a nonlinearity in the material law describing the relation between stresses and strains. This originates from a stored energy function $\psi : \mathbb{R}^{3\times 3}_{\text{sym}} \to \mathbb{R}$ which generalizes (4) and is no longer quadratic. Again, we restrict ourselves to a homogeneous material which means that ψ does not explicitly depend on the location $\mathbf{x} \in \Omega$.

Minimizing the total energy

$$I(\mathbf{v}) := \int_\Omega \psi(\boldsymbol{C}(\mathbf{v}))\,dx - \int_\Omega \mathbf{f}\cdot\mathbf{v}\,dx - \int_{\Gamma_N} \mathbf{g}\cdot\mathbf{v}\,ds \tag{36}$$

among all admissible displacements $\mathbf{v} \in \mathbf{V}$ for some suitable space $\mathbf{V}$ is again equivalent to finding a solution $\mathbf{u} \in \mathbf{V}$ of the variational problem

$$(\boldsymbol{P}(\mathbf{u}), \nabla\mathbf{v}) = (\mathbf{f}, \mathbf{v}) + \langle \mathbf{g}, \mathbf{v}\rangle_{0,\Gamma_N} \text{ for all } \mathbf{v} \in \mathbf{V}, \tag{37}$$

where $\boldsymbol{P}(\mathbf{u}) := \partial_{\boldsymbol{F}}\psi(\boldsymbol{C})(\mathbf{u})$ denotes the first Piola–Kirchhoff stress tensor. We assume that our problem is sufficiently regular so that we can choose $\mathbf{V} = W^{1,p}_{\Gamma_D}(\Omega)^3$ for $p > 2$ as our solution space for (37). In that case, we may also write (37) as a first-order system as

$$\begin{aligned} -\operatorname{div}\boldsymbol{P} &= \mathbf{f} && \text{in } \Omega \\ \boldsymbol{P} &= \partial_{\boldsymbol{F}}\psi(\boldsymbol{C}) && \text{in } \Omega \\ \boldsymbol{P}\cdot\mathbf{n} &= \mathbf{g} \text{ on } \Gamma_N, && \mathbf{u} = \mathbf{0} \text{ on } \Gamma_D. \end{aligned} \tag{38}$$

The first equation in (38) is an immediate consequence of the physically necessary conservation of linear momentum for a static problem. Conservation of angular momentum for a static problem leads additionally to the symmetry of $\boldsymbol{P}(\mathbf{u})\boldsymbol{F}(\mathbf{u})^T$ which is implicitly contained in the formulations (37) and (38).

For homogeneous isotropic materials it is possible to express the stored energy function ψ by a function $\tilde{\psi} : \mathbb{R}^3 \to \mathbb{R}$, depending on three terms $I_1, I_2, I_3 : \mathbb{R}^{3\times 3} \to \mathbb{R}$, i.e.,

$$\psi(\boldsymbol{C}) = \tilde{\psi}(I_1(\boldsymbol{C}), I_2(\boldsymbol{C}), I_3(\boldsymbol{C})), \quad \boldsymbol{C} = \boldsymbol{F}^T\boldsymbol{F}, \tag{39}$$

with the principal invariants $I_1(\boldsymbol{C}) := \operatorname{tr}(\boldsymbol{C})$, $I_2(\boldsymbol{C}) := \operatorname{tr}(\mathbf{Cof}\,\boldsymbol{C})$ and $I_3(\boldsymbol{C}) := \det \boldsymbol{C}$ (cf. Simo 1998, Theorem 31.1 and Exercise 31.2). Introducing the so-called Kirchhoff stress tensor $\boldsymbol{\tau} := \boldsymbol{P}\boldsymbol{F}^T$, a simple calculation then leads to

$$\boldsymbol{\tau} = 2\frac{\partial\tilde{\psi}}{\partial I_3} I_3(\mathbf{B})\mathbf{I} + 2\left(\frac{\partial\tilde{\psi}}{\partial I_1} + \frac{\partial\tilde{\psi}}{\partial I_2} I_1(\mathbf{B})\right)\mathbf{B} - 2\frac{\partial\tilde{\psi}}{\partial I_2}\mathbf{B}^2 =: \mathcal{G}(\mathbf{B}) \tag{40}$$

or, equivalently, for the second Piola–Kirchhoff stress tensor $\boldsymbol{\Sigma} := \boldsymbol{F}^{-1}\boldsymbol{P}$:

$$\boldsymbol{\Sigma} = 2\left(\frac{\partial\tilde{\psi}}{\partial I_1} + \frac{\partial\tilde{\psi}}{\partial I_2} I_1(\boldsymbol{C})\right)\mathbf{I} - 2\frac{\partial\tilde{\psi}}{\partial I_2}\boldsymbol{C} + 2\frac{\partial\tilde{\psi}}{\partial I_3} I_3(\boldsymbol{C})\boldsymbol{C}^{-1} := \tilde{\mathcal{G}}(\boldsymbol{C}), \tag{41}$$

where $\mathcal{G}$ and $\tilde{\mathcal{G}}$ are mappings from strains into stresses, similar as the fourth-order elasticity tensor $\mathcal{C}$ in linear elasticity.

In the following we assume that $\mathcal{G}(\mathbf{I}) = \mathbf{0} = \tilde{\mathcal{G}}(\mathbf{I})$, i.e., the reference configuration is stress-free, and that $\mathcal{G}, \tilde{\mathcal{G}}$ are continuously differentiable in the identity matrix with $\mathcal{G}'(\mathbf{I})[\boldsymbol{E}] = \frac{1}{2}\mathcal{C}\boldsymbol{E} = \tilde{\mathcal{G}}'(\mathbf{I})[\boldsymbol{E}]$, i.e., consistency of the nonlinear model with the model of linear elasticity (cf. Ciarlet 1988, Sect. 3.8). Since the elasticity tensor $\mathcal{C}$ itself is an isomorphism, the mappings $\mathcal{G}'(\mathbf{I}) = \frac{1}{2}\mathcal{C}$ and $\tilde{\mathcal{G}}'(\mathbf{I}) = \frac{1}{2}\mathcal{C}$ are also isomorphisms. Thus the local inversion theorem (cf. Ciarlet 1988, Theorem 1.2–4) is applicable and guarantees that the inverse mappings $\mathcal{G}^{-1}(\boldsymbol{\tau})$ and $\tilde{\mathcal{G}}^{-1}(\boldsymbol{\Sigma})$ are well-defined in a neighborhood of $\boldsymbol{\tau} = \mathbf{0}$ and $\boldsymbol{\Sigma} = \mathbf{0}$, respectively. Using these considerations we can modify the strong formulation (38) into

$$\begin{aligned} \operatorname{div}\boldsymbol{P} + \mathbf{f} &= \mathbf{0} \quad \text{in } \Omega, \\ \mathcal{G}^{-1}(\boldsymbol{P}\boldsymbol{F}(\mathbf{u})^T) - \mathbf{B}(\mathbf{u}) &= \mathbf{0} \quad \text{in } \Omega, \\ \boldsymbol{P}\cdot\mathbf{n} = \mathbf{g} \text{ on } \Gamma_N, &\quad \mathbf{u} = \mathbf{0} \text{ on } \Gamma_D \end{aligned} \tag{42}$$

using the representation in $\mathbf{B}$ or into

$$\begin{aligned} \operatorname{div}\boldsymbol{P} + \mathbf{f} &= \mathbf{0} \quad \text{in } \Omega, \\ \tilde{\mathcal{G}}^{-1}(\boldsymbol{F}(\mathbf{u})^{-1}\boldsymbol{P}) - \boldsymbol{C}(\mathbf{u}) &= \mathbf{0} \quad \text{in } \Omega, \\ \boldsymbol{P}\cdot\mathbf{n} = \mathbf{g} \text{ on } \Gamma_N, &\quad \mathbf{u} = \mathbf{0} \text{ on } \Gamma_D \end{aligned} \tag{43}$$

using the representation in $\boldsymbol{C}$. Both systems are at least well-defined for small stresses.

5.1 A Least Squares Finite Element Method for Isotropic Hyperelastic Materials

Since $\mathcal{G}'(\mathbf{I}) = \tilde{\mathcal{G}}'(\mathbf{I}) = \frac{1}{2}\mathcal{C}$, the implicit function theorem tells us that $(\mathcal{G}^{-1})'(\mathbf{0}) = (\tilde{\mathcal{G}}^{-1})'(\mathbf{0}) = 2\mathcal{C}^{-1}$. This means that we encounter the same problem as in the linear case, namely, that $\mathcal{G}^{-1}$ and $\tilde{\mathcal{G}}^{-1}$ are not invertible anymore in the incompressible limit. Due to this observation we use the notation $\mathcal{A}$ and $\tilde{\mathcal{A}}$ instead of $\mathcal{G}^{-1}$ and $\tilde{\mathcal{G}}^{-1}$ in (42) and (43). With this in mind we introduce for $\boldsymbol{P} = \boldsymbol{P}^N + \hat{\boldsymbol{P}} \in W^q(\text{div};\Omega)^3 +$

$W^q_{\Gamma_N}(\text{div};\Omega)^3$ (with $\boldsymbol{P}^N \cdot \mathbf{n} = \mathbf{g}$ on Γ_N), $\mathbf{u} \in W^{1,p}_{\Gamma_D}(\Omega)^3$ and $\mathbf{f} \in L^q(\Omega)^3$, $p, q \geq 4$ sufficiently large, the nonlinear operators

$$\mathcal{R}(\boldsymbol{P},\mathbf{u}) := \begin{pmatrix} \operatorname{div} \boldsymbol{P} + \mathbf{f} \\ \mathcal{A}(\boldsymbol{P}\boldsymbol{F}(\mathbf{u})^T) - \mathbf{B}(\mathbf{u}) \end{pmatrix},$$
$$\tilde{\mathcal{R}}(\boldsymbol{P},\mathbf{u}) := \begin{pmatrix} \operatorname{div} \boldsymbol{P} + \mathbf{f} \\ \tilde{\mathcal{A}}(\boldsymbol{F}(\mathbf{u})^{-1}\boldsymbol{P}) - \boldsymbol{C}(\mathbf{u}) \end{pmatrix} \tag{44}$$

for (42) and (43), respectively. Based on these operators we define nonlinear least squares functionals

$$\mathcal{F}(\boldsymbol{P},\mathbf{u}) := \|\mathcal{R}(\boldsymbol{P},\mathbf{u})\|^2 = \|\operatorname{div} \boldsymbol{P} + \mathbf{f}\|^2 + \|\mathcal{A}(\boldsymbol{P}\boldsymbol{F}(\mathbf{u})^T) - \mathbf{B}(\mathbf{u})\|^2 \tag{45}$$

for the formulation in $\mathbf{B}$ and

$$\tilde{\mathcal{F}}(\boldsymbol{P},\mathbf{u}) := \|\tilde{\mathcal{R}}(\boldsymbol{P},\mathbf{u})\|^2 = \|\operatorname{div} \boldsymbol{P} + \mathbf{f}\|^2 + \|\tilde{\mathcal{A}}(\boldsymbol{F}(\mathbf{u})^{-1}\boldsymbol{P}) - \boldsymbol{C}(\mathbf{u})\|^2 \tag{46}$$

for the formulation in $\boldsymbol{C}$.

5.2 *Gauss–Newton Iterative Method*

In the following we restrict ourselves to the minimization problem (45) corresponding to the $\mathbf{B}$-formulation. All further steps below can be handled similarly for the $\boldsymbol{C}$-formulation. The minimization of (45) is carried out iteratively solving a sequence of linearized least squares problems. Since the operator $\mathcal{R}(\boldsymbol{P},\mathbf{u})$ is continuously differentiable with respect to $(\boldsymbol{P},\mathbf{u})$, we can linearize it around a given approximation $(\boldsymbol{P}^{(k)},\mathbf{u}^{(k)}) \in \boldsymbol{P}^N + W^q_{\Gamma_N}(\text{div};\Omega)^3 \times W^{1,p}_{\Gamma_D}(\Omega)^3$. The resulting linearized least squares functional depending on $(\boldsymbol{P}^{(k)},\mathbf{u}^{(k)})$ is given by

$$\begin{aligned} \mathcal{F}^{\text{lin}}(\boldsymbol{Q},\mathbf{v}) &= \mathcal{F}^{\text{lin}}(\boldsymbol{Q},\mathbf{v};\mathcal{R}(\boldsymbol{P}^{(k)},\mathbf{u}^{(k)})) \\ &:= \|\mathcal{R}(\boldsymbol{P}^{(k)},\mathbf{u}^{(k)}) + \mathcal{R}'(\boldsymbol{P}^{(k)},\mathbf{u}^{(k)})[\boldsymbol{Q},\mathbf{v}]\|^2. \end{aligned} \tag{47}$$

The minimizer $(\boldsymbol{Q}^{(k)},\mathbf{u}^{(k)})$ of (47) is then sought in a suitable normed function space $\boldsymbol{\Pi}_{\Gamma_N} \times \mathbf{V}_{\Gamma_D}$, provided that the values q and p are chosen sufficiently large such that $\mathcal{R}(\boldsymbol{P}^{(k)},\mathbf{u}^{(k)})$ and also $\mathcal{R}'(\boldsymbol{P}^{(k)},\mathbf{u}^{(k)})[\boldsymbol{Q},\mathbf{v}]$ are contained in $L^2(\Omega)^3 \times L^2(\Omega)^{3\times 3}$. The linearized minimization problem (47) is equivalent to the variational problem of finding $(\boldsymbol{Q}^{(k)},\mathbf{v}^{(k)}) \in \boldsymbol{\Pi}_{\Gamma_N} \times \mathbf{V}_{\Gamma_D}$ such that

$$\mathcal{B}((\boldsymbol{Q}^{(k)},\mathbf{v}^{(k)}),(\hat{\boldsymbol{Q}},\hat{\mathbf{v}})) = -\left(\mathcal{R}(\boldsymbol{P}^{(k)},\mathbf{u}^{(k)}),\mathcal{R}'(\boldsymbol{P}^{(k)},\mathbf{u}^{(k)})[\hat{\boldsymbol{Q}},\hat{\mathbf{v}}]\right) \tag{48}$$

holds for all $(\hat{\boldsymbol{Q}}, \hat{\mathbf{v}}) \in \boldsymbol{\Pi}_{\Gamma_N} \times \mathbf{V}_{\Gamma_D}$. The bilinear form in (48) is defined on $\boldsymbol{\Pi}_{\Gamma_N} \times \mathbf{V}_{\Gamma_D}$ and given by

$$\mathcal{B}((\boldsymbol{Q}, \mathbf{v}), (\hat{\boldsymbol{Q}}, \hat{\mathbf{v}})) := \left(\mathcal{R}'(\boldsymbol{P}^{(k)}, \mathbf{u}^{(k)})[\boldsymbol{Q}, \mathbf{v}], \mathcal{R}'(\boldsymbol{P}^{(k)}, \mathbf{u}^{(k)})[\hat{\boldsymbol{Q}}, \hat{\mathbf{v}}]\right). \tag{49}$$

For the numerical implementation, a finite-dimensional space $\boldsymbol{\Pi}_h \times \mathbf{V}_h \subset \boldsymbol{\Pi}_{\Gamma_N} \times \mathbf{V}_{\Gamma_D}$ is chosen. Starting with an initial guess $(\boldsymbol{P}_h^{(0)}, \mathbf{u}_h^{(0)}) \in \boldsymbol{P}^N + W^q_{\Gamma_N}(\mathrm{div}; \Omega)^3 \times W^{1,p}_{\Gamma_D}(\Omega)^3$ and setting $k = 0$, the discrete analogue of (48) is then solved in $\boldsymbol{\Pi}_h \times \mathbf{V}_h$ to obtain the correction term $(\boldsymbol{Q}_h^{(k)}, \mathbf{v}_h^{(k)})$. Afterwards the new iterate is set to

$$(\boldsymbol{P}_h^{(k+1)}, \mathbf{u}_h^{(k+1)}) = (\boldsymbol{P}_h^{(k)}, \mathbf{u}_h^{(k)}) + \alpha^{(k)}(\boldsymbol{Q}_h^{(k)}, \mathbf{v}_h^{(k)})$$

where $\alpha^{(k)} > 0$ describes the step length in the direction $(\boldsymbol{Q}_h^{(k)}, \mathbf{v}_h^{(k)})$. The described approach is the well-known Gauss–Newton method combined with a line search strategy, cf. Nocedal and Wright (2006, Chap. 3 and Sect. 10.3). For instance, for the determination of a suitable step length one can use a backtracking line search approach, cf. Nocedal and Wright (2006, Algorithm 3.1). The new iterate $(\boldsymbol{P}_h^{(k+1)}, \mathbf{u}_h^{(k+1)})$ automatically satisfies $\boldsymbol{P}_h^{(k+1)} \cdot \mathbf{n} = \mathbf{g}$ on Γ_N and $\mathbf{u}_h^{(k+1)} = \mathbf{0}$ on Γ_D. An alternative approach for minimizing the linearized problems of the form (47) in finite dimensional spaces is the usage of the Levenberg–Marquardt method which replaces the line search with a trust-region method, cf. Nocedal and Wright (2006, Chap. 4 and Sect. 10.3).

Considering (48) and (49) one has to evaluate the nonlinear operator $\mathcal{R}$ locally for given $(\boldsymbol{P}_h, \mathbf{u}_h)$ at each quadrature point for a numerical implementation. Due to the representations in (44) the problematical part is the evaluation of $\mathcal{A}$. A remedy, which works independently of the used stored energy function, is to solve the problem $\mathcal{G}(\mathbf{B}) = \boldsymbol{\tau}_h$ for given $\boldsymbol{\tau}_h := \boldsymbol{P}_h \boldsymbol{F}(\mathbf{u}_h)^T$ with the help of Newton's method. Assuming a finite λ and a sufficiently small $\boldsymbol{\tau}_h$, the sequence of Newton iterations is given by

$$\mathbf{B}^{(j+1)} = \mathbf{B}^{(j)} + \boldsymbol{\Delta}^{(j)},$$

where $\boldsymbol{\Delta}^{(j)} \in \mathbb{R}^{3\times 3}$ is the solution of

$$\mathcal{G}'(\mathbf{B}^{(j)})[\boldsymbol{\Delta}^{(j)}] = \boldsymbol{\tau}_h - \mathcal{G}(\mathbf{B}^{(j)}). \tag{50}$$

The initial guess $\mathbf{B}^{(0)} = \mathbf{I} \in \mathbb{R}^{3\times 3}$ is at least for small $(\boldsymbol{P}_h, \mathbf{u}_h)$ reasonable, since for $(\boldsymbol{P}_h, \mathbf{u}_h) = (\mathbf{0}, \mathbf{0})$ the solution is given by $\mathbf{B} = \mathbf{I}$. The Eq. (50) can be solved with the help of a linear system of equations with nine unknowns, where the matrix on the left-hand side depends on the old approximation $\mathbf{B}^{(j)}$ and the right-hand side depends on $(\boldsymbol{P}_h, \mathbf{u}_h)$ and $\mathbf{B}^{(j)}$. Applying Newton's method on each quadrature point and on each element of the triangulation is numerically expensive. For a special neo-Hookean material which we consider in the next section it is possible to evaluate the operator $\mathcal{A}$ locally without using Newton's method. Moreover, one can take the limit $\lambda \to \infty$ in the operator $\mathcal{A}$ and can even set $\lambda = \infty$ in this model.

5.3 Least Squares Formulation for neo-Hookean Model

The method described in Sect. 5.1 works generally for an arbitrary isotropic stored energy function ψ, provided that the stresses are not too large such that invertibility of the operators $\mathcal{G}$ and $\tilde{\mathcal{G}}$ in (40) and (41) is ensured. In this section we consider an isotropic material of neo-Hookean type described by

$$\tilde{\psi}_{NH}(I_1, I_3) = \alpha I_1 + \beta I_3 - \frac{\gamma}{2}\ln(I_3), \quad \alpha, \beta, \gamma > 0,$$

with stored energy function

$$\psi_{NH}(\boldsymbol{C}) = \alpha\,\mathrm{tr}(\boldsymbol{C}) + \beta\det(\boldsymbol{C}) - \frac{\gamma}{2}\ln(\det\boldsymbol{C}) \tag{51}$$

via (39) (cf. Ciarlet 1988, Sect. 4.10). With the derivatives $\frac{\partial\tilde{\psi}_{NH}}{\partial I_1} = \alpha$, $\frac{\partial\tilde{\psi}_{NH}}{\partial I_2} = 0$, $\frac{\partial\tilde{\psi}_{NH}}{\partial I_3} = \beta - \frac{\gamma}{2I_3}$ and Eqs. (40) and (41) we achieve

$$\begin{aligned}\mathcal{G}_{NH}(\mathbf{B}) &= 2\alpha\mathbf{B} + (2\beta\det\mathbf{B} - \gamma)\mathbf{I},\\ \tilde{\mathcal{G}}_{NH}(\boldsymbol{C}) &= 2\alpha\mathbf{I} + (2\beta\det\boldsymbol{C} - \gamma)\boldsymbol{C}^{-1}.\end{aligned} \tag{52}$$

With $\mathcal{A}_{NH} = \mathcal{G}_{NH}^{-1}$ and $\tilde{\mathcal{A}}_{NH} = \tilde{\mathcal{G}}_{NH}^{-1}$ denoting the corresponding inverses, we end up with the nonlinear operators

$$\begin{aligned}\mathcal{R}_{NH}(\boldsymbol{P}, \mathbf{u}) &:= \begin{pmatrix}\mathrm{div}\,\boldsymbol{P} + \mathbf{f}\\ \mathcal{A}_{NH}(\boldsymbol{P}\boldsymbol{F}(\mathbf{u})^T) - \mathbf{B}(\mathbf{u})\end{pmatrix},\\ \tilde{\mathcal{R}}_{NH}(\boldsymbol{P}, \mathbf{u}) &:= \begin{pmatrix}\mathrm{div}\,\boldsymbol{P} + \mathbf{f}\\ \tilde{\mathcal{A}}_{NH}(\boldsymbol{F}(\mathbf{u})^{-1}\boldsymbol{P}) - \boldsymbol{C}(\mathbf{u})\end{pmatrix}\end{aligned} \tag{53}$$

in the neo-Hooke case. The derivatives of (52) are given by

$$\begin{aligned}\mathcal{G}'_{NH}(\mathbf{B})[\boldsymbol{E}] &= 2\alpha\boldsymbol{E} + 2\beta(\mathbf{Cof}\,\mathbf{B} : \boldsymbol{E})\mathbf{I},\\ \tilde{\mathcal{G}}'_{NH}(\boldsymbol{C})[\boldsymbol{E}] &= 2\beta(\mathbf{Cof}\,\boldsymbol{C} : \boldsymbol{E})\boldsymbol{C}^{-1} - (2\beta\det\boldsymbol{C} - \gamma)\boldsymbol{C}^{-1}\boldsymbol{E}\boldsymbol{C}^{-1}.\end{aligned} \tag{54}$$

The conditions $\mathcal{G}_{NH}(\mathbf{I}) = \mathbf{0}$ and $\mathcal{G}'_{NH}(\mathbf{I})[\boldsymbol{E}] = \frac{1}{2}\mathcal{C}\boldsymbol{E}$ (or $\tilde{\mathcal{G}}_{NH}(\mathbf{I}) = \mathbf{0}$ and $\tilde{\mathcal{G}}'_{NH}(\mathbf{I})[\boldsymbol{E}] = \frac{1}{2}\mathcal{C}\boldsymbol{E}$, respectively) lead to a linear system of equations for the determination of α, β, γ which is uniquely solvable through

$$\alpha = \frac{\mu}{2}, \quad \beta = \frac{\lambda}{4}, \quad \gamma = \mu + \frac{\lambda}{2}. \tag{55}$$

The derivatives in (54) can be directly inverted. After inserting the coefficients (55) in (54), the inverses are given by

$$\mathcal{G}'_{NH}(\mathbf{B})^{-1}[\boldsymbol{\Sigma}] = \frac{1}{\mu}\left(\boldsymbol{\Sigma} - \frac{\lambda}{2\mu + \lambda\,\mathrm{tr}(\mathbf{Cof\,B})}(\mathbf{Cof\,B} : \boldsymbol{\Sigma})\mathbf{I}\right),$$
$$\tilde{\mathcal{G}}'_{NH}(\boldsymbol{C})^{-1}[\boldsymbol{\Sigma}] = \frac{1}{\mu + \frac{\lambda}{2}(1 - \det \boldsymbol{C})} \qquad (56)$$
$$\boldsymbol{C}\left(\boldsymbol{\Sigma} - \frac{\lambda(\det \boldsymbol{C})^2}{2\mu + \lambda(1 + 2\det \boldsymbol{C})}(\mathbf{Cof}\,\boldsymbol{C}^{-1} : \boldsymbol{\Sigma})\boldsymbol{C}^{-1}\right)\boldsymbol{C}.$$

The inverses are very helpful for the direct calculation of

$$\mathcal{R}'_{NH}(\boldsymbol{P}, \mathbf{u})[\boldsymbol{Q}, \mathbf{v}]$$
$$= \begin{pmatrix} \mathrm{div}\,\boldsymbol{Q} \\ \mathcal{A}'_{NH}(\boldsymbol{P}\boldsymbol{F}(\mathbf{u})^T)[\boldsymbol{Q}\boldsymbol{F}(\mathbf{u})^T + \boldsymbol{P}(\nabla\mathbf{v})^T] - (\nabla\mathbf{v})\boldsymbol{F}(\mathbf{u})^T - \boldsymbol{F}(\mathbf{u})(\nabla\mathbf{v})^T \end{pmatrix}$$

and $\tilde{\mathcal{R}}'_{NH}(\boldsymbol{P}, \mathbf{u})[\boldsymbol{Q}, \mathbf{v}]$, respectively. Inserting $\mathbf{B} = \mathbf{I} = \boldsymbol{C}$ in (56) for finite λ leads to $\mathcal{G}'_{NH}(\mathbf{I})^{-1}[\boldsymbol{\Sigma}] = 2\mathcal{C}^{-1}\boldsymbol{\Sigma} = \tilde{\mathcal{G}}'_{NH}(\mathbf{I})^{-1}[\boldsymbol{\Sigma}]$, as expected. For $\lambda \to \infty$ the first equation in (56) becomes

$$\mathcal{G}'_{NH}(\mathbf{B})^{-1}[\boldsymbol{\Sigma}] = \frac{1}{\mu}\left(\boldsymbol{\Sigma} - \frac{1}{\mathrm{tr}(\mathbf{Cof\,B})}(\mathbf{Cof\,B} : \boldsymbol{\Sigma})\mathbf{I}\right)$$

and coincides for $\mathbf{B} = \mathbf{I}$ with $2\,\mathcal{A}\boldsymbol{\Sigma}$ from the linear elasticity case. The well-posedness of $\tilde{\mathcal{G}}'_{NH}(\boldsymbol{C})^{-1}$ for given strain $\boldsymbol{C} := \tilde{\mathcal{A}}_{NH}(\boldsymbol{\Sigma})$ and the identity $\tilde{\mathcal{G}}'_{NH}(\boldsymbol{C})^{-1} = 2\mathcal{A}$ for $\lambda \to \infty$ will be discussed later.

Local Evaluation of $\mathcal{A}_{NH}$ and $\tilde{\mathcal{A}}_{NH}$

We have seen at the end of Sect. 5.1 that we must evaluate $\mathcal{A}_{NH}(\boldsymbol{P}\boldsymbol{F}(\mathbf{u})^T)$ in the $\mathbf{B}$-formulation in each quadrature point. Analogously we have to evaluate $\tilde{\mathcal{A}}_{NH}(\boldsymbol{F}(\mathbf{u})^{-1}\boldsymbol{P})$ locally using the formulation in $\boldsymbol{C}$. For both formulations one can evaluate $\mathcal{A}_{NH}(\boldsymbol{P}\boldsymbol{F}(\mathbf{u})^T)$ and $\tilde{\mathcal{A}}_{NH}(\boldsymbol{F}(\mathbf{u})^{-1}\boldsymbol{P})$ directly without using Newton's method as described in the sequel:

For the $\mathbf{B}$-formulation on the one hand, given any stress tensor $\boldsymbol{\tau} \in \mathbb{R}^{3\times 3}$, the corresponding strain $\mathbf{B} := \mathcal{A}_{NH}(\boldsymbol{\tau}) \in \mathbb{R}^{3\times 3}$ can be determined via

$$\mathbf{B} = \mathbf{dev\,B} + \frac{1}{3}\mathrm{tr}(\mathbf{B})\mathbf{I} = \frac{\mathbf{dev}\,\boldsymbol{\tau}}{\mu} + \frac{1}{3}\mathrm{tr}(\mathbf{B})\mathbf{I} \qquad (57)$$

with $\mathrm{tr}(\mathbf{B})$ solution of

$$(\mathrm{tr}(\mathbf{B}))^3 + S\,\mathrm{tr}(\mathbf{B}) + T = 0, \qquad (58)$$

depending on the coefficients

$$\begin{aligned} S &= \frac{9}{\mu^2}\operatorname{tr}(\mathbf{Cof}\,(\mathbf{dev}\,\boldsymbol{\tau})) + \frac{18\mu}{\lambda}, \\ T &= 27\left(\frac{1}{\mu^3}\det(\mathbf{dev}\,\boldsymbol{\tau}) - 1 - \frac{2\mu}{\lambda} - \frac{2}{3\lambda}\operatorname{tr}(\boldsymbol{\tau})\right). \end{aligned} \tag{59}$$

A detailed derivation of (57) and (58) can be found in Müller et al. (2014). If the discriminant $D := \left(\frac{S}{3}\right)^3 + \left(\frac{T}{2}\right)^2$ is positive, the cubic equation (58) has only one real solution and we obtain exactly one reasonable strain which corresponds to the given stress.

For the $\boldsymbol{C}$-formulation on the other hand, given any stress tensor $\boldsymbol{\Sigma} \in \mathbb{R}^{3\times3}$, the corresponding strain $\boldsymbol{C} := \tilde{\mathcal{A}}_{NH}(\boldsymbol{\Sigma}) \in \mathbb{R}^{3\times3}$ can be determined via

$$\boldsymbol{C} = \rho(\boldsymbol{\Sigma} - \mu\mathbf{I})^{-1}, \tag{60}$$

provided that $\boldsymbol{\Sigma} - \mu\mathbf{I}$ is invertible. The parameter ρ in (60) is a solution of

$$\rho^3 + S\rho + T = 0 \tag{61}$$

with coefficients

$$S := -\frac{2}{\lambda}\det(\boldsymbol{\Sigma} - \mu\mathbf{I}), \quad T := -\left(1 + \frac{2\mu}{\lambda}\right)\det(\boldsymbol{\Sigma} - \mu\mathbf{I}), \tag{62}$$

cf. Wriggers (2008, Sect. 10.3). Provided that the discriminant of (61) is positive, we get again one real solution for ρ and a unique real strain tensor $\boldsymbol{C}$ corresponding to the given stress tensor $\boldsymbol{\Sigma}$ can be easily computed. One remarkable fact in the cubic equations (58) and (61) is that we can take also the limit $\lambda \to \infty$ here. In fact, we can also set $\lambda = \infty$. In this case all fractions with λ in the denominator in the coefficients (59) and (62) vanish. With this in mind, we can come back to the discussion of the well-posedness of $\tilde{\mathcal{G}}'_{NH}(\boldsymbol{C})^{-1}$ for given strain $\boldsymbol{C} := \tilde{\mathcal{A}}_{NH}(\boldsymbol{\Sigma})$ in the incompressible limit $\lambda \to \infty$. Inserting (62) into (61) we obtain

$$\rho^3 = \frac{2}{\lambda}\det(\boldsymbol{\Sigma} - \mu\mathbf{I})\,\rho + \left(1 + \frac{2\mu}{\lambda}\right)\det(\boldsymbol{\Sigma} - \mu\mathbf{I})$$

and with the help of (60) we can conclude that

$$\det \boldsymbol{C} = \det\left(\tilde{\mathcal{A}}_{NH}(\boldsymbol{\Sigma})\right) = \rho^3 \det\left((\boldsymbol{\Sigma} - \mu\mathbf{I})^{-1}\right) = \frac{2}{\lambda}\rho + 1 + \frac{2\mu}{\lambda}$$

holds. Due to

$$\mu + \frac{\lambda}{2}(1 - \det \boldsymbol{C}) = \mu + \frac{\lambda}{2}\left(-\frac{2}{\lambda}\rho - \frac{2\mu}{\lambda}\right) = -\rho \overset{\lambda\to\infty}{\to} -\sqrt[3]{\det(\boldsymbol{\Sigma} - \mu\mathbf{I})},$$

the second equation of (56) remains well-posed for $\lambda \to \infty$ with

$$\begin{aligned}\tilde{\mathcal{G}}'_{NH}(\tilde{\mathcal{A}}_{NH}(\boldsymbol{\Sigma}))^{-1}[\boldsymbol{Q}] &= \tilde{\mathcal{G}}'_{NH}(\boldsymbol{C})^{-1}[\boldsymbol{Q}]\\ &= \frac{1}{-\sqrt[3]{\det(\boldsymbol{\Sigma}-\mu\mathbf{I})}}\boldsymbol{C}\left(\boldsymbol{Q}-\frac{(\det\boldsymbol{C})^2}{1+2\det\boldsymbol{C}}(\mathbf{Cof}\,\boldsymbol{C}^{-1}:\boldsymbol{Q})\boldsymbol{C}^{-1}\right)\boldsymbol{C}.\end{aligned}$$

In the case $\boldsymbol{\Sigma} = \mathbf{0}$ (corresponding to $\boldsymbol{C} = \tilde{\mathcal{A}}_{NH}(\mathbf{0}) = \mathbf{I}$) this leads to $-\sqrt[3]{\det(\mathbf{0}-\mu\mathbf{I})} = \mu$ and hence $\tilde{\mathcal{G}}'_{NH}(\mathbf{I})^{-1} = 2\mathcal{A}$ from the linear elasticity case for $\lambda \to \infty$. Combining the neo-Hookean model (51) with the first-order systems (42) and (43) we have thus established a formulation which allows us to consider fully incompressible materials.

Analysis of the Formulation in B

Based on the convex sets

$$\begin{aligned}\boldsymbol{\Pi}^\infty &:= \{\boldsymbol{Q} \in L^\infty(\Omega)^3 : \|\boldsymbol{Q}\|_{L^\infty(\Omega)} \le \delta\} \cap (\boldsymbol{P}^N + W^4_{\Gamma_N}(\text{div};\Omega)^3),\\ \mathbf{V}^\infty &:= \{\mathbf{u} \in W^{1,\infty}(\Omega)^3 : \|\nabla\mathbf{u}\|_{L^\infty(\Omega)} \le \delta\} \cap W^{1,4}_{\Gamma_D}(\Omega)^3\end{aligned} \tag{63}$$

again with $\boldsymbol{P}^N \in W^\infty(\text{div};\Omega)^3$ satisfying $\boldsymbol{P}^N\cdot\mathbf{n} = \mathbf{g}$ on Γ_N, one can prove for the first-order system operator $\mathcal{R}_{NH}$ (cf. (44) with $\mathcal{A} = \mathcal{A}_{NH}$ locally defined by (57), (58) and (59)) the estimates

$$\begin{aligned}\|\mathcal{R}_{NH}(\hat{\boldsymbol{Q}},\hat{\mathbf{v}}) - \mathcal{R}_{NH}(\boldsymbol{Q},\mathbf{v})\|^2 &\lesssim \|\hat{\boldsymbol{Q}}-\boldsymbol{Q}\|^2_{H(\text{div};\Omega)} + \|\hat{\mathbf{v}}-\mathbf{v}\|^2_{H^1(\Omega)}\\ \|\mathcal{R}_{NH}(\hat{\boldsymbol{Q}},\hat{\mathbf{v}}) - \mathcal{R}_{NH}(\boldsymbol{Q},\mathbf{v})\|^2 &\gtrsim \|\hat{\boldsymbol{Q}}-\boldsymbol{Q}\|^2_{H(\text{div};\Omega)} + \|\hat{\mathbf{v}}-\mathbf{v}\|^2_{H^1(\Omega)},\end{aligned} \tag{64}$$

provided that $(\hat{\boldsymbol{Q}},\hat{\mathbf{v}}), (\boldsymbol{Q},\mathbf{v}) \in \boldsymbol{\Pi}^\infty \times \mathbf{V}^\infty$ with sufficient small δ, cf. Müller et al. (2014, Theorem 4.4). In particular, (64) holds uniformly for $\lambda \to \infty$. Inserting the exact solution $(\boldsymbol{P},\mathbf{u}) \in \boldsymbol{\Pi}^\infty \times \mathbf{V}^\infty$ with $\mathcal{R}_{NH}(\boldsymbol{P},\mathbf{u}) = \mathbf{0}$ as $(\boldsymbol{Q},\mathbf{v})$ and an approximation $(\boldsymbol{P}_h,\mathbf{u}_h) \in \boldsymbol{\Pi}^\infty \times \mathbf{V}^\infty$ as $(\hat{\boldsymbol{Q}},\hat{\mathbf{v}})$ in (64) directly leads to

$$\|(\boldsymbol{P}-\boldsymbol{P}_h,\mathbf{u}-\mathbf{u}_h)\|^2_{\mathcal{V}} \lesssim \mathcal{F}_{NH}(\boldsymbol{P}_h,\mathbf{u}_h) \lesssim \|(\boldsymbol{P}-\boldsymbol{P}_h,\mathbf{u}-\mathbf{u}_h)\|^2_{\mathcal{V}} \tag{65}$$

with $\mathcal{V} := H_{\Gamma_N}(\text{div};\Omega)^3 \times H^1_{\Gamma_D}(\Omega)^3$. The coercivity and continuity of the nonlinear least-squares functional $\mathcal{F}_{NH}(\boldsymbol{P}_h,\mathbf{u}_h) = \|\mathcal{R}_{NH}(\boldsymbol{P}_h,\mathbf{u}_h)\|^2$ in (65) justifies its use as an a posteriori error estimator. The left inequality in (65) implies the reliability while the right inequality stands for the efficiency. Since the constants in (65) are independent of λ, the approach (45) combined with the neo-Hookean stored energy function (51) is (Poisson) locking-free.

Equation (65) also leads to a priori error estimates. For instance, if we combine, for some $l \ge 1$, Raviart–Thomas elements $\boldsymbol{\Pi}^l_h := (\mathcal{RT}_{l-1}(\mathcal{T}_h))^3 \subset \boldsymbol{\Pi}^\infty \subset H(\text{div};\Omega)^3$ for the approximation of $\boldsymbol{P}_h$ with continuous elements $\mathbf{V}^l_h := (\mathcal{P}_l(\mathcal{T}_h))^3 \subset \mathbf{V}^\infty \subset H^1(\Omega)^3$ for the approximation of $\mathbf{u}_h$ and let $(\boldsymbol{P}_h,\mathbf{u}_h)$ be the minimizer of $\mathcal{F}_{NH}(\boldsymbol{Q}_h,\mathbf{v}_h)$ among all $(\boldsymbol{Q}_h,\mathbf{v}_h) \in \boldsymbol{\Pi}^l_h \times \mathbf{V}^l_h \subset H(\text{div};\Omega)^3 \times H^1(\Omega)^3$, then we obtain

$$
\begin{aligned}
\|(\boldsymbol{P}-\boldsymbol{P}_h,\mathbf{u}-\mathbf{u}_h)\|_{\mathcal{V}} &\lesssim (\mathcal{F}_{NH}(\boldsymbol{P}_h,\mathbf{u}_h))^{\frac{1}{2}}\\
&= \inf\left\{(\mathcal{F}_{NH}(\boldsymbol{Q}_h,\mathbf{v}_h))^{\frac{1}{2}} : (\boldsymbol{Q}_h,\mathbf{v}_h)\in \boldsymbol{\Pi}_h^l\times \mathbf{V}_h^l\right\}\\
&\lesssim \|(\boldsymbol{P}-\rho_h\boldsymbol{P},\mathbf{u}-r_h\mathbf{u})\|_{\mathcal{V}}\\
&\lesssim h^l\left(\|\boldsymbol{P}\|^2_{H^l(\Omega)}+\|\operatorname{div}\boldsymbol{P}\|^2_{H^l(\Omega)}+\|\mathbf{u}\|^2_{H^{l+1}(\Omega)}\right)^{\frac{1}{2}}
\end{aligned}
\tag{66}
$$

under the assumption that $\boldsymbol{P}\in\boldsymbol{\Pi}^\infty\cap H^l(\Omega)^{3\times 3}$ with $\operatorname{div}\boldsymbol{P}\in H^l(\Omega)^3$, $\mathbf{u}\in\mathbf{V}^\infty\cap H^{l+1}(\Omega)^3$ and $(\boldsymbol{P}_h,\mathbf{u}_h)\in\boldsymbol{\Pi}^\infty\times\mathbf{V}^\infty$ holds. Here ρ_h and r_h denote the usual interpolation operators for the Raviart–Thomas and the standard conforming finite elements, respectively, cf. Boffi et al. (2013, Sects. 2.2 and 2.5). Under these assumptions the square of the error $\|(\boldsymbol{P}-\boldsymbol{P}_h,\mathbf{u}-\mathbf{u}_h)\|^2_{\mathcal{V}}$ and the least-squares functional $\mathcal{F}_{NH}$ both are proportional to $h^{2l}\sim n_t^{-\frac{2l}{d}}$ as $h\to 0$, where n_t denotes the number of elements in the triangulation $\mathcal{T}_h$.

Accuracy of Balance of Momentum in the Nonlinear Case

We investigate the generalization of Theorem 3.2 to the hyperelastic situation. To this end, we assume that the linearized problem

$$
\begin{aligned}
\operatorname{div}\boldsymbol{\sigma}+\mathbf{f}&=\mathbf{0},\\
\mathcal{R}_2'(\boldsymbol{P},\mathbf{u})[\boldsymbol{\sigma},\boldsymbol{v}]&=\mathbf{0}(=\mathcal{R}_2(\boldsymbol{P},\mathbf{u}))
\end{aligned}
\tag{67}
$$

has the following regularity properties, similar to those stated at the end of Sect. 1 for the linear elasticity problem: For any $\mathbf{f}\in L^2(\Omega)^3$, the solution $(\boldsymbol{\sigma},\boldsymbol{v})\in H_{\Gamma_N}(\operatorname{div},\Omega)^d\times H^1_{\Gamma_D}(\Omega)^d$ of (67) satisfies $(\boldsymbol{\sigma},\boldsymbol{v})\in H^\alpha(\Omega)^{d\times d}\times H^{1+\alpha}(\Omega)^d$ with

$$
\|\boldsymbol{\sigma}\|_{H^\alpha(\Omega)}+\|\boldsymbol{v}\|_{H^{1+\alpha}(\Omega)}\le C_R\|\mathbf{f}\| \tag{68}
$$

for some constant $C_R>0$ and $\alpha>0$.

Theorem 5.1 *Under our regularity assumptions, the momentum balance accuracy associated with the first-order system least-squares approximation for the neo-Hooke model satisfies*

$$
\|\operatorname{div}\boldsymbol{P}_h^{LS}+\mathbf{f}\|\lesssim h^\alpha\left(\|\boldsymbol{P}-\boldsymbol{P}_h^{LS}\|+\|\nabla\mathbf{u}-\nabla\mathbf{u}_h^{LS}\|\right)+\inf_{\mathbf{z}_h\in\mathbf{Z}_h}\|\mathbf{f}-\mathbf{z}_h\|. \tag{69}
$$

Proof Starting as in the proof of Theorem 3.2, we arrive at

$$
\|\operatorname{div}\boldsymbol{P}_h^{LS}+\mathbf{f}\|\le\sup_{\mathbf{z}_h\in\mathbf{Z}_h}\frac{(\operatorname{div}(\boldsymbol{P}_h^{LS}-\boldsymbol{P}),\mathbf{z}_h)}{\|\mathbf{z}_h\|}+\|\mathbf{f}-\pi_h\mathbf{f}\|. \tag{70}
$$

Recalling that $(\boldsymbol{P}_h^{LS},\mathbf{u}_h^{LS})\in\boldsymbol{\Pi}_h\times\mathbf{V}_h$ minimizes $\|\mathcal{R}(\boldsymbol{P}_h,\mathbf{u}_h)\|^2$ (and that $\mathcal{R}(\boldsymbol{P},\mathbf{u})=\mathbf{0}$), we have

$$(\mathcal{R}(\boldsymbol{P},\mathbf{u}) - \mathcal{R}(\boldsymbol{P}_h^{LS},\mathbf{u}_h^{LS}), \mathcal{R}'(\boldsymbol{P}_h^{LS},\mathbf{u}_h^{LS})[\boldsymbol{Q}_h,\mathbf{v}_h]) = 0 \tag{71}$$

for all $(\boldsymbol{Q}_h, \mathbf{v}_h) \in \mathbf{\Pi}_h \times \mathbf{V}_h$, where

$$\begin{aligned}
&\mathcal{R}'(\boldsymbol{P},\mathbf{u})[\boldsymbol{Q},\mathbf{v}] \\
&= \begin{pmatrix} \operatorname{div} \boldsymbol{Q} \\ \mathcal{A}'(\boldsymbol{P}\boldsymbol{F}(\mathbf{u})^T)[\boldsymbol{Q}\boldsymbol{F}(\mathbf{u})^T + \boldsymbol{P}(\nabla\mathbf{v})^T] - \boldsymbol{F}(\mathbf{u})(\nabla\mathbf{v})^T - \nabla\mathbf{v}\boldsymbol{F}(\mathbf{u})^T \end{pmatrix} \\
&=: \begin{pmatrix} \operatorname{div} \boldsymbol{Q} \\ \mathcal{R}_2'(\boldsymbol{P},\mathbf{u})[\boldsymbol{Q},\mathbf{v}] \end{pmatrix}.
\end{aligned}$$

We replace the auxiliary boundary value problem (23) by the following: Find $\Xi \in H_{\Gamma_N}(\operatorname{div},\Omega)^3$ and $\boldsymbol{\eta} \in H^1_{\Gamma_D}(\Omega)^3$ such that

$$\begin{aligned}
\operatorname{div} \Xi &= \mathbf{z}_h, \\
\mathcal{R}_2'(\boldsymbol{P},\mathbf{u})[\Xi,\boldsymbol{\eta}] &= \mathbf{0}
\end{aligned} \tag{72}$$

holds. For arbitrary $\Xi_h \in \mathbf{\Pi}_h$ with $\operatorname{div} \Xi_h = \mathbf{z}_h$ and $\boldsymbol{\eta}_h \in \mathbf{V}_h$, we obtain from (71) that

$$\begin{aligned}
&(\operatorname{div}(\boldsymbol{P} - \boldsymbol{P}_h^{LS}), \mathbf{z}_h) \\
&= (\operatorname{div}(\boldsymbol{P} - \boldsymbol{P}_h^{LS}), \operatorname{div} \Xi) \\
&= (\operatorname{div}(\boldsymbol{P} - \boldsymbol{P}_h^{LS}), \operatorname{div} \Xi) \\
&\quad + (\mathcal{R}_2(\boldsymbol{P},\mathbf{u}) - \mathcal{R}_2(\boldsymbol{P}_h^{LS},\mathbf{u}_h^{LS}), \mathcal{R}_2'(\boldsymbol{P},\mathbf{u})[\Xi,\boldsymbol{\eta}]) \\
&= (\operatorname{div}(\boldsymbol{P} - \boldsymbol{P}_h^{LS}), \operatorname{div} \Xi - \operatorname{div} \Xi_h) \\
&\quad + (\mathcal{R}_2(\boldsymbol{P},\mathbf{u}) - \mathcal{R}_2(\boldsymbol{P}_h^{LS},\mathbf{u}_h^{LS}), \mathcal{R}_2'(\boldsymbol{P},\mathbf{u})[\Xi,\boldsymbol{\eta}] - \mathcal{R}_2'(\boldsymbol{P}_h^{LS},\mathbf{u}_h^{LS})[\Xi_h,\boldsymbol{\eta}_h]) \\
&= (\mathcal{R}_2(\boldsymbol{P},\mathbf{u}) - \mathcal{R}_2(\boldsymbol{P}_h^{LS},\mathbf{u}_h^{LS}), \mathcal{R}_2'(\boldsymbol{P},\mathbf{u})[\Xi,\boldsymbol{\eta}] - \mathcal{R}_2'(\boldsymbol{P}_h^{LS},\mathbf{u}_h^{LS})[\Xi_h,\boldsymbol{\eta}_h])
\end{aligned}$$

holds. Combining this with (70) leads to

$$\begin{aligned}
\|\operatorname{div} \boldsymbol{P}_h^{LS} + \mathbf{f}_h\| &\le \|\mathcal{R}_2(\boldsymbol{P},\mathbf{u}) - \mathcal{R}_2(\boldsymbol{P}_h^{LS},\mathbf{u}_h^{LS})\| \\
&\quad \sup_{\Xi} \frac{\|\mathcal{R}_2'(\boldsymbol{P},\mathbf{u})[\Xi,\boldsymbol{\eta}] - \mathcal{R}_2'(\boldsymbol{P}_h^{LS},\mathbf{u}_h^{LS})[\Xi_h,\boldsymbol{\eta}_h]\|}{\|\operatorname{div} \Xi\|}.
\end{aligned} \tag{73}$$

The second term in (73) can be bounded further as

$$\begin{aligned}
&\|\mathcal{R}_2'(\boldsymbol{P},\mathbf{u})[\Xi,\boldsymbol{\eta}] - \mathcal{R}_2'(\boldsymbol{P}_h^{LS},\mathbf{u}_h^{LS})[\Xi_h,\boldsymbol{\eta}_h]\| \\
&\le \|\mathcal{R}_2'(\boldsymbol{P},\mathbf{u})[\Xi - \Xi_h, \boldsymbol{\eta} - \boldsymbol{\eta}_h]\| \\
&\quad + \|(\mathcal{R}_2'(\boldsymbol{P},\mathbf{u}) - \mathcal{R}_2'(\boldsymbol{P}_h^{LS},\mathbf{u}_h^{LS}))[\Xi_h,\boldsymbol{\eta}_h]\|.
\end{aligned} \tag{74}$$

The first term in (74) may be bounded using (Müller et al. 2014, Lemma 4.3) to get

$$\|\mathcal{R}_2'(\boldsymbol{P},\mathbf{u})[\Xi - \Xi_h, \boldsymbol{\eta} - \boldsymbol{\eta}_h]\| \lesssim \|\Xi - \Xi_h\| + \|\boldsymbol{\eta} - \boldsymbol{\eta}_h\| \lesssim h^\alpha \|\mathrm{div}\, \boldsymbol{\eta}\|$$

using our regularity assumption. For the second term in (74),

$$\begin{aligned}
&\|(\mathcal{R}_2'(\boldsymbol{P},\mathbf{u}) - \mathcal{R}_2'(\boldsymbol{P}_h^{LS},\mathbf{u}_h^{LS}))[\Xi_h, \boldsymbol{\eta}_h]\| \\
&\quad\lesssim \left(\|\boldsymbol{P} - \boldsymbol{P}_h^{LS}\|_{L^\infty(\Omega)} + \|\nabla(\mathbf{u} - \mathbf{u}_h^{LS})\|_{L^\infty(\Omega)}\right)\left(\|\Xi_h\| + \|\nabla \boldsymbol{\eta}_h\|\right) \\
&\quad\lesssim h^\alpha \|\mathrm{div}\, \boldsymbol{\eta}\|
\end{aligned}$$

can be shown in a similar way using the expression for $\mathcal{R}_{NH}'$ below (56). This finishes the proof of (69). □

Stress reconstruction in the nonlinear case In analogy to (28) one may consider a displacement-pressure finite element approach to the nonlinear variational problem (37) associated with hyperelasticity. In the case of a neo-Hookean model with

$$\boldsymbol{P}(\mathbf{u}) = \mu\left(\boldsymbol{F}(\mathbf{u}) - \boldsymbol{F}(\mathbf{u})^{-T}\right) + \frac{\lambda}{2}\left((\det \boldsymbol{F}(\mathbf{u}))^2 - 1\right)\boldsymbol{F}(\mathbf{u})^{-T}, \tag{75}$$

a pressure-like variable

$$p = \frac{\lambda}{2}\left((\det \boldsymbol{F}(\mathbf{u}))^2 - 1\right)$$

may be introduced leading to the system

$$\begin{aligned}
\mu\left(\boldsymbol{F}(\mathbf{u}) - \boldsymbol{F}(\mathbf{u})^{-T}, \nabla \mathbf{v}\right) + \left(p\,\boldsymbol{F}(\mathbf{u})^{-T}, \nabla \mathbf{v}\right) &= (\mathbf{f},\mathbf{v}) + \langle \mathbf{g},\mathbf{v}\rangle_{0,\Gamma_N} \\
\left(\frac{1}{2}\left((\det \boldsymbol{F}(\mathbf{u}))^2 - 1\right), q\right) - \frac{1}{\lambda}(p,q) &= 0
\end{aligned} \tag{76}$$

to hold for all $\mathbf{v}$ and q in suitable test spaces. In order to allow for general $\mathbf{u} \in H_{\Gamma_D}^1(\Omega)^3$ and to keep the pressure space as big as possible, one may rewrite (75) slightly as

$$\begin{aligned}
\boldsymbol{P}(\mathbf{u}) = \mu&\left(\boldsymbol{F}(\mathbf{u}) - \frac{\mathbf{Cof}\ \boldsymbol{F}(\mathbf{u})}{\det \boldsymbol{F}(\mathbf{u})}\right) \\
&+ \frac{\lambda}{2}\left(\det \boldsymbol{F}(\mathbf{u}) - \frac{1}{\det \boldsymbol{F}(\mathbf{u})}\right)\mathbf{Cof}\ \boldsymbol{F}(\mathbf{u})
\end{aligned} \tag{77}$$

and define the pressure variable as

$$p = \frac{\lambda}{2}\left(\det \boldsymbol{F}(\mathbf{u}) - \frac{1}{\det \boldsymbol{F}(\mathbf{u})}\right).$$

This is motivated by the fact that, for $\mathbf{u} \in H^1(\Omega)^3$ and under the additional assumption that $\mathbf{Cof}\ \boldsymbol{F}(\mathbf{u}) \in L^2(\Omega)^{3\times 3}$, we have $\det\ \boldsymbol{F}(\mathbf{u}) \in L^1(\Omega)$, see (Ciarlet 1988,

Theorem 7.6-1). The assumption on the cofactor is not as restrictive as it seems since it is usually fulfilled for the solutions associated with edge singularities at reentrant corners. For the same reason, one may also assume that $(\det \boldsymbol{F}(\mathbf{u}))^{-1} \in L^\infty(\Omega)$ which suggests the well-posedness of the following system: Find $\mathbf{u} \in H^1_{\Gamma_D}(\Omega)^3$ and $p \in L^1(\Omega)$ such that

$$\mu\left(\boldsymbol{F}(\mathbf{u}) - \frac{\textbf{Cof } \boldsymbol{F}(\mathbf{u})}{\det \boldsymbol{F}(\mathbf{u})}, \nabla\mathbf{v}\right) + (p\,\textbf{Cof } \boldsymbol{F}(\mathbf{u}), \nabla\mathbf{v}) = (\mathbf{f}, \mathbf{v}) + \langle \mathbf{g}, \mathbf{v}\rangle_{0,\Gamma_N}$$
$$\left(\frac{1}{2}\left(\det \boldsymbol{F}(\mathbf{u}) - \frac{1}{\det \boldsymbol{F}(\mathbf{u})}\right), q\right) - \frac{1}{\lambda}(p, q) = 0 \tag{78}$$

for all $\mathbf{v} \in H^1_{\Gamma_D}(\Omega)^3$ and $q \in L^\infty(\Omega)$. In the small-strain limit, (79) (as well as (76)) turns into the displacement-pressure formulation of linear elasticity (28).

The discrete problem consists in finding $\mathbf{u}_h \in \mathbf{V}_h$ and $p_h \in Q_h$ such that

$$\mu\left(\boldsymbol{F}(\mathbf{u}_h) - \frac{\textbf{Cof } \boldsymbol{F}(\mathbf{u}_h)}{\det \boldsymbol{F}(\mathbf{u}_h)}, \nabla\mathbf{v}_h\right) + (p_h\,\textbf{Cof } \boldsymbol{F}(\mathbf{u}_h), \nabla\mathbf{v}_h) = (\mathbf{f}, \mathbf{v}_h) + \langle \mathbf{g}, \mathbf{v}_h\rangle_{0,\Gamma_N}$$
$$\left(\frac{1}{2}\left(\det \boldsymbol{F}(\mathbf{u}_h) - \frac{1}{\det \boldsymbol{F}(\mathbf{u}_h)}\right), q_h\right) - \frac{1}{\lambda}(p_h, q_h) = 0 \tag{79}$$

for all $\mathbf{v}_h \in \mathbf{V}_h$ and $q_h \in Q_h$. It is natural to use the same combination of spaces $\mathbf{V}_h$ and Q_h as in the linear case although the compatibility is not guaranteed. However, the investigations in Auricchio et al. (2010, 2013) indicate that at least some of these finite element spaces (like Taylor–Hood) are also safe to use in the hyperelastic case for incompressible materials. The stability and approximation properties of the Fortin–Soulie elements (in combination with piecewise linear pressure) for hyperelastic models with incompressible materials needs still to be investigated.

But even if the approximation quality is similar to the linear elasticity case, the stress reconstruction procedure is more involved. The stress

$$\widetilde{\boldsymbol{P}}_h(\mathbf{u}_h, p_h) = \mu\left(\boldsymbol{F}(\mathbf{u}_h) - \boldsymbol{F}(\mathbf{u}_h)^{-T}\right) + p_h\,\boldsymbol{F}(\mathbf{u}_h)^{-T}$$

associated with a piecewise quadratic $\mathbf{u}_h$ and piecewise linear p_h is certainly not piecewise linear and therefore does not satisfy the properties of Proposition 4.1. The element-wise stress reconstruction procedure by Kim (2012) is therefore not immediately applicable in the nonlinear case. The stress reconstruction can be expected to be more involved for hyperelastic material models speaking in favor of our first-order system approach which produces accurate stress approximations simultaneously.

5.4 Computational Results

For the confirmation of our theoretical results, we consider some two- and three-dimensional examples. Although the three-dimensional situation is, of course, more realistic from an application point of view, two-dimensional computations have the advantage that the asymptotic behavior is more accessible.

Two-Dimensional Numerical Tests

We start with the two-dimensional case and use a plane strain configuration, meaning that the displacement components u_1 and u_2 of $\mathbf{u}$ depend only on x_1 and x_2 and the component u_3 is constant. In this situation the deformation gradient becomes

$$\boldsymbol{F} = \boldsymbol{F}(\mathbf{u}) = \mathbf{I} + \nabla \mathbf{u} = \begin{pmatrix} 1+\partial_1 u_1 & \partial_2 u_1 & 0 \\ \partial_1 u_2 & 1+\partial_2 u_2 & 0 \\ 0 & 0 & 1 \end{pmatrix}.$$

Due to the definition of the Cauchy–Green strain tensors $\mathbf{B} = \boldsymbol{F}\boldsymbol{F}^T$ and $\boldsymbol{C} = \boldsymbol{F}^T\boldsymbol{F}$, they have the same structure as $\boldsymbol{F}$. Moreover, also the stress tensor $\boldsymbol{P}$ has this structure (see (40) and (41)). Our numerical simulations in the plane strain situation may therefore be based on a two-dimensional domain with planar Raviart–Thomas elements for the first two rows of $\boldsymbol{P}$ and piecewise polynomial functions without any continuity requirement for the remaining nonzero entry P_{33} of $\boldsymbol{P}$. For the displacement components u_1 and u_2 we can use continuous piecewise polynomial functions. In fact, we use next-to-lowest-order Raviart–Thomas elements $RT_1(\mathcal{T}_h)$ for the plane stress field, piecewise linear discontinuous elements $DP_1(\mathcal{T}_h)$ for P_{33} and conforming, piecewise quadratic, finite elements $P_2(\mathcal{T}_h)$ for each displacement component. Thus our finite dimensional space is altogether given by $\mathbf{\Pi}_h \times \mathbf{V}_h := \left(RT_1(\mathcal{T}_h)^2 \times DP_1(\mathcal{T}_h)\right) \times P_2(\mathcal{T}_h)^2$.

Example 2 In this example we consider again the polygonal Cook's membrane domain as in Sect. 4. With respect to the vertices (0, 0), (48, 44), (48, 60) and (0, 44), the left part of the boundary is again used as $\Gamma_D := \{(0, x_2) : 0 < x_2 < 44\}$ and the remaining boundary as Γ_N. The volume force is set to $\mathbf{f} = \mathbf{0}$ and for the surface force, $\mathbf{g} = \mathbf{0}$ is prescribed on the upper and lower part of Γ_N and $\mathbf{g} = (0, \gamma^{\text{load}})^T$ with $\gamma^{\text{load}} \in \mathbb{R}$ on the right part $\Gamma_R := \{(48, x_2) : 44 < x_2 < 60\}$ of Γ_N. We choose $\mu = 1$ and $\lambda = \infty$ as Lamé constants, i.e., we simulate a fully incompressible material which is the numerical most challenging case. We use the formulation in $\mathbf{B}$ and as load value $\gamma^{\text{load}} = 0.05$. Before we present some numerical results, note that the exact solution $(\boldsymbol{P}, \mathbf{u})$ of this problem is not in $\mathbf{\Pi}^\infty \times \mathbf{V}^\infty$, since at the point $(0, 44)$ the boundary condition changes from hard clamped ($\mathbf{u} = \mathbf{0}$) to stress-free ($\boldsymbol{P} \cdot \mathbf{n} = \mathbf{0}$) and the interior angle is larger than the critical one, cf. Rössle (2000). Although the regularity assumptions in (63) are not satisfied for the exact solution, we see in Table 3 that the nonlinear least-squares functional $\mathcal{F}_{NH}$ works reasonable as a posteriori error estimator.

In the last column of Table 3 the approximation order

$$\frac{\log\left(\mathcal{F}_{NH}\left(\boldsymbol{P}_h^{(l+1)}, \mathbf{u}_h^{(l+1)}\right)\right) - \log\left(\mathcal{F}_{NH}\left(\boldsymbol{P}_h^{(l)}, \mathbf{u}_h^{(l)}\right)\right)}{\log\left(n_t^{(l)}\right) - \log\left(n_t^{(l+1)}\right)}$$

Table 3 Convergence rates of $\mathcal{F}_{NH}$ with adaptive refinement (2d)

n_t	dim $\boldsymbol{\Pi}_h$	dim $\mathbf{V}_h$	$\mathcal{F}_{NH}(\boldsymbol{P}_h, \mathbf{u}_h)$	(Order)
186	2378	784	2.9972×10^{-2}	
275	3525	1150	1.4042×10^{-2}	(1.939)
390	5010	1620	6.7178×10^{-3}	(2.110)
559	7189	2314	3.2427×10^{-3}	(2.023)
821	10583	3374	1.5525×10^{-3}	(1.916)
1211	15633	4954	7.3322×10^{-4}	(1.930)
1796	23208	7324	3.3695×10^{-4}	(1.973)
2622	33918	10656	1.4855×10^{-4}	(2.165)

of $\mathcal{F}_{NH}$ is listed. Here $(\boldsymbol{P}_h, \mathbf{u}_h) := (\boldsymbol{P}_h^{(l)}, \mathbf{u}_h^{(l)})$ denotes the approximated solution and $n_t := n_t^{(l)}$ the number of elements on level $l \in \mathbb{N} \cup \{0\}$. One observes that the optimal convergence rate of 2 is achieved using adaptive refinement. Figure 4 shows the mesh after four adaptive refinement steps resulting in a triangulation with 821 triangles.

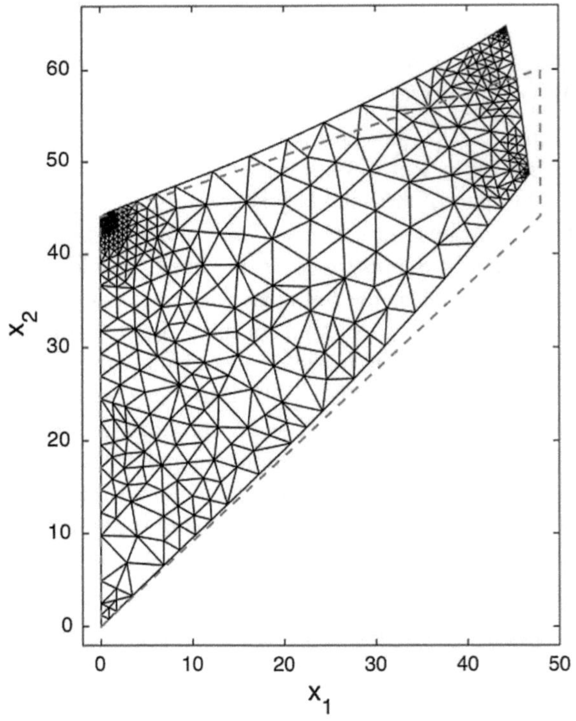

Fig. 4 Adaptively refined triangulation for Cook's membrane

Table 4 Convergence rates of $\mathcal{F}_{NH}$ with uniform refinement (2d)

n_t	$\dim \boldsymbol{\Pi}_h$	$\dim \mathbf{V}_h$	$\mathcal{F}_{NH}(\boldsymbol{P}_h, \mathbf{u}_h)$	(Order)
186	2378	784	2.9972×10^{-2}	
744	9592	3056	1.3800×10^{-2}	(0.559)
2976	38528	12064	6.4895×10^{-3}	(0.544)
11904	154432	47936	3.0743×10^{-3}	(0.539)
47616	618368	191104	1.4538×10^{-3}	(0.540)

In Table 4 the approximation order using uniform refinement is illustrated. Obviously the optimal convergence rate is not reached and adaptive refinement is superior. This is as expected due to the singularity at (0, 44).

In Table 5 we can confirm numerically that the convergence rate of the term $\|\operatorname{div}(\boldsymbol{P} - \boldsymbol{P}_h)\|^2 = \|\operatorname{div} \boldsymbol{P}_h + \mathbf{f}\|^2$ is approximately doubled, regardless of using uniform or adaptive refinement. Furthermore, the values itself are close to zero which means that the approximations satisfy the conservation of linear momentum quite well.

Besides the convergence rates, we are interested in the quality of the surface traction forces resulting from our stress approximations. The distribution of the normal component of the traction force acting at the left boundary is shown in Fig. 5. For a closer investigation of the accuracy of these quantities we focus on the integral $\int_{\Gamma_D} \boldsymbol{P} \cdot \mathbf{n}\, ds$ which constitutes the resultant force acting on the left-hand boundary segment. Due to $\mathbf{f} = \mathbf{0}$, the divergence theorem implies

$$\int_{\Gamma_D} \boldsymbol{P} \cdot \mathbf{n}\, ds = -\int_{\Gamma_N} \boldsymbol{P} \cdot \mathbf{n}\, ds = -\int_{\Gamma_R} \mathbf{g}\, ds = \begin{pmatrix} 0 \\ -\gamma^{\text{load}} |\Gamma_R| \end{pmatrix}. \tag{80}$$

With the outward normal $\mathbf{n} = (-1, 0)^T$ of Γ_D, the load values $\gamma^{\text{load}} \in \{0.0005, 0.05\}$ and $|\Gamma_R| = 16$ it follows immediately that

Table 5 Improved convergence rates for balance of momentum (2d)

Adaptive refinement			Uniform refinement		
n_t	$\|\operatorname{div}(\boldsymbol{P} - \boldsymbol{P}_h)\|^2$	(Order)	n_t	$\|\operatorname{div}(\boldsymbol{P} - \boldsymbol{P}_h)\|^2$	(Order)
186	8.3534×10^{-9}		186	8.3534×10^{-9}	
275	1.9602×10^{-9}	(3.707)	744	1.9315×10^{-9}	(1.056)
390	4.5544×10^{-10}	(4.177)	2976	4.4487×10^{-10}	(1.059)
559	1.0488×10^{-10}	(4.079)	11904	1.0116×10^{-10}	(1.068)
821	2.3596×10^{-11}	(3.881)	47616	2.2323×10^{-11}	(1.090)
1211	4.9448×10^{-12}	(4.021)			
1796	9.4024×10^{-13}	(4.212)			
2622	1.4687×10^{-13}	(4.907)			

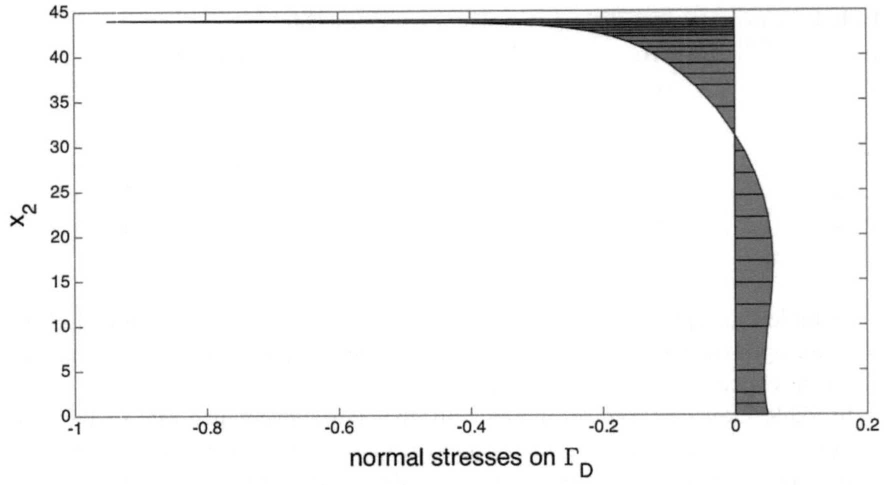

Fig. 5 Normal traction at left boundary for Cook's membrane

$$\int_{\Gamma_D} P_{21}\, ds = 16\gamma^{\text{load}} = \begin{cases} 8 \times 10^{-3}, & \gamma^{\text{load}} = 0.0005 \\ 8 \times 10^{-1}, & \gamma^{\text{load}} = 0.05 \end{cases}$$

holds for the second entry in (80). One observes in Table 6 that the least-squares approach does in both cases produce quite satisfactory approximations to the resultant force.

Three-Dimensional Numerical Tests

For fully three-dimensional examples we use the finite-dimensional spaces $\mathbf{\Pi}_h \times \mathbf{V}_h := (RT_1(\mathcal{T}_h))^3 \times (P_2(\mathcal{T}_h))^3$ on a tetrahedral decomposition of the given domain.

Example 3 We consider a three-dimensional Cook membrane problem. For this purpose we expand the two-dimensional domain of Example 2 in x_3-direction with thickness 5. Thus the three-dimensional polyhedral domain is defined through the vertices $(0,0,0)$, $(48,44,0)$, $(48,60,0)$, $(0,44,0)$, $(0,0,5)$, $(48,44,5)$, $(48,60,5)$, and $(0,44,5)$. We split the boundary $\Gamma = \partial\Omega$ into the left lateral face $\Gamma_D := \{(0, x_2, x_3) : 0 < x_2 < 44, 0 < x_3 < 5\}$ and Γ_N consisting of the remaining five lateral faces.

Table 6 Comparison of $\int_{\Gamma_D} P_{21}\, ds$ for different load values

n_t	$\gamma^{\text{load}} = 0.0005$	$\gamma^{\text{load}} = 0.05$
186	7.9750×10^{-3}	7.9764×10^{-1}
744	7.9881×10^{-3}	7.9886×10^{-1}
2976	7.9944×10^{-3}	7.9945×10^{-1}
11904	7.9973×10^{-3}	7.9974×10^{-1}
47616	7.9987×10^{-3}	7.9988×10^{-1}

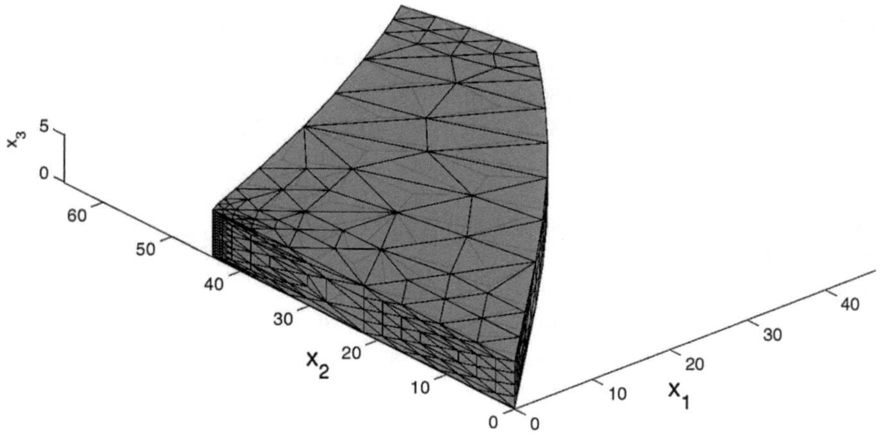

Fig. 6 Adaptively refined triangulation for the 3D Cook's membrane

We clamp the body on Γ_D and apply a surface force $\mathbf{g} = (0, \gamma^{\text{load}}, 0)^T$ with load value $\gamma^{\text{load}} \in \mathbb{R}$ on the right part of the boundary $\Gamma_R := \{(48, x_2, x_3) : 44 < x_2 < 60, 0 < x_3 < 5\}$. On the other parts of Γ_N no surface forces act ($\mathbf{g} = \mathbf{0}$). As body force density we use $\mathbf{f} = \mathbf{0}$, choose $\gamma^{\text{load}} = 0.05$ and Lamé constants $\mu = 1, \lambda = \infty$, i.e., we consider again a fully incompressible material.

Figure 6 shows the mesh after three adaptive refinement steps resulting in a triangulation with 2892 tetrahedra. The concentration of the refinement in the vicinity of the singularity at the edge at $x_1 = 0$ and $x_2 = 44$ is clearly visible. In Tables 7 and 8 the numerically obtained convergence rates corresponding to $\mathcal{F}_{NH}(\boldsymbol{P}_h, \mathbf{u}_h)$ and $\|\text{div}(\boldsymbol{P} - \boldsymbol{P}_h)\|^2$ using adaptive and uniform refinement, respectively, can be compared. One observes in Table 7 that we obtain good convergence rates, close to the optimal value $\frac{4}{3}$, for the nonlinear functional using adaptive refinement. Moreover we see, similar as in the two-dimensional example, that the convergence for the balance of momentum is significantly faster than for the overall functional. Moreover, the value $\|\text{div}(\boldsymbol{P} - \boldsymbol{P}_h)\|^2$ on each considered level is again close to zero, i.e., linear momentum is conserved quite well.

Similar as in the two-dimensional example we consider again the boundary integral values $\int_{\Gamma_D} \boldsymbol{P} \cdot \mathbf{n}\, ds$. Due to $|\Gamma_R| = 16 \cdot 5 = 80$ the exact values are

Table 7 Convergence rates of $\mathcal{F}_{NH}$ and $\|\text{div}(\boldsymbol{P} - \boldsymbol{P}_h)\|^2$ with adaptive refinement (3d)

n_t	$\dim \boldsymbol{\Pi}_h$	$\dim \mathbf{V}_h$	$\mathcal{F}_{NH}(\boldsymbol{P}_h, \mathbf{u}_h)$	(Order)	$\|\text{div}(\boldsymbol{P} - \boldsymbol{P}_h)\|^2$	(Order)
880	22968	4104	3.8682×10^{-1}		2.3313×10^{-7}	
1410	37161	6321	2.0062×10^{-1}	(1.393)	4.8949×10^{-8}	(3.311)
1928	50859	8607	1.3179×10^{-1}	(1.343)	1.8969×10^{-8}	(3.030)
2892	76734	12576	8.1998×10^{-2}	(1.170)	5.7679×10^{-9}	(2.936)

Table 8 Convergence rates of $\mathcal{F}_{NH}$ and $\|\mathrm{div}(\boldsymbol{P} - \boldsymbol{P}_h)\|^2$ with uniform refinement (3d)

n_t	dim $\boldsymbol{\Pi}_h$	dim $\mathbf{V}_h$	$\mathcal{F}_{NH}(\boldsymbol{P}_h, \mathbf{u}_h)$	(Order)	$\|\mathrm{div}(\boldsymbol{P} - \boldsymbol{P}_h)\|^2$	(Order)
880	22968	4104	3.8682×10^{-1}		2.3313×10^{-7}	
7040	186912	30384	1.3719×10^{-1}	(0.498)	3.3031×10^{-8}	(0.940)

Table 9 Values of boundary integrals on Γ_D (3d)

Adaptive refinement			
n_t	Val_1	Val_2	Val_3
880	1.7462×10^{-2}	3.9723×10^{0}	-1.1473×10^{-4}
1410	6.6751×10^{-3}	3.9872×10^{0}	8.6541×10^{-6}
1928	3.0716×10^{-3}	3.9921×10^{0}	5.3635×10^{-6}
2892	1.8159×10^{-3}	3.9959×10^{0}	-6.7773×10^{-5}
Uniform refinement			
n_t	Val_1	Val_2	Val_3
880	1.7462×10^{-2}	3.9723×10^{0}	-1.1473×10^{-4}
7040	6.7975×10^{-3}	3.9895×10^{0}	-3.8141×10^{-6}

$$\begin{pmatrix} \mathrm{Val}_1 \\ \mathrm{Val}_2 \\ \mathrm{Val}_3 \end{pmatrix} := -\int_{\Gamma_D} \boldsymbol{P} \cdot \begin{pmatrix} -1 \\ 0 \\ 0 \end{pmatrix} ds = \begin{pmatrix} 0 \\ 80\gamma^{\mathrm{load}} \\ 0 \end{pmatrix} = \begin{pmatrix} 0 \\ 4 \\ 0 \end{pmatrix}$$

following the same calculations as in the two-dimensional derivation. We can observe in Table 9 that our least-squares approach yields already on a coarse mesh good approximations to the resultant forces and converges to the correct values.

Acknowledgments The work reported here was supported by the German Research Foundation (DFG) under grant STA 402/11-1. The authors would also like to thank Jörg Schröder and Alexander Schwarz for many discussions on the subject in the past years, especially related to the topic of Sect. 5.

References

Arnold, D. N., Brezzi, F., & Douglas, J. (1984a). PEERS: A new mixed finite element for plane elasticity. *Japan Journal of Industrial and Applied Mathematics*, *1*, 347–367.

Arnold, D. N., Douglas, J., & Gupta, C. P. (1984b). A family of higher order mixed finite element methods for plane elasticity. *Numerische Mathematik*, *45*, 1–22.

Auricchio, F., Beirão da Veiga, L., Lovadina, C., & Reali, A. (2010). The importance of the exact satisfaction of the incompressibility constraint in nonlinear elasticity: Mixed FEMs versus NURBS-based approximations. *Computer Methods in Applied Mechanics and Engineering*, *199*, 314–323.

Auricchio, F., Beirão da Veiga, L., Lovadina, C., Reali, A., Taylor, R., & Wriggers, P. (2013). Approximation of incompressible large deformation elastic problems: Some unresolved issues. *Computational Mechanics*, *52*, 1153–1167.

Bertrand, F., Münzenmaier, S., & Starke, G. (2014). First-order system least squares on curved boundaries: Higher-order Raviart-Thomas elements. *SIAM Journal on Numerical Analysis*, *52*, 3165–3180.

Bochev, P., & Gunzburger, M. (2009). *Least-squares finite element methods*. New York: Springer.

Boffi, D., Brezzi, F., & Fortin, M. (2009). Reduced symmetry elements in linear elasticity. *Communications on Pure and Applied Analysis*, *8*, 95–121.

Boffi, D., Brezzi, F., & Fortin, M. (2013). *Mixed finite element methods and applications*. Heidelberg: Springer.

Braess, D., Pillwein, V., & Schöberl, J. (2009). Equilibrated residual error estimates are p-robust. *Computer Methods in Applied Mechanics and Engineering*, *198*, 1189–1197.

Brandts, J., Chen, Y., & Yang, J. (2006). A note on least-squares mixed finite elements in relation to standard and mixed finite elements. *IMA Journal of Numerical Analysis*, *26*, 779–789.

Brenner, S. C. (2003). Korn's inequalities for piecewise H^1 vector fields. *Mathematics of Computation*, *73*, 1067–1087.

Brenner, S. C., & Scott, L. R. (2008). *The mathematical theory of finite element methods* (3rd ed.). New York: Springer.

Cai, Z., & Starke, G. (2004). Least squares methods for linear elasticity. *SIAM Journal on Numerical Analysis*, *42*, 826–842.

Cai, Z., & Zhang, S. (2012). Robust equilibrated residual error estimator for diffusion problems: Conforming elements. *SIAM Journal on Numerical Analysis*, *50*, 151–170.

Cai, Z., Korsawe, J., & Starke, G. (2005). An adaptive least squares mixed finite element method for the stress-displacement formulation of linear elasticity. *Numerical Methods for Partial Differential Equations*, *21*, 132–148.

Carstensen, C., & Dolzmann, G. (1998). A posteriori error estimates for mixed FEM in elasticity. *Numerische Mathematik*, *81*, 187–209.

Carstensen, C., & Dolzmann, G. (2004). An a priori error estimate for finite element discretizations in nonlinear elasticity for polyconvex materials under small loads. *Numerische Mathematik*, *97*, 67–80.

Ciarlet, P. G. (1988). *Mathematical elasticity volume I: Three-dimensional elasticity*. Amsterdam: North-Holland.

Ern, A., & Vohralík, M. (2015). Polynomial-degree-robust a posteriori error estimates in a unified setting for conforming, nonconforming, discontinuous Galerkin, and mixed discretizations. *SIAM Journal on Numerical Analysis*, *53*, 1058–1081.

Fortin, M. (1985). A three-dimensional quadratic nonconforming element. *Numerische Mathematik*, *46*, 269–279.

Fortin, M., & Soulie, M. (1983). A non-conforming piecewise quadratic finite element on triangles. *International Journal for Numerical Methods in Engineering*, *19*, 505–520.

Girault, V., & Raviart, P.-A. (1986). *Finite element methods for Navier-stokes equations*. New York: Springer.

Kim, K.-Y. (2012). Flux reconstruction for the P2 nonconforming finite element method with application to a posteriori error estimation. *Applied Numerical Mathematics*, *62*, 1701–1717.

Klaas, O., Schröder, J., Stein, E., & Miehe, C. (1995). A regularized dual mixed element for plane elasticity: Implementation and performance of the BDM element. *Computer Methods in Applied Mechanics and Engineering*, *121*, 201–209.

LeTallec, P. (1994). In Ciarlet, P. G. & Lions, J. L. (Eds.), *Numerical methods for nonlinear three-dimensional elasticity* (pp. 465–662). Handbook of numerical analysis III. Amsterdam: North-Holland.

Luce, R., & Wohlmuth, B. (2004). A local a posteriori error estimator based on equilibrated fluxes. *SIAM Journal on Numerical Analysis*, *42*, 1394–1414.

Müller, B., Starke, G., Schwarz, A., & Schröder, J. (2014). A first-order system least squares method for hyperelasticity. *SIAM Journal on Scientific Computing*, *36*, B795–B816.

Nicaise, S., Witowski, K., & Wohlmuth, B. (2008). An a posteriori error estimator for the Lamé equation based on equilibrated fluxes. *IMA Journal of Numerical Analysis*, *28*, 331–353.

Nocedal, J., & Wright, S. J. (2006). *Numerical optimization* (2nd ed.). New York: Springer.
Parés, N., Bonet, J., Huerta, A., & Peraire, J. (2006). The computation of bounds for linear-functional outputs of weak solutions to the two-dimensional elasticity equations. *Computer Methods in Applied Mechanics and Engineering*, *195*, 406–429.
Reddy, B. D. (1992). Mixed variational inequalities arising in elastoplasticity. *Nonlinear Analysis*, *19*, 1071–1089.
Rössle, A. (2000). Corner singularities and regularity of weak solutions for the two-dimensional Lamé equations on domains with angular corners. *Journal of Elasticity*, *60*, 57–75.
Schwarz, A., Schröder, J., & Starke, G. (2009). Least-squares mixed finite elements for small strain elasto-viscoplasticity. *International Journal for Numerical Methods in Engineering*, *77*, 1351–1370.
Simo, J . C. (1998). In Ciarlet, P. G. & Lions, J. L. (Eds.), *Numerical analysis and simulation of plasticity* (pp. 183–499). Handbook of numerical analysis VI. Amsterdam: North-Holland.
Starke, G. (2007). An adaptive least-squares mixed finite element method for elasto-plasticity. *SIAM Journal on Numerical Analysis*, *45*, 371–388.
Stenberg, R. (1988). A family of mixed finite elements for the elasticity problem. *Numerische Mathematik*, *53*, 513–538.
Wriggers, P. (2008). *Nonlinear finite element methods*. Berlin: Springer.

Tutorial on Hybridizable Discontinuous Galerkin (HDG) for Second-Order Elliptic Problems

Ruben Sevilla and Antonio Huerta

Abstract The HDG is a new class of discontinuous Galerkin (DG) methods that shares favorable properties with classical mixed methods such as the well known Raviart–Thomas methods. In particular, HDG provides optimal convergence of both the primal and the dual variables of the mixed formulation. This property enables the construction of superconvergent solutions, contrary to other popular DG methods. In addition, its reduced computational cost, compared to other DG methods, has made HDG an attractive alternative for solving problems governed by partial differential equations. A tutorial on HDG for the numerical solution of second-order elliptic problems is presented. Particular emphasis is placed on providing all the necessary details for the implementation of HDG methods.

1 Introduction

Efficient and robust solution of equations of mathematical physics has been and still is a major concern for numerical analysts. In the last decades, discontinuous Galerkin (DG) techniques, originally introduced in Reed and Hill (1973), have become popular beyond their original applications in fluid dynamics or electromagnetic problems. DG methods provide a natural stabilization to the solution due to the inter-element fluxes. In recent years, hybrid DG methods have become more popular. According to Ciarlet (2002), a hybrid method is "any finite element method based on a formulation where one unknown is a function, or some of its derivatives, on the set Ω, and the

R. Sevilla
Zienkiewicz Centre for Computational Engineering, College of Engineering,
Swansea University, Bay Campus, Wales SA1 8EN, UK
e-mail: r.sevilla@swansea.ac.uk

A. Huerta (✉)
Laboratori de Calcul Numeric (LaCaN). ETS de Ingenieros de Caminos,
Canales y Puertos, Universitat Politecnica de Catalunya·BarcelonaTech, Barcelona, Spain
e-mail: antonio.huerta@upc.es

J. Schröder and P. Wriggers (eds.), *Advanced Finite Element Technologies*,
CISM International Centre for Mechanical Sciences 566,
DOI 10.1007/978-3-319-31925-4_5

other unknown is the trace of some of the derivatives of the same function, or the trace of the function itself, along the boundaries of the set." In fact, as noted by Arnold and Brezzi (1985), hybridization of DG methods derives from the mixed methods of Raviart and Thomas (1977), where the continuity constrain is eliminated from the finite element space and imposed by means of Lagrange multipliers on the inter-element boundaries. The idea was exploited by Cockburn and Gopalakrishnan (2004, 2005a, b) and Cockburn et al. (2009b) to formally develop the HDG method for second-order elliptic problems.

The HDG method is able to provide the optimal approximation properties that are characteristic of mixed methods, including the possibility to build a superconvergent solution, while retaining the advantages of DG methods. In addition, HDG methods are known to reduce the globally coupled degrees of freedom, when compared to other DG methods analyzed in Arnold et al. (2002). Recently, the comparisons between the HDG method and the traditional continuous Galerkin (CG) method, performed by Cockburn et al. (2009a) and Kirby et al. (2011), indicate that HDG methods are competitive, both in terms of the nonzero entries of the resulting matrix and the actual computing time. Huerta et al. (2013) evaluate the floating point operation counts for CG, DG, and HDG schemes in 2D and 3D and for direct and iterative solvers. They conclude that HDG has comparable costs (in terms of floating point operations) to CG and that other advantages of HDG, such as block structured information and element-by-element operations, must be exploited to improve its performance compared to CG because parallelism and memory access are crucial for the final runtime. In fact, Yakovlev et al. (2015) compare CG and HDG from a practical perspective using the same object-oriented spectral element framework. They show that HDG can outperform the traditional CG approach when direct solvers for the linear system are used. Different conclusions are obtained when iterative solvers are employed, suggesting the need for tailormade preconditioners in an HDG framework. It is also worth emphasizing that the superconvergent properties of HDG enable the definition of efficient and inexpensive p-adaptive procedures not feasible in a standard CG approach, see for instance Giorgiani et al. (2013, 2014).

Since its introduction, the HDG has been objective of intensive research and has been applied to a large number of problems in different areas, including fluid mechanics (Nguyen et al. 2010, 2011a; Peraire et al. 2010), wave propagation (Nguyen et al. 2011c, b; Giorgiani et al. 2013) and solid mechanics (Soon et al. 2009; Kabaria et al. 2015), to name but a few.

This work presents a tutorial on the HDG method for the numerical solution of second-order elliptic problems. Section 2 presents the model second-order elliptic problem and its mixed formulation. The necessary notation is introduced in Sect. 3. The HDG method is presented in detail in Sect. 4, including the strong, weak and discrete forms and the corresponding equations. A new formulation, consisting on a variation of the standard HDG method is presented and its advantages are discussed. Special emphasis is placed on the computational aspects, providing an easy guide for the implementation of the HDG method. Additionally, numerical examples are used to illustrate the performance and the optimal approximation properties of the two HDG formulations. Section 5 presents the postprocessing technique that enables the computation of a superconvergent solution. Numerical examples are also included to show the benefits of the postprocessing technique and to illustrate its optimal approximation properties. Finally, Appendix A provides detailed expression of all the elemental matrices and vectors appearing in the discrete form of the HDG method.

2 Problem Statement

Let $\Omega \in \mathbb{R}^{\mathtt{n_{sd}}}$ be an open bounded domain with boundary $\partial\Omega$ and $\mathtt{n_{sd}}$ the number of spatial dimensions. The strong from for the second-order elliptic problem can be written as

$$\begin{cases} -\nabla \cdot \nabla u = f & \text{in } \Omega, \\ u = u_D & \text{on } \Gamma_D, \\ \boldsymbol{n} \cdot \nabla u = t & \text{on } \Gamma_N, \end{cases} \tag{1}$$

where $\partial\Omega = \overline{\Gamma}_D \cup \overline{\Gamma}_N$, $\overline{\Gamma}_D \cap \overline{\Gamma}_N = \emptyset$, $f \in \mathcal{L}_2(\Omega)$ is a source term and $\boldsymbol{n}$ is the outward unit normal vector to $\partial\Omega$. Note that standard Dirichlet and Neumann boundary conditions are considered. Of course, other mixed (i.e., Robin) boundary conditions can also be imposed but here, for clarity, they will not be detailed.

Moreover, assume that Ω is partitioned in $\mathtt{n_{el}}$ disjoint subdomains Ω_i

$$\overline{\Omega} = \bigcup_{i=1}^{\mathtt{n_{el}}} \overline{\Omega}_i, \quad \Omega_i \cap \Omega_j = \emptyset \quad \text{for } i \neq j,$$

with boundaries $\partial\Omega_i$, which define an internal interface Γ

$$\Gamma := \Big[\bigcup_{i=1}^{\mathtt{n_{el}}} \partial\Omega_i \Big] \backslash \partial\Omega. \tag{2}$$

An equivalent strong form of the second-order elliptic problem can be written in the *broken* computational domain as

$$\begin{cases} -\nabla \cdot \nabla u = f & \text{in } \Omega_i, \text{ and for } i = 1, \dots, \mathtt{n_{el}}, \\ u = u_D & \text{on } \Gamma_D, \\ \boldsymbol{n} \cdot \nabla u = t & \text{on } \Gamma_N, \\ [\![u\boldsymbol{n}]\!] = \boldsymbol{0} & \text{on } \Gamma, \\ [\![\boldsymbol{n} \cdot \nabla u]\!] = 0 & \text{on } \Gamma, \end{cases} \tag{3}$$

where the two last equations correspond to the imposition of the continuity of the primal variable u and the normal fluxes respectively along the internal interface Γ.

Note that the *jump* $[\![\cdot]\!]$ operator has been introduced following the definition by Montlaur et al. (2008). That is, along each portion of the interface Γ it sums the values from the left and right of say, Ω_i and Ω_j, namely

$$[\![\odot]\!] = \odot_i + \odot_j.$$

It is important to observe that this definition always requires the normal vector $\boldsymbol{n}$ in the argument and always produces functions in the same space as the argument.

Finally, the strong form is written in mixed form as a system of first-order equations over the *broken* computational domain, namely

$$\begin{cases} \nabla \cdot \boldsymbol{q} = f & \text{in } \Omega_i, \text{ and for } i = 1, \dots, \mathtt{n_{el}}, \\ \boldsymbol{q} + \nabla u = \boldsymbol{0} & \text{in } \Omega_i, \text{ and for } i = 1, \dots, \mathtt{n_{el}}, \\ u = u_D & \text{on } \Gamma_D, \\ \boldsymbol{n} \cdot \boldsymbol{q} = -t & \text{on } \Gamma_N, \\ [\![u\boldsymbol{n}]\!] = \boldsymbol{0} & \text{on } \Gamma, \\ [\![\boldsymbol{n} \cdot \boldsymbol{q}]\!] = 0 & \text{on } \Gamma. \end{cases} \tag{4}$$

3 Functional and Interpolation Setting

In what follows, as usual, $(\cdot, \cdot)_D$ denotes the $\mathcal{L}_2$ scalar product in a generic subdomain D, that is

$$(u, v)_D = \int_D u\, v\, d\Omega \text{ and } (\boldsymbol{u}, \boldsymbol{v})_D = \int_D \boldsymbol{u} \cdot \boldsymbol{v}\, d\Omega$$

for scalars and vectors respectively.

Analogously, $< \cdot, \cdot >_S$ denotes the $\mathcal{L}_2$ scalar product in any domain $S \subset \Gamma \cup \partial\Omega$, that is

$$< u, v >_S = \int_S u\, v\, d\Gamma \text{ and } < \boldsymbol{u}, \boldsymbol{v} >_S = \int_S \boldsymbol{u} \cdot \boldsymbol{v}\, d\Gamma$$

for scalars and vectors, respectively.

In the subsequent formulation the following scalar and vector spaces are used:

$$\begin{aligned} \mathcal{W}(D) &= \{\boldsymbol{w} \in [\mathcal{H}^1(D)]^{\mathtt{n_{sd}}}, D \subseteq \Omega\}, \\ \mathcal{V}(D) &= \{v \in \mathcal{H}^1(D), D \subseteq \Omega\}, \\ \mathcal{M}(S) &= \{\mu \in \mathcal{L}_2(S), S \subseteq \Gamma \cup \partial\Omega\}. \end{aligned}$$

Moreover, the following discrete finite element spaces are introduced

$$\begin{aligned} \mathcal{W}^h(\Omega) &= \{\boldsymbol{w} \in [\mathcal{L}_2(\Omega)]^{\mathtt{n_{sd}}}; \boldsymbol{w}|_{\Omega_i} \in [\mathcal{P}^p(\Omega_i)]^{\mathtt{n_{sd}}}\ \forall \Omega_i\} && \subset \mathcal{W}(\Omega), \\ \mathcal{V}^h(\Omega) &= \{v \in \mathcal{L}_2(\Omega); v|_{\Omega_i} \in \mathcal{P}^p(\Omega_i)\ \forall \Omega_i\} && \subset \mathcal{V}(\Omega), \\ \mathcal{M}^h(S) &= \{\mu \in \mathcal{L}_2(S); \mu|_{\Gamma_i} \in \mathcal{P}^p(\Gamma_i)\ \forall \Gamma_i \subset S \subseteq \Gamma \cup \partial\Omega\} && \subset \mathcal{M}(S), \end{aligned}$$

where $\mathcal{P}^p(\Omega_i)$ and $\mathcal{P}^p(\Gamma_i)$ are the spaces of polynomial functions of degree at most $p \geq 1$ in Ω_i and Γ_i respectively. Note that $\mathcal{M}^h$ can be defined over all the mesh skeleton interior and exterior faces (or edges in two dimensions).

These spaces give rise to an element-by-element nodal interpolation of the corresponding variables, namely

$$\boldsymbol{q} \approx \boldsymbol{q}^h = \sum_{j=1}^{\mathtt{n_{en}}} N_j \, \mathbf{q}_j \qquad \in \mathcal{W}^h, \tag{5a}$$

$$u \approx u^h = \sum_{j=1}^{\mathtt{n_{en}}} N_j \, \mathrm{u}_j \qquad \in \mathcal{V}^h, \tag{5b}$$

$$\hat{u} \approx \hat{u}^h = \sum_{j=1}^{\mathtt{n_{fn}}} \widehat{N}_j \, \hat{\mathrm{u}}_j \qquad \in \mathcal{M}^h(\Gamma \cup \Gamma_N) \text{ or } \mathcal{M}^h(\Gamma), \tag{5c}$$

where $\mathbf{q}_j$, u_j, and $\hat{\mathrm{u}}_j$ are nodal values, N_j are polynomial shape functions of order p in each element, $\mathtt{n_{en}}$ is the number of nodes per element, $\widehat{N}_j$ are polynomial shape functions of order p in each element face/edge, and $\mathtt{n_{fn}}$ is the corresponding number of nodes per face/edge.

Given the element-by-element formulation, the vectors $\mathbf{u}_i$ and $\mathbf{q}_i$ are defined for each element $i = 1, \ldots, \mathtt{n_{el}}$. They include the corresponding nodal values described previously and are of dimension $\mathtt{n_{en}}$ and $\mathtt{n_{sd}}\mathtt{n_{en}}$, respectively. The vector $\hat{\mathbf{u}}$ is defined globally over the mesh skeleton (faces/edges). Its dimension depends on the formulation and corresponds to the number of nodes on $\Gamma \cup \Gamma_N$ or on Γ. More precisely,

$$\dim(\hat{\mathbf{u}}) = \sum_{k=1}^{\mathtt{n_{ef}}} \mathtt{n}^k_{\mathtt{fn}},$$

where $\mathtt{n_{ef}}$ is the number of element faces/edges in the mesh skeleton and $\mathtt{n}^k_{\mathtt{fn}}$ is the number of nodes in the kth face. The number of element faces/edges in the mesh skeleton always includes those on the interior, i.e., those belonging to Γ. But, depending on the formulation used, $\mathtt{n_{ef}}$ also includes the faces/edges on the Neumann boundary, Γ_N.

The HDG formulation solves problem (4) in two phases, see the seminal contribution by Cockburn et al. (2009b) and the subsequent papers by Cockburn et al. (2008) and Nguyen et al. (2009a, b, 2010, 2011a).

First, an element-by-element problem is defined with $(\boldsymbol{q}, u)$ as unknowns, and then a global problem is setup to determine the traces of u, denoted by $\hat{u}$, on the element boundaries. The local problem determines $\boldsymbol{q}_i := \boldsymbol{q}|_{\Omega_i}$ and $u_i := u|_{\Omega_i}$ for $i = 1, \ldots, \mathtt{n_{el}}$ with a new variable $\hat{u}$ along the interface Γ acting as a Dirichlet boundary condition.

There are however several options for the detailed implementation. They are presented and discussed in the following sections.

4 The Hybridizable Discontinuous Galerkin

4.1 The Strong Forms

This is the classical formulation, it can be found in the series of papers by Nguyen et al. (2009a, b, 2010, 2011a) and rewrites (4) as two equivalent problems. First, the local–element-by-element–problem with Dirichlet boundary conditions is defined, namely

$$\begin{cases} \nabla \cdot \boldsymbol{q}_i = f & \text{in } \Omega_i, \\ \boldsymbol{q}_i + \nabla u_i = \boldsymbol{0} & \text{in } \Omega_i, \\ u_i = u_D & \text{on } \partial\Omega_i \cap \Gamma_D, \\ u_i = \hat{u} & \text{on } \partial\Omega_i \backslash \Gamma_D, \end{cases} \tag{6}$$

for $i = 1, \ldots, \mathtt{n_{el}}$. Note that this approach assumes $\hat{u} \in \mathcal{L}_2(\Gamma \cup \Gamma_N)$ given. In each element Ω_i this problem produces an element-by-element solution $\boldsymbol{q}_i$ and u_i as a function of the unknown $\hat{u} \in \mathcal{L}_2(\Gamma \cup \Gamma_N)$. Note that these problems can be solved independently element by element.

Second, a global problem is defined to determine $\hat{u}$. It corresponds to the imposition of the Neumann boundary condition and the so-called *transmission conditions*, see Cockburn et al. (2009b). These transmission conditions were already introduced in (4) to ensure inter-element continuity when the broken computational domain formulation was presented,

$$\begin{cases} [\![u\boldsymbol{n}]\!] = \boldsymbol{0} & \text{on } \Gamma, \\ [\![\boldsymbol{n} \cdot \boldsymbol{q}]\!] = 0 & \text{on } \Gamma, \\ \boldsymbol{n} \cdot \boldsymbol{q} = -t & \text{on } \Gamma_N. \end{cases}$$

Note that the first equation in the previous global problem imposes continuity of u along Γ. But $u = \hat{u}$ on Γ as imposed by the local problems (6). Hence, continuity of the primal variable, $[\![\hat{u}\boldsymbol{n}]\!] = \boldsymbol{0}$, is imposed automatically because $\hat{u}$ is unique for adjacent elements. In summary, the transmission conditions are simply

$$\begin{cases} [\![\boldsymbol{n} \cdot \boldsymbol{q}]\!] = 0 & \text{on } \Gamma, \\ \boldsymbol{n} \cdot \boldsymbol{q} = -t & \text{on } \Gamma_N. \end{cases} \tag{7}$$

4.2 The Weak Forms

The weak formulation for each element equivalent to (6) is as follows: for $i = 1, \ldots, \mathtt{n_{el}}$, given u_D on Γ_D and $\hat{u}$ on $\Gamma \cup \Gamma_N$, find $(\boldsymbol{q}_i, u_i) \in \boldsymbol{\mathcal{W}}(\Omega_i) \times \mathcal{V}(\Omega_i)$ that satisfies

$$
\begin{aligned}
-(\nabla v, \boldsymbol{q}_i)_{\Omega_i} + < v, \boldsymbol{n}_i \cdot \widehat{\boldsymbol{q}}_i >_{\partial\Omega_i} &= (v, f)_{\Omega_i} \\
-(\boldsymbol{w}, \boldsymbol{q}_i)_{\Omega_i} + (\nabla \cdot \boldsymbol{w}, u_i)_{\Omega_i} &= < \boldsymbol{n}_i \cdot \boldsymbol{w}, u_D >_{\partial\Omega_i \cap \Gamma_D} + < \boldsymbol{n}_i \cdot \boldsymbol{w}, \hat{u} >_{\partial\Omega_i \setminus \Gamma_D},
\end{aligned}
$$

for all $(\boldsymbol{w}, v) \in \boldsymbol{\mathcal{W}}(\Omega_i) \times \mathcal{V}(\Omega_i)$, where the numerical traces of the fluxes $\widehat{\boldsymbol{q}}_i$ must be defined. Note that this problem imposes the Dirichlet boundary conditions weakly.

The numerical traces of the fluxes are formally $\boldsymbol{n}_i \cdot \widehat{\boldsymbol{q}}_i = \boldsymbol{n}_i \cdot \boldsymbol{q}_i$ but, in practice, for stability, they are defined element-by-element (i.e. for $i = 1, \ldots, \mathtt{n_{el}}$) as

$$
\boldsymbol{n}_i \cdot \widehat{\boldsymbol{q}}_i := \begin{cases} \boldsymbol{n}_i \cdot \boldsymbol{q}_i + \tau_i (u_i - u_D) & \text{on } \partial\Omega_i \cap \Gamma_D, \\ \boldsymbol{n}_i \cdot \boldsymbol{q}_i + \tau_i (u_i - \hat{u}) & \text{elsewhere,} \end{cases} \tag{8}
$$

with τ_i being a stabilization parameter defined element-by-element, whose selection has an important effect on the stability, accuracy, and convergence properties of the resulting HDG method. The influence of the stabilization parameter has been studied extensively by Cockburn and co-workers, see for instance Cockburn et al. (2009b, 2008) and Nguyen et al. (2009a, b, 2010, 2011a). Choosing the correct stabilization parameter provides sufficient stabilization to the solution. Note that such a definition for the numerical trace is consistent, i.e., $\boldsymbol{n}_i \cdot \widehat{\boldsymbol{q}}_i = \boldsymbol{n}_i \cdot \boldsymbol{q}_i$ when $u_i = \hat{u}$ (and $u_i = u_D$). With the definition of the numerical fluxes given by (8), the weak problem becomes: for $i = 1, \ldots, \mathtt{n_{el}}$, find $(\boldsymbol{q}_i, u_i) \in \boldsymbol{\mathcal{W}}(\Omega_i) \times \mathcal{V}(\Omega_i)$ that satisfies

$$
\begin{aligned}
&< v, \tau_i u_i >_{\partial\Omega_i} - (\nabla v, \boldsymbol{q}_i)_{\Omega_i} + < v, \boldsymbol{n}_i \cdot \boldsymbol{q}_i >_{\partial\Omega_i} \\
&\quad = (v, f)_{\Omega_i} + < v, \tau_i u_D >_{\partial\Omega_i \cap \Gamma_D} + < v, \tau_i \hat{u} >_{\partial\Omega_i \setminus \Gamma_D}, \qquad \text{(9a)} \\
&-(\boldsymbol{w}, \boldsymbol{q}_i)_{\Omega_i} + (\nabla \cdot \boldsymbol{w}, u_i)_{\Omega_i} \\
&\quad = < \boldsymbol{n}_i \cdot \boldsymbol{w}, u_D >_{\partial\Omega_i \cap \Gamma_D} + < \boldsymbol{n}_i \cdot \boldsymbol{w}, \hat{u} >_{\partial\Omega_i \setminus \Gamma_D}, \qquad \text{(9b)}
\end{aligned}
$$

for all $(\boldsymbol{w}, v) \in \boldsymbol{\mathcal{W}}(\Omega_i) \times \mathcal{V}(\Omega_i)$. The weak form (9) for the *local problem* is equivalent to the strong form described by (6).

Once the weak form for the *local problem* is presented, the *global problem* (7) is of interest. The weak form equivalent to (7) is simply: find $\hat{u} \in \mathcal{M}(\Gamma \cup \Gamma_N)$ for all $\mu \in \mathcal{M}(\Gamma \cup \Gamma_N)$ such that

$$
\sum_{i=1}^{\mathtt{n_{el}}} < \mu, \boldsymbol{n}_i \cdot \widehat{\boldsymbol{q}}_i >_{\partial\Omega_i \setminus \partial\Omega} + \sum_{i=1}^{\mathtt{n_{el}}} < \mu, \boldsymbol{n}_i \cdot \widehat{\boldsymbol{q}}_i + t >_{\partial\Omega_i \cap \Gamma_N} = 0,
$$

where it is important to recall the definition of internal interface Γ given by (2).

Then, replacing (8) in the previous equation results in the global weak problem: find $\hat{u} \in \mathcal{M}(\Gamma \cup \Gamma_N)$ for all $\mu \in \mathcal{M}(\Gamma \cup \Gamma_N)$ such that

$$
\begin{aligned}
&\sum_{i=1}^{\mathtt{n_{el}}} \Big\{ < \mu, \tau_i u_i >_{\partial\Omega_i \setminus \Gamma_D} + < \mu, \boldsymbol{n}_i \cdot \boldsymbol{q}_i >_{\partial\Omega_i \setminus \Gamma_D} - < \mu, \tau_i \hat{u} >_{\partial\Omega_i \setminus \Gamma_D} \Big\} \\
&\quad = -\sum_{i=1}^{\mathtt{n_{el}}} < \mu, t >_{\partial\Omega_i \cap \Gamma_N} . \qquad \text{(10)}
\end{aligned}
$$

Note that both u_i and $\boldsymbol{q}_i$ are known functions of $\hat{u}$ once the local problems (9) are solved.

Remark 4.1 (*Symmetric Dirichlet local problem*) There are two alternatives to symmetrize the local problem. The first one consists of integrating by parts the second term of the l.h.s. in (9a) leaving on the boundary of the element the values of the flux $\boldsymbol{q}_i$ in the interior. This strategy produces the following local problem:

$$
\begin{aligned}
&< v, \tau_i u_i >_{\partial\Omega_i} + (v, \nabla \cdot \boldsymbol{q}_i)_{\Omega_i} \\
&\quad = (v, f)_{\Omega_i} + < v, \tau_i u_D >_{\partial\Omega_i \cap \Gamma_D} + < v, \tau_i \hat{u} >_{\partial\Omega_i \setminus \Gamma_D}, && \text{(11a)} \\
&(\nabla \cdot \boldsymbol{w}, u_i)_{\Omega_i} - (\boldsymbol{w}, \boldsymbol{q}_i)_{\Omega_i} \\
&\quad = < \boldsymbol{n}_i \cdot \boldsymbol{w}, u_D >_{\partial\Omega_i \cap \Gamma_D} + < \boldsymbol{n}_i \cdot \boldsymbol{w}, \hat{u} >_{\partial\Omega_i \setminus \Gamma_D} . && \text{(11b)}
\end{aligned}
$$

The second alternative consists of integrating by parts the second term on the l.h.s. of (9b) and change the sign of (9a), namely

$$
\begin{aligned}
&< v, \tau_i u_i >_{\partial\Omega_i} - (\nabla v, \boldsymbol{q}_i)_{\Omega_i} + < v, \boldsymbol{n}_i \cdot \boldsymbol{q}_i >_{\partial\Omega_i} \\
&\quad = (v, f)_{\Omega_i} + < v, \tau_i u_D >_{\partial\Omega_i \cap \Gamma_D} + < v, \tau_i \hat{u} >_{\partial\Omega_i \setminus \Gamma_D}, \\
&-(\boldsymbol{w}, \nabla u_i)_{\Omega_i} + < \boldsymbol{n}_i \cdot \boldsymbol{w}, u >_{\partial\Omega_i} - (\boldsymbol{w}, \boldsymbol{q}_i)_{\Omega_i} \\
&\quad = < \boldsymbol{n}_i \cdot \boldsymbol{w}, u_D >_{\partial\Omega_i \cap \Gamma_D} + < \boldsymbol{n}_i \cdot \boldsymbol{w}, \hat{u} >_{\partial\Omega_i \setminus \Gamma_D} .
\end{aligned}
$$

The first alternative is retained because it requires less computational effort (during the loop on faces/edges) than the second one.

4.3 The Discrete Forms and the Corresponding Equations

Section 3 introduced the necessary discrete spaces in order to prescribe the discrete weak forms for the local (11) and global (10) problems. The local problems are: for $i = 1, \dots \mathtt{n_{el}}$, find $(\boldsymbol{q}_i^h, u_i^h) \in \boldsymbol{\mathcal{W}}^h \times \mathcal{V}^h$ for all $(\boldsymbol{w}, v) \in \boldsymbol{\mathcal{W}}^h \times \mathcal{V}^h$ such that

$$
\begin{aligned}
&< v, \tau_i u_i^h >_{\partial\Omega_i} + (v, \nabla \cdot \boldsymbol{q}_i^h)_{\Omega_i} \\
&\quad = (v, f)_{\Omega_i} + < v, \tau_i u_D >_{\partial\Omega_i \cap \Gamma_D} + < v, \tau_i \hat{u}^h >_{\partial\Omega_i \setminus \Gamma_D}, && \text{(12a)} \\
&(\nabla \cdot \boldsymbol{w}, u_i^h)_{\Omega_i} - (\boldsymbol{w}, \boldsymbol{q}_i^h)_{\Omega_i} \\
&\quad = < \boldsymbol{n}_i \cdot \boldsymbol{w}, u_D >_{\partial\Omega_i \cap \Gamma_D} + < \boldsymbol{n}_i \cdot \boldsymbol{w}, \hat{u}^h >_{\partial\Omega_i \setminus \Gamma_D}, && \text{(12b)}
\end{aligned}
$$

whereas the global problem is: find $\hat{u}^h \in \mathcal{M}^h(\Gamma \cup \Gamma_N)$ for all $\mu \in \mathcal{M}^h(\Gamma \cup \Gamma_N)$ such that

$$\sum_{i=1}^{\mathtt{n_{el}}}\left\{<\mu,\tau_i u_i^h>_{\partial\Omega_i\setminus\Gamma_D}+<\mu,\boldsymbol{n}_i\cdot\boldsymbol{q}_i^h>_{\partial\Omega_i\setminus\Gamma_D}-<\mu,\tau_i\hat{u}^h>_{\partial\Omega_i\setminus\Gamma_D}\right\}$$
$$=-\sum_{i=1}^{\mathtt{n_{el}}}<\mu,t>_{\partial\Omega_i\cap\Gamma_N}. \tag{13}$$

At this point, it is important to notice that (12) is well defined, see Theorem 4.2. Thus, with the interpolation chosen by (5), Eq. (12) give rise to the following system of equations for each element Ω_i (i.e., for $i=1,\ldots,\mathtt{n_{el}}$)

$$\begin{bmatrix}\mathbf{A}_{uu} & \mathbf{A}_{uq}\\ \mathbf{A}_{uq}^T & \mathbf{A}_{qq}\end{bmatrix}_i\begin{Bmatrix}\mathbf{u}_i\\ \mathbf{q}_i\end{Bmatrix}=\begin{Bmatrix}\mathbf{f}_u\\ \mathbf{f}_q\end{Bmatrix}_i+\begin{bmatrix}\mathbf{A}_{u\hat{u}}\\ \mathbf{A}_{q\hat{u}}\end{bmatrix}_i\hat{\mathbf{u}}_i. \tag{14a}$$

Recalling the dimensions of the different vectors presented in Sect. 3, this system requires inverting a dense matrix of dimension $(\mathtt{n_{sd}}+1)^2\mathtt{n_{en}^2}$.

Similarly, the interpolation defined by (5) applied to (13) produce the following system of equations

$$\sum_{i=1}^{\mathtt{n_{el}}}\left\{\begin{bmatrix}\mathbf{A}_{u\hat{u}}^T & \mathbf{A}_{q\hat{u}}^T\end{bmatrix}_i\begin{Bmatrix}\mathbf{u}_i\\ \mathbf{q}_i\end{Bmatrix}+[\mathbf{A}_{\hat{u}\hat{u}}]_i\,\hat{\mathbf{u}}_i\right\}=\sum_{i=1}^{\mathtt{n_{el}}}[\mathbf{f}_{\hat{u}}]_i. \tag{14b}$$

A detailed description of the matrices and vectors appearing in (14) is given in Appendix A.

After replacing the solution of the local problem (14a) in (14b), the global problem becomes

$$\widehat{\mathbf{K}}\hat{\mathbf{u}}=\hat{\mathbf{f}}, \tag{15}$$

with

$$\widehat{\mathbf{K}}=\mathop{\mathbf{A}}_{i=1}^{\mathtt{n_{el}}}\begin{bmatrix}\mathbf{A}_{u\hat{u}}^T & \mathbf{A}_{q\hat{u}}^T\end{bmatrix}_i\begin{bmatrix}\mathbf{A}_{uu} & \mathbf{A}_{uq}\\ \mathbf{A}_{uq}^T & \mathbf{A}_{qq}\end{bmatrix}_i^{-1}\begin{bmatrix}\mathbf{A}_{u\hat{u}}\\ \mathbf{A}_{q\hat{u}}\end{bmatrix}_i+[\mathbf{A}_{\hat{u}\hat{u}}]_i \tag{16a}$$

and

$$\hat{\mathbf{f}}=\mathop{\mathbf{A}}_{i=1}^{\mathtt{n_{el}}}[\mathbf{f}_{\hat{u}}]_i-\begin{bmatrix}\mathbf{A}_{u\hat{u}}^T & \mathbf{A}_{q\hat{u}}^T\end{bmatrix}_i\begin{bmatrix}\mathbf{A}_{uu} & \mathbf{A}_{uq}\\ \mathbf{A}_{uq}^T & \mathbf{A}_{qq}\end{bmatrix}_i^{-1}\begin{Bmatrix}\mathbf{f}_u\\ \mathbf{f}_q\end{Bmatrix}_i. \tag{16b}$$

Note the symmetry of the (local and) global problem.

Theorem 4.2 (Well posedness of the local problem (Cockburn et al. 2009b)) *The local solver defined by (12) on Ω_i for each element $i=1,\ldots,\mathtt{n_{el}}$ is well defined if $\tau_i>0$ on $\partial\Omega_i$ and $\nabla\mathcal{V}^h(\Omega_i)\subset\mathcal{W}^h(\Omega_i)$.*

Proof For homogeneous conditions, i.e., $\hat{u}^h=0$, $f=0$ and $u_D=0$, and for $(\boldsymbol{w},v):=(\boldsymbol{q}_i^h,u_i^h)$, Eq. (12) read

$$< u_i^h, \tau_i u_i^h >_{\partial\Omega_i} + (u_i^h, \nabla \cdot \boldsymbol{q}_i^h)_{\Omega_i} = 0,$$
$$(\nabla \cdot \boldsymbol{q}_i^h, u_i^h)_{\Omega_i} - (\boldsymbol{q}_i^h, \boldsymbol{q}_i^h)_{\Omega_i} = 0.$$

Hence, subtracting both equations

$$< u_i^h, \tau_i u_i^h >_{\partial\Omega_i} + (\boldsymbol{q}_i^h, \boldsymbol{q}_i^h)_{\Omega_i} = 0,$$

which implies, for $\tau_i > 0$ on $\partial\Omega_i$, that $\boldsymbol{q}_i^h = \mathbf{0}$ in Ω_i and $u_i^h = 0$ on $\partial\Omega_i$. Then, since $\boldsymbol{q}_i^h = \mathbf{0}$ in Ω_i Eq. (12b) becomes

$$(u_i^h, \nabla \cdot \boldsymbol{w})_{\Omega_i} = 0 \quad \forall \boldsymbol{w} \in \mathcal{W}^h$$

or, equivalently,

$$(\nabla u_i^h, \boldsymbol{w})_{\Omega_i} = 0 \quad \forall \boldsymbol{w} \in \mathcal{W}^h,$$

which implies $\nabla u_i^h = \mathbf{0}$ in Ω_i and proves the result. □

Remark 4.3 (Computational aspect) For implementation purposes, some auxiliary vectors are defined. As noticed by Theorem 4.2, problem (12) is well posed thus, Eq. (14a) can be solved and written as

$$\begin{Bmatrix} \mathbf{u}_i \\ \mathbf{q}_i \end{Bmatrix} = \begin{Bmatrix} \mathbf{z}_u^f \\ \mathbf{z}_q^f \end{Bmatrix}_i + \begin{bmatrix} \mathbf{Z}_u^{\hat{u}} \\ \mathbf{Z}_q^{\hat{u}} \end{bmatrix}_i \hat{\mathbf{u}}_i \tag{17}$$

where

$$\begin{Bmatrix} \mathbf{z}_u^f \\ \mathbf{z}_q^f \end{Bmatrix}_i = \begin{bmatrix} \mathbf{A}_{uu} & \mathbf{A}_{uq} \\ \mathbf{A}_{uq}^T & \mathbf{A}_{qq} \end{bmatrix}_i^{-1} \begin{Bmatrix} \mathbf{f}_u \\ \mathbf{f}_q \end{Bmatrix}_i \text{ and } \begin{bmatrix} \mathbf{Z}_u^{\hat{u}} \\ \mathbf{Z}_q^{\hat{u}} \end{bmatrix}_i = \begin{bmatrix} \mathbf{A}_{uu} & \mathbf{A}_{uq} \\ \mathbf{A}_{uq}^T & \mathbf{A}_{qq} \end{bmatrix}_i^{-1} \begin{bmatrix} \mathbf{A}_{u\hat{u}} \\ \mathbf{A}_{q\hat{u}} \end{bmatrix}_i.$$

Then, (17) is replaced in (14b), which induces the same system of Eq. (15) but the matrix and vector defined by (16) are computed as follows:

$$\widehat{\mathbf{K}} = \mathop{\mathbf{A}}_{i=1}^{\mathtt{n_{el}}} [\mathbf{A}_{u\hat{u}}^T \; \mathbf{A}_{q\hat{u}}^T]_i \begin{bmatrix} \mathbf{Z}_u^{\hat{u}} \\ \mathbf{Z}_q^{\hat{u}} \end{bmatrix}_i + [\mathbf{A}_{\hat{u}\hat{u}}]_i \quad \text{and} \quad \hat{\mathbf{f}} = \mathop{\mathbf{A}}_{i=1}^{\mathtt{n_{el}}} [\mathbf{f}_{\hat{u}}]_i - [\mathbf{A}_{u\hat{u}}^T \; \mathbf{A}_{q\hat{u}}^T]_i \begin{Bmatrix} \mathbf{z}_u^f \\ \mathbf{z}_q^f \end{Bmatrix}_i.$$

4.4 Numerical Example

In order to illustrate the results of HDG, the model problem (1) is solved in $\Omega :=]0,1[\times]0,1[$ with $\Gamma_N = \{(x,y) \in \partial\Omega \mid y = 0\}$ and $\Gamma_D = \partial\Omega \backslash \Gamma_N$. The source and boundary conditions are taken such that the analytical solution is given by

$$u(x, y) = 4y^2 - 4\lambda^2 y \exp(-\lambda y) \cos(6\pi x) + \lambda \exp(-2\lambda y),$$

where λ is a parameter that enables to control the strength of the solution gradient near the Neumann boundary. The effect of this parameter is illustrated in Fig. 1, where the analytical solution is represented for two values of λ, namely 4 and 10.

The first example involves the solution of the model problem with a value of $\lambda = 4$. An extremely coarse mesh, with only eight elements, is considered, as shown in the left plot of Fig. 2. The right plot of Fig. 2 depicts the degrees of freedom used in an HDG computation with approximation order $p = 6$. The black dots on the triangles denote the nodes used to build the polynomial approximation of the primal and dual solutions, u^h and $\boldsymbol{q}^h$, respectively. The red lines are the set of edges $\Gamma \cup \Gamma_N$ where the trace of the solution is approximated and the dots over these lines are the nodes used to build the polynomial approximation of $\hat{u}$. Note that there are no $\hat{u}^h$ unknowns along the Dirichlet boundary $\Gamma_D = \partial\Omega \backslash \Gamma_N$. The nodal distributions in elements and edges correspond to approximated optimal points presented in Chen and Babuška (1995) that are known to have better approximation properties than traditional equally spaced nodal distributions.

The numerical solution computed with a polynomial approximation of degree $p = 6$ is depicted in Fig. 3, showing both the approximation of the solution in the element interiors and the approximation of the trace of the solution on $\Gamma \cup \Gamma_N$. It can be clearly observed that the numerical solution, u^h, is obviously discontinuous. More important, the numerical solution u^h and the numerical trace, $\hat{u}^h$, do not coincide on $\Gamma \cup \Gamma_N$ because the condition $u = \hat{u}$ in problem (6) is imposed in a weak sense.

Next, the model problem is considered with a value of $\lambda = 10$. Figure 4 shows the numerical solution computed on a finer mesh, with 32 elements, and with a degree of approximation $p = 4$ and $p = 5$. It is worth noting how the jump of the solution on the element interfaces decreases as the degree of the approximation increases, suggesting the higher accuracy of the solution computed with $p = 5$.

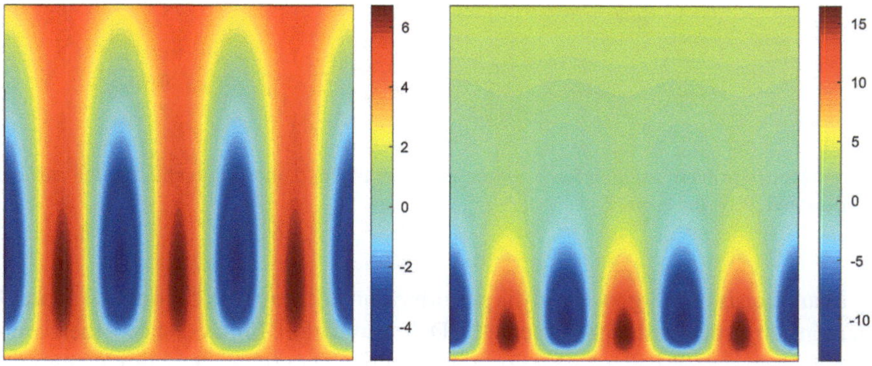

Fig. 1 Model problem analytical solution: $\lambda = 4$ (*left*) and $\lambda = 10$ (*right*)

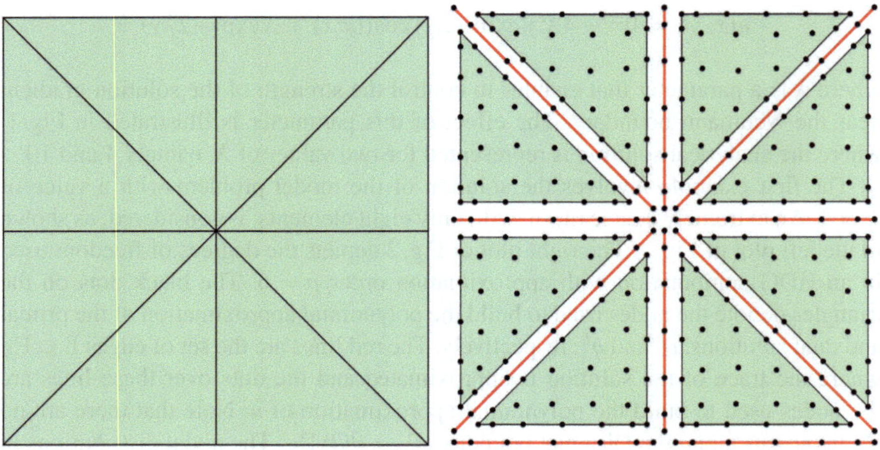

Fig. 2 Coarse mesh of the domain Ω (*left*) and illustration of the degrees of freedom employed in an HDG computation with $p = 6$ (*right*)

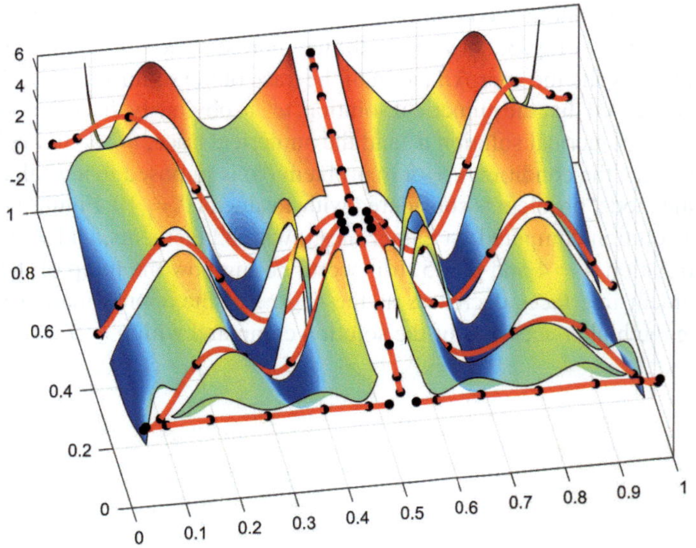

Fig. 3 Model problem solution for $p = 6$ on the mesh of Fig. 2 showing both u^h in the element interiors and $\hat{u}^h$ on the edges $\Gamma \cup \Gamma_N$

Finally, an h-convergence study is performed in order to check the optimal approximation properties of the implemented HDG formulation. Figure 5 shows the evolution of the error of u^h in the $\mathcal{L}_2(\Omega)$ norm as a function of the characteristic element size h for a degree of approximation p ranging from 1 to 5. For all the degrees of approximation considered, the optimal rate of convergence (i.e., $p + 1$) is obtained. The results also illustrate the benefits of using high-order approximations. For instance,

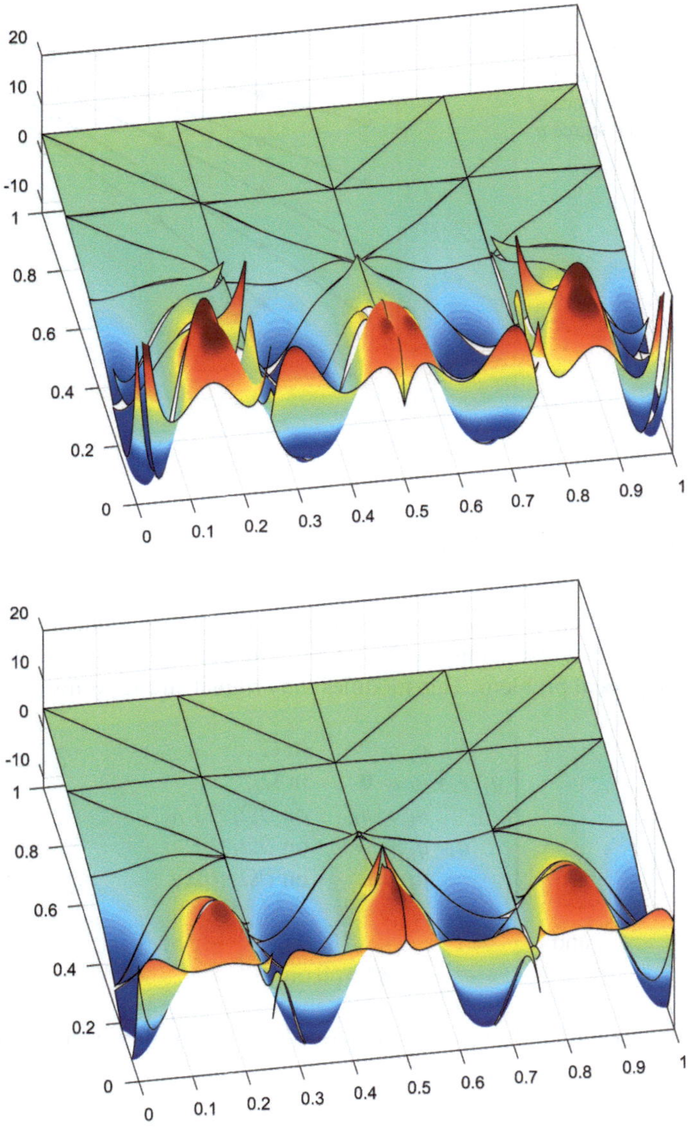

Fig. 4 Model problem solution for $p = 4$ (*top*) and $p = 5$ (*bottom*) showing the improvement induced by an increase on the degree of approximation

similar accuracy is obtained with a quartic approximation in a the mesh with 32 elements and with a linear approximation in a mesh with 2 048 elements. This implies that, in order to obtain a similar accuracy, linear elements require the solution of a system of equations ten times larger than the one induced by a quartic approximation.

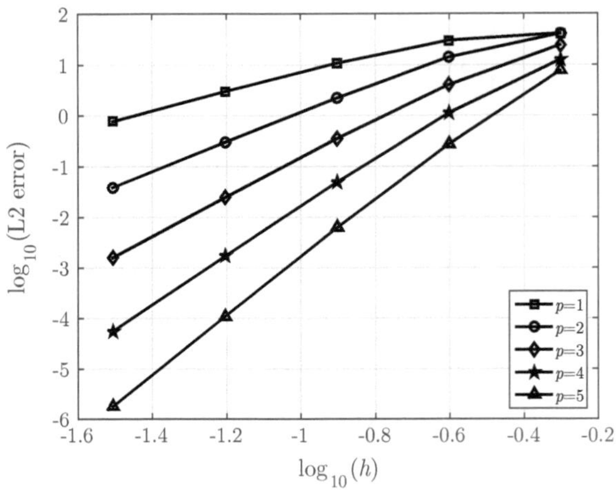

Fig. 5 Error of u^h in the $\mathcal{L}_2(\Omega)$ norm as a function of the characteristic element size h for different values of the approximation degree p

4.5 Neumann Local Problems

A minor modification of the previous formulation can induce a smaller global problem. It consists of prescribing the Neumann boundary conditions already in the corresponding local problem. This modifies the original strong forms (6) and (7) as

$$\begin{cases} \nabla \cdot \boldsymbol{q}_i = f & \text{in } \Omega_i, \\ \boldsymbol{q}_i + \nabla u_i = \boldsymbol{0} & \text{in } \Omega_i, \\ u_i = u_D & \text{on } \partial\Omega_i \cap \Gamma_D, \\ \boldsymbol{n}_i \cdot \boldsymbol{q}_i = -t & \text{on } \partial\Omega_i \cap \Gamma_N, \\ u_i = \hat{u} & \text{on } \partial\Omega_i \backslash \partial\Omega, \end{cases} \tag{18}$$

for $i = 1, \ldots \mathtt{n_{el}}$, and

$$[\![\boldsymbol{n} \cdot \boldsymbol{q}]\!] = 0 \text{ on } \Gamma. \tag{19}$$

It also implies a new definition for the numerical traces of the fluxes, thus (8) becomes, for $i = 1, \ldots \mathtt{n_{el}}$,

$$\boldsymbol{n}_i \cdot \widehat{\boldsymbol{q}}_i := \begin{cases} \boldsymbol{n}_i \cdot \boldsymbol{q}_i + \tau_i(u_i - u_D) & \text{on } \partial\Omega_i \cap \Gamma_D, \\ \boldsymbol{n}_i \cdot \boldsymbol{q}_i + \tau_i(u_i - \hat{u}) & \text{on } \partial\Omega_i \cap \Gamma, \\ -t & \text{on } \partial\Omega_i \cap \Gamma_N. \end{cases} \tag{20}$$

Consequently, the weak form for the local problem, originally defined by (9) is now: for $i = 1, \ldots \mathtt{n_{el}}$, find $(\boldsymbol{q}_i, u_i) \in \boldsymbol{\mathcal{W}}(\Omega_i) \times \mathcal{V}(\Omega_i)$ that satisfies

$$
\begin{aligned}
&< v, \tau_i\, u_i >_{\partial\Omega_i \setminus \Gamma_N} - (\nabla v, \boldsymbol{q}_i)_{\Omega_i} + < v, \boldsymbol{n}_i \cdot \boldsymbol{q}_i >_{\partial\Omega_i \setminus \Gamma_N} \\
&\quad = (v, f)_{\Omega_i} + < v, t >_{\partial\Omega_i \cap \Gamma_N} + < v, \tau_i\, u_D >_{\partial\Omega_i \cap \Gamma_D} + < v, \tau_i\, \hat{u} >_{\partial\Omega_i \setminus \partial\Omega}, \\
&-(\boldsymbol{w}, \boldsymbol{q}_i)_{\Omega_i} + (\nabla \cdot \boldsymbol{w}, u_i)_{\Omega_i} - < \boldsymbol{n} \cdot \boldsymbol{w}, u_i >_{\partial\Omega_i \cap \Gamma_N} \\
&\quad = < \boldsymbol{n} \cdot \boldsymbol{w}, u_D >_{\partial\Omega_i \cap \Gamma_D} + < \boldsymbol{n} \cdot \boldsymbol{w}, \hat{u} >_{\partial\Omega_i \setminus \partial\Omega} .
\end{aligned}
$$

For all $(\boldsymbol{w}, v) \in \boldsymbol{\mathcal{W}}(\Omega_i) \times \mathcal{V}(\Omega_i)$. To obtain the second equation above, it is important to recall that $\hat{u} \in \mathcal{L}_2(\Gamma)$ is not defined along Γ_N and, consequently, u_i is left along $\partial\Omega_i \cap \Gamma_N$. Following Remark 4.1, a symmetric version can also be obtained, namely

$$
\begin{aligned}
&< v, \tau_i\, u_i >_{\partial\Omega_i \setminus \Gamma_N} + (v, \nabla \cdot \boldsymbol{q}_i)_{\Omega_i} - < v, \boldsymbol{n}_i \cdot \boldsymbol{q}_i >_{\partial\Omega_i \cap \Gamma_N} \\
&\quad = (v, f)_{\Omega_i} + < v, t >_{\partial\Omega_i \cap \Gamma_N} + < v, \tau_i\, u_D >_{\partial\Omega_i \cap \Gamma_D} + < v, \tau_i\, \hat{u} >_{\partial\Omega_i \setminus \partial\Omega}, \qquad (21a) \\
&(\nabla \cdot \boldsymbol{w}, u_i)_{\Omega_i} - < \boldsymbol{n} \cdot \boldsymbol{w}, u_i >_{\partial\Omega_i \cap \Gamma_N} - (\boldsymbol{w}, \boldsymbol{q}_i)_{\Omega_i} \\
&\quad = < \boldsymbol{n} \cdot \boldsymbol{w}, u_D >_{\partial\Omega_i \cap \Gamma_D} + < \boldsymbol{n} \cdot \boldsymbol{w}, \hat{u} >_{\partial\Omega_i \setminus \partial\Omega} . \qquad (21b)
\end{aligned}
$$

For the global problem, originally (10), continuity of fluxes is now only imposed along the internal faces, see (19). Hence, the global weak problem is: find $\hat{u} \in \mathcal{L}_2(\Gamma)$ for all $\mu \in \mathcal{L}_2(\Gamma)$ such that

$$
\sum_{i=1}^{\mathtt{n_{el}}} < \mu, \left[\boldsymbol{n}_i \cdot \boldsymbol{q}_i + \tau(u_i - \hat{u})\right] >_{\partial\Omega_i \setminus \partial\Omega} = 0, \tag{22}
$$

where the definition of the numerical flux, see (20), has already been used.

The discrete versions of these weak problem (21) and (22) are automatically determined as: for $i = 1, \ldots \mathtt{n_{el}}$, find $(\boldsymbol{q}_i^h, u_i^h) \in \boldsymbol{\mathcal{W}}^h \times \mathcal{V}^h$ for all $(\boldsymbol{w}, v) \in \boldsymbol{\mathcal{W}}^h \times \mathcal{V}^h$ such that

$$
\begin{aligned}
&< v, \tau_i\, u_i^h >_{\partial\Omega_i \setminus \Gamma_N} + (v, \nabla \cdot \boldsymbol{q}_i^h)_{\Omega_i} - < v, \boldsymbol{n}_i \cdot \boldsymbol{q}_i^h >_{\partial\Omega_i \cap \Gamma_N} \\
&\quad = (v, f)_{\Omega_i} + < v, t >_{\partial\Omega_i \cap \Gamma_N} + < v, \tau_i\, u_D >_{\partial\Omega_i \cap \Gamma_D} + < v, \tau_i\, \hat{u}^h >_{\partial\Omega_i \setminus \partial\Omega}, \qquad (23a) \\
&(\nabla \cdot \boldsymbol{w}, u_i^h)_{\Omega_i} - < \boldsymbol{n} \cdot \boldsymbol{w}, u_i^h >_{\partial\Omega_i \cap \Gamma_N} - (\boldsymbol{w}, \boldsymbol{q}_i^h)_{\Omega_i} \\
&\quad = < \boldsymbol{n} \cdot \boldsymbol{w}, u_D >_{\partial\Omega_i \cap \Gamma_D} + < \boldsymbol{n} \cdot \boldsymbol{w}, \hat{u}^h >_{\partial\Omega_i \setminus \partial\Omega}, \qquad (23b)
\end{aligned}
$$

and find $\hat{u}^h \in \mathcal{M}^h(\Gamma)$ for all $\mu \in \mathcal{M}^h(\Gamma)$ such that

$$
\sum_{i=1}^{\mathtt{n_{el}}} \left\{ < \mu, \tau\, u_i^h >_{\partial\Omega_i \setminus \partial\Omega} + < \mu, \boldsymbol{n}_i \cdot \boldsymbol{q}_i^h >_{\partial\Omega_i \setminus \partial\Omega} - < \mu, \tau\, \hat{u}^h >_{\partial\Omega_i \setminus \partial\Omega} \right\} = 0, \tag{24}
$$

where, again, $(\boldsymbol{q}_i^h, u_i^h)$ are directly functions of $\hat{u}_i^h$ as determined by (23).

Finally, the following system of equations is obtained for the local problem, for each element $i = 1, \ldots, \mathtt{n_{el}}$,

$$\begin{bmatrix} \mathbf{A}^{\star}_{uu} & \mathbf{A}^{\star}_{uq} \\ \mathbf{A}^{\star T}_{uq} & \mathbf{A}_{qq} \end{bmatrix}_i \begin{Bmatrix} \mathbf{u}_i \\ \mathbf{q}_i \end{Bmatrix} = \begin{Bmatrix} \mathbf{f}^{\star}_u \\ \mathbf{f}_q \end{Bmatrix}_i + \begin{bmatrix} \mathbf{A}^{\star}_{u\hat{u}} \\ \mathbf{A}^{\star}_{q\hat{u}} \end{bmatrix}_i \hat{\mathbf{u}}_i. \tag{25a}$$

whereas the global system of equations is simply

$$\sum_{i=1}^{\mathtt{n_{el}}} \left\{ \begin{bmatrix} \mathbf{A}^{\star T}_{u\hat{u}} & \mathbf{A}^{\star T}_{q\hat{u}} \end{bmatrix}_i \begin{Bmatrix} \mathbf{u}_i \\ \mathbf{q}_i \end{Bmatrix} + [\mathbf{A}^{\star}_{\hat{u}\hat{u}}]_i \, \hat{\mathbf{u}}_i \right\} = 0. \tag{25b}$$

A detailed description of the matrices and vectors appearing in (25) is given in Appendix A.

The final global system, which retains all the symmetries, becomes

$$\widehat{\mathbf{K}}^{\star} \hat{\mathbf{u}} = \hat{\mathbf{f}}^{\star}, \tag{26a}$$

with

$$\widehat{\mathbf{K}}^{\star} = \mathop{\mathbf{A}}_{i=1}^{\mathtt{n_{el}}} \begin{bmatrix} \mathbf{A}^{\star T}_{u\hat{u}} & \mathbf{A}^{\star T}_{q\hat{u}} \end{bmatrix}_i \begin{bmatrix} \mathbf{A}^{\star}_{uu} & \mathbf{A}^{\star}_{uq} \\ \mathbf{A}^{\star T}_{uq} & \mathbf{A}_{qq} \end{bmatrix}_i^{-1} \begin{bmatrix} \mathbf{A}^{\star}_{u\hat{u}} \\ \mathbf{A}^{\star}_{q\hat{u}} \end{bmatrix}_i + [\mathbf{A}^{\star}_{\hat{u}\hat{u}}]_i \tag{26b}$$

and

$$\hat{\mathbf{f}}^{\star} = \mathop{\mathbf{A}}_{i=1}^{\mathtt{n_{el}}} - \begin{bmatrix} \mathbf{A}^{\star T}_{u\hat{u}} & \mathbf{A}^{\star T}_{q\hat{u}} \end{bmatrix}_i \begin{bmatrix} \mathbf{A}^{\star}_{uu} & \mathbf{A}^{\star}_{uq} \\ \mathbf{A}^{\star T}_{uq} & \mathbf{A}_{qq} \end{bmatrix}_i^{-1} \begin{Bmatrix} \mathbf{f}^{\star}_u \\ \mathbf{f}_q \end{Bmatrix}_i. \tag{26c}$$

Note that, in this case, the dimension of $\hat{\mathbf{u}}$ corresponds only to the degrees of freedom along the interior skeleton Γ, which is slightly smaller than in the previous case where unknowns had also to be determined along the Neumann boundary.

Theorem 4.4 (Well posedness of the local Neumann problem) *The local solver defined by (23) on Ω_i for each element $i = 1, \ldots, \mathtt{n_{el}}$ is well defined if $\tau_i > 0$ on $\partial\Omega_i$ and $\nabla\mathcal{V}^h(\Omega_i) \subset \mathcal{W}^h(\Omega_i)$.*

Proof For homogeneous conditions, i.e., $\hat{u}^h = 0$, $f = 0$ and $u_D = 0$, and for $(\boldsymbol{w}, v) := (\boldsymbol{q}_i^h, u_i^h)$, Eq. (23) read

$$< u_i^h, \tau_i u_i^h >_{\partial\Omega_i \setminus \Gamma_N} + (u_i^h, \nabla \cdot \boldsymbol{q}_i^h)_{\Omega_i} - < u_i^h, \boldsymbol{n}_i \cdot \boldsymbol{q}_i^h >_{\partial\Omega_i \cap \Gamma_N} = 0, \tag{27}$$

$$(\nabla \cdot \boldsymbol{q}_i^h, u_i^h)_{\Omega_i} - < \boldsymbol{n} \cdot \boldsymbol{q}_i^h, u_i^h >_{\partial\Omega_i \cap \Gamma_N} - (\boldsymbol{q}_i^h, \boldsymbol{q}_i^h)_{\Omega_i} = 0. \tag{28}$$

Hence, subtracting both equations

$$< u_i^h, \tau_i u_i^h >_{\partial\Omega_i \setminus \Gamma_N} + (\boldsymbol{q}_i^h, \boldsymbol{q}_i^h)_{\Omega_i} = 0,$$

which implies, for $\tau_i > 0$ on $\partial\Omega_i$, that $\boldsymbol{q}_i^h = \boldsymbol{0}$ in Ω_i and $u_i^h = 0$ on $\partial\Omega_i \backslash \Gamma_N$. Then, since $\boldsymbol{q}_i^h = \boldsymbol{0}$ in Ω_i Eq. (23b) becomes

$$(\nabla \cdot \boldsymbol{w}, u_i^h)_{\Omega_i} - < \boldsymbol{n} \cdot \boldsymbol{w}, u_i^h >_{\partial\Omega_i \cap \Gamma_N} = 0 \quad \forall \boldsymbol{w} \in \boldsymbol{\mathcal{W}}^h$$

or, equivalently,

$$-(\nabla u_i^h, \boldsymbol{w})_{\Omega_i} + < \boldsymbol{n} \cdot \boldsymbol{w}, u_i^h >_{\partial\Omega_i \backslash \Gamma_N} = 0 \quad \forall \boldsymbol{w} \in \boldsymbol{\mathcal{W}}^h.$$

But $u_i^h = 0$ on $\partial\Omega_i \backslash \Gamma_N$. Thus,

$$(\nabla u_i^h, \boldsymbol{w})_{\Omega_i} = 0 \quad \forall \boldsymbol{w} \in \boldsymbol{\mathcal{W}}^h,$$

which implies $\nabla u_i^h = \boldsymbol{0}$ in Ω_i and proves the result. □

4.6 Numerical Example

In order to illustrate the results of HDG by using the formulation with Neumann local problems, the model problem of Sect. 4.4 is considered with a value of $\lambda = 10$. Figure 6 shows the numerical solution computed with a degree of approximation $p = 5$. A visual comparison of the bottom plot in Figs. 4 and 6 suggests that the

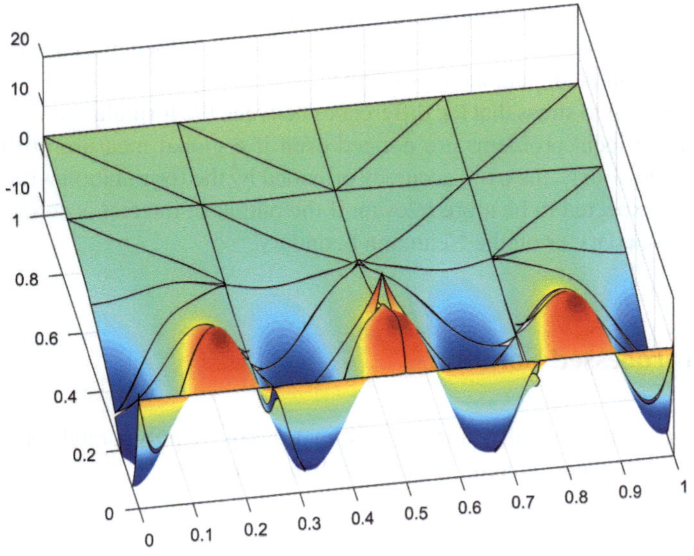

Fig. 6 Model problem solution for $p = 5$ using the formulation with Neumann local problems

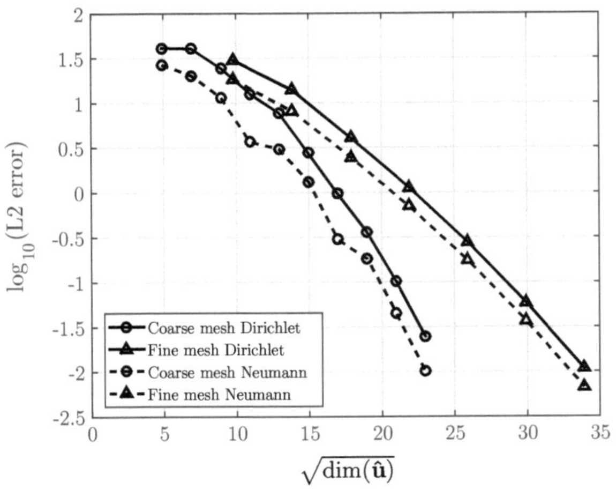

Fig. 7 *p*-refinement: error of u^h for $p = 1, 2, 3, \ldots$ in the $\mathcal{L}_2(\Omega)$ norm as a function of $\sqrt{\dim(\hat{\mathbf{u}})}$. The results are displayed for the HDG formulations with Dirichlet and Neumann local problems and for two different meshes

formulation with Neumann local problems provides a better accuracy of the solution near Neumann boundaries.

Next, an *p*-convergence study is performed in order to check the optimality of the approximation using the formulation with Neumann local problems and to compare the accuracy of the two HDG formulations considered in this work. Figure 7 shows the evolution of the error of u^h in the $\mathcal{L}_2(\Omega)$ norm as a function of the square root of the number of degrees of freedom of the global system of equations, i.e., $\mathtt{n_{dof}} = \dim(\hat{\mathbf{u}})$. Two meshes with 8 and 32 elements are considered and the degree of approximation is increased in each mesh from $p = 1$. The exponential rate of convergence is observed in all cases and the results reveal the advantage of using the formulation with Neumann local problems.

It is important to stress that the differences between the formulation with Dirichlet and Neumann local problems are noticed even if a global measure of the error is employed. Obviously, the extra accuracy provided by the formulation with Neumann problems is expected to be more relevant if the output of interest is defined near the Neumann boundary or on the Neumann boundary.

5 Postprocessed Solution

The following well-known a priori error estimate holds if a polynomial approximation of degree $p \geq 0$ is considered for the primal variable, u,

$$\|e_u\|_{\mathcal{L}_2(\Omega)} \leq Ch^{p+1}|u|_{\mathcal{H}^{p+1}(\Omega)},$$

where e_u denotes the error of the primal variable, h is the characteristic mesh size and $\|\cdot\|$ and $|\cdot|$ denote the norm and the semi-norm, respectively, induced by the scalar product defined in Sect. 3, see for instance (Szabó and Babuška 1991; Brenner and Scott 1994).

Optimal convergence of the dual variable $\mathbf{q}$ is strongly dependent on the definition of the numerical flux. For a variety of DG methods, only convergence with order p was proved, see the unified analysis by Arnold et al. (2002). The first DG method with optimal convergence for the dual variable was introduced by Cockburn et al. (2009b). For a given element, assuming that the stabilization parameter τ is equal to zero except on an arbitrary chosen element face, it was proved that the following a priori error estimate holds if a polynomial approximation of degree $p \geq 0$ is considered for the dual variable, $\mathbf{q}$,

$$\|e_{\mathbf{q}}\|_{\mathcal{L}_2(\Omega)} \leq Ch^{p+1}|\mathbf{q}|_{\mathcal{H}^{p+1}(\Omega)},$$

where $e_{\mathbf{q}}$ denotes the error of the dual variable, see Cockburn et al. (2008, 2009b, c) for more details.

Using the similarities of the HDG method and the Raviart–Thomas and Brezzi–Douglas–Marini mixed methods, see Raviart and Thomas (1977), Brezzi et al. (1985), it is possible to devise a *superconvergent* solution, $u_\star$, such that the following a priori error estimate holds

$$\|e_{u_\star}\|_{\mathcal{L}_2(\Omega)} \leq Ch^{p+2}|u|_{\mathcal{H}^{p+2}(\Omega)}.$$

for $p \geq 1$, see for instance (Cockburn et al. 2008, 2009b, c).

The postprocessed solution is computed by performing a postprocessing similar to the projection traditionally employed in the mixed method by Raviart and Thomas (1977), see also Arnold and Brezzi (1985). More precisely, the superconvergent postprocessed solution is obtained by solving the following problem in each element

$$\begin{cases} -\boldsymbol{\nabla}\cdot\boldsymbol{\nabla}u_\star = -\boldsymbol{\nabla}\cdot\mathbf{q}_h & \text{in } \Omega_i, \\ \boldsymbol{n}\cdot\boldsymbol{\nabla}u_\star = \boldsymbol{n}\cdot\mathbf{q}_h & \text{on } \partial\Omega_i, \end{cases} \tag{29}$$

with the additional solvability constraint

$$\int_{\Omega_i} u_\star = \int_{\Omega_i} u_h,$$

for $i = 1, \ldots, \mathtt{n_{el}}$.

If the approximation to the postprocessed solution, namely $u_\star^h$, is sought in a space $\mathcal{V}_\star^h(\Omega)$ that contains $\mathcal{V}^h(\Omega)$, asymptotic convergence of order $p+2$ can be proved, as shown by Cockburn et al. (2008). A typical choice for the richer space where $u_\star^h$ belongs is

$$\mathcal{V}_\star^h(\Omega) = \{v \in \mathcal{L}_2(\Omega); v|_{\Omega_i} \in \mathcal{P}^{p+1}(\Omega_i)\ \forall\Omega_i\}.$$

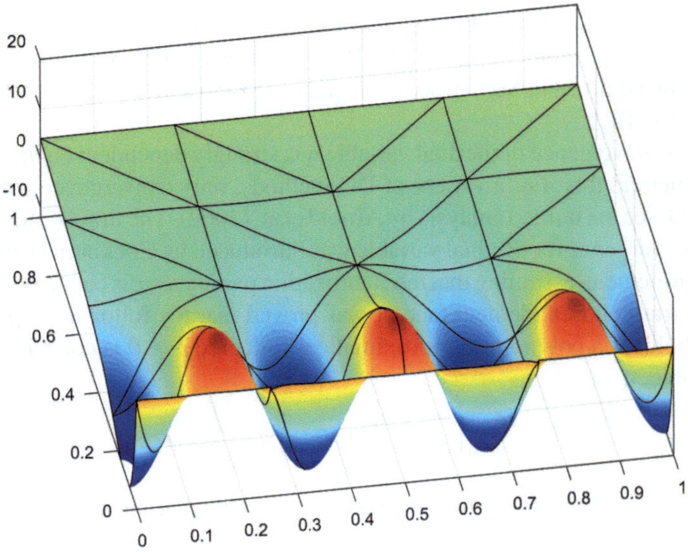

Fig. 8 Model problem postprocessed solution for $p = 5$ using the formulation with Neumann local problems

Figure 8 shows the postprocessed solution corresponding to an HDG computation with $p = 5$ for the formulation with Neumann local problems. The gain in accuracy induced by the postprocessing is clearly observed by comparing the postprocessed solution in Fig. 8 with the solution shown in Fig. 6. In this example, the postprocessed solution $u_\star^h$ has an error in the $\mathcal{L}_2(\Omega)$ norm, one order of magnitude lower than the error of the solution u^h.

It is important to remark that the significant extra accuracy provided by the postprocessing technique only requires the solution of the element-by-element problem (29), having a marginal cost compared to the cost of computing the solution u^h.

Next, an h-convergence study of the error of the postprocessed solution is performed. Figure 9 compares the evolution of the error of the solution u^h and the postprocessed solution $u_\star^h$ in the $\mathcal{L}_2(\Omega)$ norm as a function of the characteristic element size h and for a degree of approximation p ranging from 1 to 5. All the simulations correspond to the formulation with Neumann local problems.

The results show that the optimal (i.e., $p+1$ for the solution and $p+2$ for the postprocessed solution) rate of convergence is obtained in all cases. The substantial gain in accuracy introduced by the postprocessing technique is clearly illustrated. As an example, the postprocessing of the solution computed in the finer mesh with $p = 5$ reduces the error by two orders of magnitude.

As expected, the same rate of convergence is obtained for the postprocessed solution $u_\star^h$ that results from a computation with degree of approximation p and the solution u^h computed with a degree of approximation $p+1$. However, it is worth

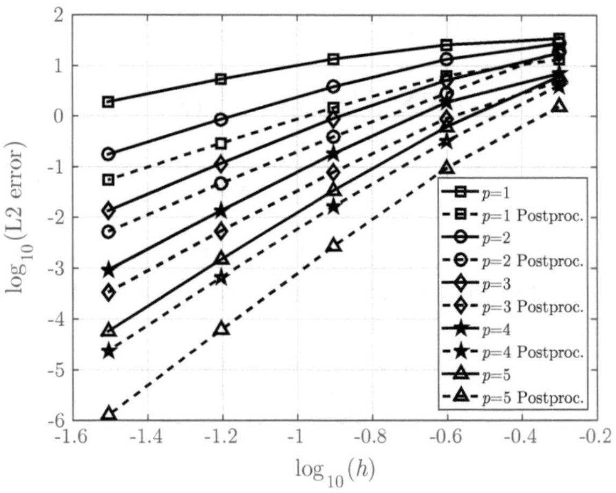

Fig. 9 Error of the solution and the postprocessed solution in the $\mathcal{L}_2(\Omega)$ norm as a function of the characteristic element size h for different values of the approximation degree p

emphasizing that the postprocessed solution derived from a computation with degree of approximation p is always more accurate than the solution computed with a degree of approximation $p+1$. For instance, the postprocessed solution computed in the finer mesh with $p=4$ is two times more accurate than the solution computed in the finer mesh with $p=5$.

The extra accuracy of the postprocessed solution has been recently exploited by Giorgiani et al. (2013, 2014) to define a simple and inexpensive error estimator than can be used to develop highly efficient p-adaptive procedures.

Appendix A: Implementation Details

This appendix is devoted to the detailed presentation of the matrices and vectors appearing in the discrete version of both the local and global problems induced by the HDG method.

The interpolation functions and their derivatives, used in (5), are defined in a reference element, with local coordinates $\boldsymbol{\xi}$. The isoparametric transformation is used to relate local and Cartesian coordinates, namely

$$\boldsymbol{x}(\boldsymbol{\xi}) = \sum_{i=1}^{\mathtt{n_{en}}} \boldsymbol{x}_i \mathbf{N}_i(\boldsymbol{\xi}),$$

where $\boldsymbol{x}_i$ denote the elemental nodal coordinates.

The following compact form of the interpolation functions is introduced

$$\mathbf{N} = \left[N_1\, N_2\, \ldots\, N_{\mathtt{n_{en}}}\right]^T, \quad \widehat{\mathbf{N}} = \left[\widehat{N}_1\, \widehat{N}_2\, \ldots\, \widehat{N}_{\mathtt{n_{fn}}}\right]^T,$$

$$\mathbf{N}_{\boldsymbol{n}} = \left[N_1\boldsymbol{n}\; N_2\boldsymbol{n}\; \ldots\; N_{\mathtt{n_{en}}}\boldsymbol{n}\right]^T, \quad \widehat{\mathbf{N}}_{\boldsymbol{n}} = \left[\widehat{N}_1\boldsymbol{n}\; \widehat{N}_2\boldsymbol{n}\; \ldots\; \widehat{N}_{\mathtt{n_{fn}}}\boldsymbol{n}\right]^T,$$

$$\nabla\mathbf{N} = \left[\left(\boldsymbol{J}^{-1}\nabla N_1\right)^T \left(\boldsymbol{J}^{-1}\nabla N_2\right)^T \ldots \left(\boldsymbol{J}^{-1}\nabla N_{\mathtt{n_{en}}}\right)^T\right]^T,$$

$$\mathbf{N}_{\mathtt{n_{sd}}} = \left[N_1\boldsymbol{I}_{\mathtt{n_{sd}}}\; N_2\boldsymbol{I}_{\mathtt{n_{sd}}}\; \ldots\; N_{\mathtt{n_{en}}}\boldsymbol{I}_{\mathtt{n_{sd}}}\right]^T,$$

where $\boldsymbol{n} = (n_1, \ldots, n_{\mathtt{n_{sd}}})$ denotes the outward unit normal vector to an edge/face, $\boldsymbol{J}$ is the Jacobian of the isoparametric transformation and $\boldsymbol{I}_{\mathtt{n_{sd}}}$ is the identity matrix of dimension $\mathtt{n_{sd}}$.

The different matrices appearing in (14), computed for each element $i = 1, \ldots, \mathtt{n_{el}}$, can be expressed as

$$[\mathbf{A}_{uu}]_i = \sum_{\partial\Omega_i} \tau_i \sum_{\mathtt{g}=1}^{\mathtt{n_{ip}^f}} \mathbf{N}(\boldsymbol{\xi}_{\mathtt{g}}^{\mathtt{f}})\, \mathbf{N}^T(\boldsymbol{\xi}_{\mathtt{g}}^{\mathtt{f}}) w_{\mathtt{g}}^{\mathtt{f}},$$

$$[\mathbf{A}_{qq}]_i = -\sum_{\mathtt{g}=1}^{\mathtt{n_{ip}^e}} \mathbf{N}_{\mathtt{n_{sd}}}(\boldsymbol{\xi}_{\mathtt{g}}^{\mathtt{e}})\, \mathbf{N}_{\mathtt{n_{sd}}}^T(\boldsymbol{\xi}_{\mathtt{g}}^{\mathtt{e}}) w_{\mathtt{g}}^{\mathtt{e}},$$

$$[\mathbf{A}_{uq}]_i = \sum_{\mathtt{g}=1}^{\mathtt{n_{ip}^e}} \mathbf{N}(\boldsymbol{\xi}_{\mathtt{g}}^{\mathtt{e}})\, \nabla\mathbf{N}^T(\boldsymbol{\xi}_{\mathtt{g}}^{\mathtt{e}}) w_{\mathtt{g}}^{\mathtt{e}},$$

$$[\mathbf{f}_u]_i = \sum_{\mathtt{g}=1}^{\mathtt{n_{ip}^e}} \mathbf{N}(\boldsymbol{\xi}_{\mathtt{g}}^{\mathtt{e}}) f\big(\boldsymbol{x}(\boldsymbol{\xi}_{\mathtt{g}}^{\mathtt{e}})\big) w_{\mathtt{g}}^{\mathtt{e}} + \sum_{\partial\Omega_i\cap\Gamma_D} \tau_i \sum_{\mathtt{g}=1}^{\mathtt{n_{ip}^f}} \mathbf{N}(\boldsymbol{\xi}_{\mathtt{g}}^{\mathtt{f}})\, u_D\big(\boldsymbol{x}(\boldsymbol{\xi}_{\mathtt{g}}^{\mathtt{e}})\big) w_{\mathtt{g}}^{\mathtt{f}},$$

$$[\mathbf{f}_q]_i = \sum_{\partial\Omega_i\cap\Gamma_D} \sum_{\mathtt{g}=1}^{\mathtt{n_{ip}^f}} \mathbf{N}_{\boldsymbol{n}}(\boldsymbol{\xi}_{\mathtt{g}}^{\mathtt{f}})\, u_D\big(\boldsymbol{x}(\boldsymbol{\xi}_{\mathtt{g}}^{\mathtt{f}})\big) w_{\mathtt{g}}^{\mathtt{f}},$$

$$[\mathbf{A}_{u\hat{u}}]_i = \sum_{\partial\Omega_i\setminus\Gamma_D} \tau_i \sum_{\mathtt{g}=1}^{\mathtt{n_{ip}^f}} \mathbf{N}(\boldsymbol{\xi}_{\mathtt{g}}^{\mathtt{f}})\, \widehat{\mathbf{N}}^T(\boldsymbol{\xi}_{\mathtt{g}}^{\mathtt{f}}) w_{\mathtt{g}}^{\mathtt{f}},$$

$$[\mathbf{A}_{q\hat{u}}]_i = \sum_{\partial\Omega_i\setminus\Gamma_D} \sum_{\mathtt{g}=1}^{\mathtt{n_{ip}^f}} \mathbf{N}_{\boldsymbol{n}}(\boldsymbol{\xi}_{\mathtt{g}}^{\mathtt{f}})\, \widehat{\mathbf{N}}^T(\boldsymbol{\xi}_{\mathtt{g}}^{\mathtt{f}}) w_{\mathtt{g}}^{\mathtt{f}},$$

$$[\mathbf{A}_{\hat{u}\hat{u}}]_i = -\sum_{\partial\Omega_i\setminus\Gamma_D} \tau_i \sum_{\mathtt{g}=1}^{\mathtt{n_{ip}^f}} \widehat{\mathbf{N}}_{\boldsymbol{n}}(\boldsymbol{\xi}_{\mathtt{g}}^{\mathtt{f}})\, \widehat{\mathbf{N}}^T(\boldsymbol{\xi}_{\mathtt{g}}^{\mathtt{f}}) w_{\mathtt{g}}^{\mathtt{f}}$$

and

$$[\mathbf{f}_{\hat{u}}]_i = -\sum_{\partial\Omega_i\cap\Gamma_N}\sum_{\mathtt{g}=1}^{\mathtt{n}^{\mathtt{f}}_{\mathtt{ip}}} \mathbf{N}(\boldsymbol{\xi}^{\mathtt{f}}_{\mathtt{g}})\, t\big(\boldsymbol{x}(\boldsymbol{\xi}^{\mathtt{f}}_{\mathtt{g}})\big)\, w^{\mathtt{f}}_{\mathtt{g}}.$$

In the above expressions, $\boldsymbol{\xi}^{\mathtt{e}}_{\mathtt{g}}$ and $w^{\mathtt{e}}_{\mathtt{g}}$ are the $\mathtt{n}^{\mathtt{e}}_{\mathtt{ip}}$ integration points and weights defined on the reference element and $\boldsymbol{\xi}^{\mathtt{f}}_{\mathtt{g}}$ and $w^{\mathtt{f}}_{\mathtt{g}}$ are the $\mathtt{n}^{\mathtt{f}}_{\mathtt{ip}}$ integration points and weights defined on the reference edge/face. The implementation considered here adopts the numerical quadratures recently proposed by Witherden and Vincent (2015).

Similarly, the different matrices appearing in (25), computed for each element $i = 1, \dots, \mathtt{n}_{\mathtt{el}}$, can be expressed as

$$[\mathbf{A}^{\star}_{uu}]_i = \sum_{\partial\Omega_i\setminus\Gamma_N} \tau_i \sum_{\mathtt{g}=1}^{\mathtt{n}^{\mathtt{f}}_{\mathtt{ip}}} \mathbf{N}(\boldsymbol{\xi}^{\mathtt{f}}_{\mathtt{g}})\, \mathbf{N}^T(\boldsymbol{\xi}^{\mathtt{f}}_{\mathtt{g}}) w^{\mathtt{f}}_{\mathtt{g}},$$

$$[\mathbf{A}^{\star}_{uq}]_i = \sum_{\mathtt{g}=1}^{\mathtt{n}^{\mathtt{e}}_{\mathtt{ip}}} \mathbf{N}(\boldsymbol{\xi}^{\mathtt{e}}_{\mathtt{g}})\, \boldsymbol{\nabla}\mathbf{N}^T(\boldsymbol{\xi}^{\mathtt{e}}_{\mathtt{g}}) w^{\mathtt{e}}_{\mathtt{g}} - \sum_{\partial\Omega_i\cap\Gamma_N}\sum_{\mathtt{g}=1}^{\mathtt{n}^{\mathtt{f}}_{\mathtt{ip}}} \mathbf{N}(\boldsymbol{\xi}^{\mathtt{f}}_{\mathtt{g}})\, \mathbf{N}_{\boldsymbol{n}T}(\boldsymbol{\xi}^{\mathtt{f}}_{\mathtt{g}}) w^{\mathtt{f}}_{\mathtt{g}},$$

$$\begin{aligned}[\mathbf{f}^{\star}_{u}]_i = &\sum_{\mathtt{g}=1}^{\mathtt{n}^{\mathtt{e}}_{\mathtt{ip}}} \mathbf{N}(\boldsymbol{x}^{\mathtt{e}}_{\mathtt{g}}) f\big(\boldsymbol{x}(\boldsymbol{\xi}^{\mathtt{f}}_{\mathtt{g}})\big)\, w^{\mathtt{e}}_{\mathtt{g}} + \sum_{\partial\Omega_i\cap\Gamma_N}\sum_{\mathtt{g}=1}^{\mathtt{n}^{\mathtt{f}}_{\mathtt{ip}}} \mathbf{N}(\boldsymbol{x}^{\mathtt{f}}_{\mathtt{g}})\, t\big(\boldsymbol{x}(\boldsymbol{\xi}^{\mathtt{f}}_{\mathtt{g}})\big)\, w^{\mathtt{f}}_{\mathtt{g}} \\ &+ \sum_{\partial\Omega_i\cap\Gamma_D} \tau_i \sum_{\mathtt{g}=1}^{\mathtt{n}^{\mathtt{f}}_{\mathtt{ip}}} \mathbf{N}(\boldsymbol{\xi}^{\mathtt{f}}_{\mathtt{g}})\, u_D\big(\boldsymbol{x}(\boldsymbol{\xi}^{\mathtt{f}}_{\mathtt{g}})\big)\, w^{\mathtt{f}}_{\mathtt{g}},\end{aligned}$$

$$[\mathbf{A}^{\star}_{u\hat{u}}]_i = \sum_{\partial\Omega_i\setminus\partial\Omega} \tau_i \sum_{\mathtt{g}=1}^{\mathtt{n}^{\mathtt{f}}_{\mathtt{ip}}} \mathbf{N}(\boldsymbol{\xi}^{\mathtt{f}}_{\mathtt{g}})\, \widehat{\mathbf{N}}^T(\boldsymbol{\xi}^{\mathtt{f}}_{\mathtt{g}}) w^{\mathtt{f}}_{\mathtt{g}},$$

$$[\mathbf{A}^{\star}_{q\hat{u}}]_i = \sum_{\partial\Omega_i\setminus\partial\Omega}\sum_{\mathtt{g}=1}^{\mathtt{n}^{\mathtt{f}}_{\mathtt{ip}}} \mathbf{N}_{\boldsymbol{n}}(\boldsymbol{\xi}^{\mathtt{f}}_{\mathtt{g}})\, \widehat{\mathbf{N}}^T(\boldsymbol{\xi}^{\mathtt{f}}_{\mathtt{g}}) w^{\mathtt{f}}_{\mathtt{g}}$$

and

$$[\mathbf{A}^{\star}_{\hat{u}\hat{u}}]_i = -\sum_{\partial\Omega_i\setminus\partial\Omega} \tau_i \sum_{\mathtt{g}=1}^{\mathtt{n}^{\mathtt{f}}_{\mathtt{ip}}} \widehat{\mathbf{N}}_{\boldsymbol{n}}(\boldsymbol{\xi}^{\mathtt{f}}_{\mathtt{g}})\, \widehat{\mathbf{N}}^T(\boldsymbol{\xi}^{\mathtt{f}}_{\mathtt{g}}) w^{\mathtt{f}}_{\mathtt{g}}.$$

References

Arnold, D. N., & Brezzi, F. (1985). Mixed and nonconforming finite element methods: implementation, postprocessing and error estimates. *RAIRO Modélisation Mathématique et Analyse Numérique*, *19*(1), 7–32.

Arnold, D. N., Brezzi, F., Cockburn, B., & Marini, L. D. (2002). Unified analysis of discontinuous Galerkin methods for elliptic problems. *SIAM Journal on Numerical Analysis*, *39*(5), 1749–1779.

Brenner, S. C., & Scott, L. R. (1994). *The mathematical theory of finite element methods*. New York: Springer.

Brezzi, F., Douglas, J, Jr., & Marini, L. D. (1985). Two families of mixed finite elements for second order elliptic problems. *Numerische Mathematik*, *47*(2), 217–235.

Chen, Q., & Babuška, I. (1995). Approximate optimal points for polynomial interpolation of real functions in an interval and in a triangle. *Computer Methods in Applied Mechanics and Engineering*, *128*(3–4), 405–417.

Ciarlet, P. G. (2002). The finite element method for elliptic problems. *Classics in applied mathematics* (Vol. 40). Society for Industrial and Applied Mathematics (SIAM), Philadelphia, PA. Amsterdam: North-Holland. Reprint of the 1978 original.

Cockburn, B., & Gopalakrishnan, J. (2004). A characterization of hybridized mixed methods for second order elliptic problems. *SIAM Journal on Numerical Analysis*, *42*(1), 283–301.

Cockburn, B., & Gopalakrishnan, J. (2005a). Incompressible finite elements via hybridization. I. The Stokes system in two space dimensions. *SIAM Journal on Numerical Analysis*, *43*(4), 1627–1650.

Cockburn, B., & Gopalakrishnan, J. (2005b). New hybridization techniques. *GAMM-Mitt.*, *28*(2), 154–182.

Cockburn, B., Dong, B., & Guzmán, J. (2008). A superconvergent LDG-hybridizable Galerkin method for second-order elliptic problems. *Mathematics of Computation*, *77*(264), 1887–1916.

Cockburn, B., Dong, B., Guzmán, J., Restelli, M., & Sacco, R. (2009a). A hybridizable discontinuous Galerkin method for steady-state convection-diffusion-reaction problems. *SIAM Journal on Scientific Computing*, *31*(5), 3827–3846.

Cockburn, B., Gopalakrishnan, J., & Lazarov, R. (2009b). Unified hybridization of discontinuous Galerkin, mixed, and continuous Galerkin methods for second order elliptic problems. *SIAM Journal on Numerical Analysis*, *47*(2), 1319–1365.

Cockburn, B., Guzmán, J., & Wang, H. (2009c). Superconvergent discontinuous Galerkin methods for second-order elliptic problems. *Mathematics of Computation*, *78*(265), 1–24.

Giorgiani, G., Fernández-Méndez, S., & Huerta, A. (2013). Hybridizable discontinuous Galerkin p-adaptivity for wave propagation problems. *International Journal for Numerical Methods in Fluids*, *72*(12), 1244–1262.

Giorgiani, G., Fernández-Méndez, S., & Huerta, A. (2014). Hybridizable discontinuous Galerkin with degree adaptivity for the incompressible Navier-Stokes equations. *Computers and Fluids*, *98*, 196–208.

Huerta, A., Angeloski, A., Roca, X., & Peraire, J. (2013). Efficiency of high-order elements for continuous and discontinuous Galerkin methods. *International Journal for Numerical Methods in Engineering*, *96*(9), 529–560.

Kabaria, H., Lew, A. J., & Cockburn, B. (2015). A hybridizable discontinuous Galerkin formulation for non-linear elasticity. *Computer Methods in Applied Mechanics and Engineering*, *283*, 303–329.

Kirby, R., Sherwin, S. J., & Cockburn, B. (2011). To CG or to HDG: A comparative study. *Journal of Scientific Computing*, *51*(1), 183–212.

Montlaur, A., Fernández-Méndez, S., & Huerta, A. (2008). Discontinuous Galerkin methods for the Stokes equations using divergence-free approximations. *International Journal for Numerical Methods in Fluids*, *57*(9), 1071–1092.

Nguyen, N. C., Peraire, J., & Cockburn, B. (2009a). An implicit high-order hybridizable discontinuous Galerkin method for linear convection-diffusion equations. *Journal of Computational Physics, 228*(9), 3232–3254.

Nguyen, N. C., Peraire, J., & Cockburn, B. (2009b). An implicit high-order hybridizable discontinuous Galerkin method for nonlinear convection-diffusion equations. *Journal of Computational Physics, 228*(23), 8841–8855.

Nguyen, N., Peraire, J., & Cockburn, B. (2010). A hybridizable discontinuous Galerkin method for Stokes flow. *Computer Methods in Applied Mechanics and Engineering, 199*(9–12), 582–597.

Nguyen, N. C., Peraire, J., & Cockburn, B. (2011a). An implicit high-order hybridizable discontinuous Galerkin method for the incompressible Navier-Stokes equations. *Journal of Computational Physics, 230*(4), 1147–1170.

Nguyen, N. C., Peraire, J., & Cockburn, B. (2011b). High-order implicit hybridizable discontinuous Galerkin methods for acoustics and elastodynamics. *Journal of Computational Physics, 230*(10), 3695–3718.

Nguyen, N. C., Peraire, J., & Cockburn, B. (2011c). Hybridizable discontinuous Galerkin methods for the time-harmonic Maxwell's equations. *Journal of Computational Physics, 230*(19), 7151–7175.

Peraire, J., Nguyen, N., & Cockburn, B. (2010). A hybridizable discontinuous Galerkin method for the compressible Euler and Navier-Stokes equations. *AIAA paper, 363*, 2010.

Raviart, P.-A., & Thomas, J. M. (1977). A mixed finite element method for 2nd order elliptic problems. *Mathematical aspects of finite element methods (Proceedings of the Conference, Consiglio Nazionale delle Ricerche (C.N.R.), Rome, 1975)* (Vol. 606, pp. 292–315). Lecture Notes in Mathematics. Berlin: Springer.

Reed, W., & Hill, T. (1973). *Triangular mesh methods for the neutron transport equation*. Technical report, Los Alamos Scientific Laboratory.

Soon, S.-C., Cockburn, B., & Stolarski, H. K. (2009). A hybridizable discontinuous Galerkin method for linear elasticity. *International Journal for Numerical Methods in Engineering, 80*(8), 1058–1092.

Szabó, B., & Babuška, I. (1991). *Finite element analysis*. New York: wiley.

Witherden, F. D., & Vincent, P. E. (2015). On the identification of symmetric quadrature rules for finite element methods. *Computers and Mathematics with Applications, 69*(10), 1232–1241.

Yakovlev, S., Moxey, D., Kirby, R. M., & Sherwin, S. J. (2015). To CG or to HDG: A comparative study in 3D. *Journal of Scientific Computing*, 1–29.

Least-Squares Mixed Finite Element Formulations for Isotropic and Anisotropic Elasticity at Small and Large Strains

Jörg Schröder, Alexander Schwarz and Karl Steeger

Abstract The performance of least-squares finite element formulations for geometrically linear and nonlinear problems is investigated in this work. We consider different elastic material behaviors as, e.g., quasi-incompressibility and transverse isotropy. Basis for the provided element formulations is a first-order system of differential equations consisting of the residual forms of the balance of momentum, a constitutive relation, and a (redundant) residual enforcing a stronger control of the balance of moment of momentum. The sum of the squared $L^2(\mathcal{B})$-norms of the residuals leads to a functional, which is the basis for the related minimization problem. As unknown fields the displacements (approximated in $W^{1,p}(\mathcal{B})$) and the stresses (approximated in $W^q(\text{div}, \mathcal{B})$) are chosen. Here, the choice of the polynomial orders of the interpolation functions for the displacements and stresses is not restricted by the so-called LBB condition; they can be chosen independently. Numerical examples for the proposed formulations are presented and compared to standard and mixed Galerkin formulations.

1 Introduction

In the last two decades least-squares variational principles have increasingly gained attention and provide the basis for several mixed finite element formulations. Although the research in these methods began in the early 1970s, see Lynn and Arya (1973), Zienkiewicz et al. (1974) and for an overview Eason (1976), mixed least-squares finite element methods (LSFEMs) become not that popular as, for example, mixed Galerkin methods. A possible reason for this could be the relatively poor approximation quality of the (at that time) widely used lower order elements, which is also addressed in the present contribution. However, not at least with rising available computer performance, the LSFEM encountered a revival in the 1990s

J. Schröder (✉) · A. Schwarz · K. Steeger
Faculty of Engineering, Department of Civil Engineering, Institute of Mechanics, University of Duisburg-Essen, Universitätsstr. 15, 45141 Essen, Germany
e-mail: j.schroeder@uni-due.de

J. Schröder and P. Wriggers (eds.), *Advanced Finite Element Technologies*, CISM International Centre for Mechanical Sciences 566,
DOI 10.1007/978-3-319-31925-4_6

and many approaches were proposed and analyzed for different problems in fluid dynamics and solid mechanics. The monographs by Jiang (1998) and Bochev and Gunzburger (2009) give a comprehensive survey of the scientific advancement in the field of LSFEM.

The main focus of research in the LSFEM can be found in fields where standard finite element methods do not yield satisfactory results, e.g., materials or fluid flow with incompressible behavior. Least-squares methods constitute a different approach to construct mixed finite element formulations. In general, mixed methods must fulfill several mathematical requirements, especially the LBB condition (Ladyzhenskaya–Babuška–Brezzi condition), see Ladyzhenskaya (1969), Babuška (1973), and Brezzi (1974), in this context we also refer to Bathe (1995, 2001) and Ern and Guermond (2013). This condition (also called inf–sup condition) demands to balance out the polynomial orders of the chosen interpolations for the different field variables. The difference of the LSFEM to conventional mixed variational formulations lies in the fact that it replaces a constrained minimization problem (with saddle point structure) by a least-squares formulation without constraints. The growing interest in developing LSFEMs is due to the fact that the method exhibits some inherent advantages, as for instance:

- The flexibility to design suited functionals directly approximating the unknown field variables of interest, e.g., stresses and displacements.
- The method provides an a posteriori error estimator without additional costs, applicable to adaptive mesh refinement algorithms.
- The resulting positive definite system matrices can be solved by using robust and fast iterative methods even for problems with non-self-adjoint operators.
- There are no restrictions by the LBB condition regarding the choice of the polynomial degree of the finite element spaces.

Besides that, there are also some disadvantages of least-squares approaches. First, as mentioned before, lower order least-squares formulations provide only a moderate performance. An illustrative example of this effect was given in Pontaza (2003) for the driven cavity problem and in Schwarz et al. (2010) for bending dominated problems in solid mechanics. In both works it was shown that higher order elements are the better choice in terms of efficiency and accuracy. A similar result was obtained for the mass loss problem in fluid dynamics where the application of higher order (spectral) finite elements demonstrates also an improvement, see, e.g., Pontaza and Reddy (2003). Second, there is a crucial impact of residual weighting, i.e., a balancing between the different (physical) residuals could be necessary. The influence of weighting has been addressed in the context of linear elasticity recently in Schwarz et al. (2014) and for Stokes flow in Deang and Gunzburger (1998).

In the field of solid mechanics different types of first-order systems were proposed for several material models ranging from quasi-incompressible over transversely isotropic hyperelasticity to elasto-viscoplasticity, only to name a few.

On the one hand there are div–curl–grad or div–curl systems, where all unknowns are approximated in H^1. An early contribution with respect to mathematical analysis was presented by Cai et al. (1995). The authors adopted the well-known VVP formulation with velocity, vorticity, and pressure as unknowns for the Stokes equations and used a similar div–curl–grad first-order system for the investigation of linear elasticity. In Cai et al. (1997, 1998) the authors proposed div–curl first-order systems with the displacements and the displacement gradient as unknowns. Further developments in this direction can be found in Cai et al. (2000a, b), where a so-called two-stage algorithm was proposed which, in a first step computes the displacement gradient followed by the recovery of the displacements. This formulation has been extended to geometrically nonlinear elasticity in Manteuffel et al. (2006) for a St.Venant–Kirchhoff material model. In the work of Jiang and Wu (2002) a displacement–stress–rotation first-order system was introduced for plane elasticity. Here, the drilling rotation about the plane normal is used as an additional independent unknown variable.

On the other hand so-called div–grad first-order systems were proposed. Here, the stress approximations are row-wise in $H(\text{div})$ and the displacements are componentwise in H^1. In most of the papers as well as in this contribution, $H(\text{div})$-conforming Raviart–Thomas elements were used for the approximation of the stress tensor, while for the displacement approximation standard Lagrange polynomials are applied. The origin of this kind of approaches for solid mechanics could be seen in the work of Cai and Starke (2003). Here, the authors presented and analyzed a formulation for linear elasticity and established ellipticity in $H(\text{div}) \times H^1$. Cai and Starke (2004) investigated a different scaling with respect to the incompressible limit. Based on the research results, the stress–displacement least-squares formulation was first extended to transversely isotropic elasticity in Schwarz and Schröder (2007). Apart from the mentioned vector-valued stress approximation, in Tchonkova and Sture (1997, 2002) stress–displacement formulations are presented and analyzed using constant and bilinear standard interpolations on quadrilaterals. Rate-dependent models for $H(\text{div}) \times H^1$ formulations are presented in Schwarz et al. (2009) for viscoplasticity and in Cai and Westphal (2009) for viscoelastic fluids. Recently, a weighted overconstrained LSFEM for static and dynamic linear elasticity was developed in Schwarz et al. (2014). Here, the expression "overconstrained" implies that a (mathematical) redundant residual is added to the functional in order to strengthen special physical relations. In the aforementioned work the authors proposed to add the symmetry condition for the stresses as third residual (besides balance of momentum and material law) with an additional weighting factor. Furthermore, the influence of a normalized weighting scheme with respect to the physical units of all residuals was investigated. Later on the idea of overconstraining with special weights was extended to hyperelasticity in Schwarz et al. (2015) and will be also discussed in the present work. Investigations to incompressible hyperelastic material models can be found in Müller et al. (2014).

2 Mechanical Foundations

In this chapter, we summarize the basic mechanical quantities: deformation of line-, area-, and volume elements, stress tensors, and balance laws. Furthermore, we discuss the principle of material frame indifference, the principle of material symmetry, and give some remarks concerning isotropic and anisotropic tensor functions.

2.1 *Placements, Deformation, and Stress Tensors*

Let $\mathcal{B} \subset \mathbb{R}^3$, parametrized in $\boldsymbol{X}$, be the body of interest in the reference placement and $\mathcal{B}_t \subset \mathbb{R}^3$, parametrized in $\boldsymbol{x}$, the body in the current placement. The nonlinear deformation map $\varphi_t : \mathcal{B} \to \mathcal{B}_t$ at time $t \in \mathbb{R}_+$ maps points $\boldsymbol{X} \in \mathcal{B}$ onto points $\boldsymbol{x} \in \mathcal{B}_t$, i.e., $\varphi_t : \boldsymbol{X} \mapsto \boldsymbol{x}$. Let $\boldsymbol{X} = \hat{\boldsymbol{X}}(\theta^\alpha)$ and $\boldsymbol{x} = \hat{\boldsymbol{x}}(\theta^\alpha)$ be the position vector from the origin to an arbitrary point in space, both parametrized in general coordinates $\{\theta^1, \theta^2, \theta^3\}$. Then the covariant base vectors in the reference and actual placement are given by

$$\boldsymbol{G}_A = \frac{\partial \boldsymbol{X}}{\partial \theta^A} \quad \text{and} \quad \boldsymbol{g}_a = \frac{\partial \boldsymbol{x}}{\partial \theta^a}\,, \tag{1}$$

respectively. The associated contravariant base vectors are defined via

$$\boldsymbol{G}_A \cdot \boldsymbol{G}^B = \delta_A{}^B \quad \text{and} \quad \boldsymbol{g}_a \cdot \boldsymbol{g}^b = \delta_a{}^b\,. \tag{2}$$

Let $\mathrm{d}\boldsymbol{x}$ denote an infinitesimal line element in $\mathbb{R}^3$ then $\{\mathrm{d}x^1\boldsymbol{g}_1,\ \mathrm{d}x^2\boldsymbol{g}_2,\ \mathrm{d}x^3\boldsymbol{g}_3\}$ denote the components of the vector and $\{\mathrm{d}x^1,\ \mathrm{d}x^2,\ \mathrm{d}x^3\}$ characterize the coordinates of $\mathrm{d}\boldsymbol{x}$. The square of the length of an infinitesimal line element

$$\mathrm{d}s^2 = \mathrm{d}\boldsymbol{x} \cdot \mathrm{d}\boldsymbol{x} = (\mathrm{d}x^a\, \boldsymbol{g}_a) \cdot (\mathrm{d}x^b\, \boldsymbol{g}_b) = \mathrm{d}x^a\, g_{ab}\, \mathrm{d}x^b, \tag{3}$$

inherently defines the metric of this space. The terms $g_{ab} := \boldsymbol{g}_a \cdot \boldsymbol{g}_b$ denote the covariant coefficients of the covariant metric tensor $g_{ab}\, \boldsymbol{g}^a \otimes \boldsymbol{g}^b$. Like any second-order tensor, the identity tensor $\boldsymbol{I}$ has the tensorial representations

$$\boldsymbol{I} = g_{ab}\, \boldsymbol{g}^a \otimes \boldsymbol{g}^b = g^{ab}\, \boldsymbol{g}_a \otimes \boldsymbol{g}_b = g^a{}_b\, \boldsymbol{g}_a \otimes \boldsymbol{g}^b = g_a{}^b\, \boldsymbol{g}^a \otimes \boldsymbol{g}_b, \tag{4}$$

with $g^{ab} := \boldsymbol{g}^a \cdot \boldsymbol{g}^b$, $g^a{}_b := \boldsymbol{g}^a \cdot \boldsymbol{g}_b = \delta^a{}_b$ and $g_a{}^b := \boldsymbol{g}_a \cdot \boldsymbol{g}^b = \delta_a{}^b$. Rules for lowering and raising of indices follow directly from

$$\boldsymbol{g}_a = \boldsymbol{g}_a \cdot \boldsymbol{I} = \boldsymbol{g}_a \cdot (g_{bc}\, \boldsymbol{g}^b \otimes \boldsymbol{g}^c) = g_{bc}\, \delta_a{}^b\, \boldsymbol{g}^c = g_{ac}\, \boldsymbol{g}^c \tag{5}$$

and

$$\boldsymbol{g}^a = \boldsymbol{g}^a \cdot \boldsymbol{I} = \boldsymbol{g}^a \cdot (g^{bc}\, \boldsymbol{g}_b \otimes \boldsymbol{g}_c) = g^{bc}\, \delta^a{}_b\, \boldsymbol{g}_c = g^{ac}\, \boldsymbol{g}_c. \tag{6}$$

Exploiting the orthogonality condition (2) we obtain

$$\delta_a{}^b = \boldsymbol{g}_a \cdot \boldsymbol{g}^b = (g_{ac}\, \boldsymbol{g}^c) \cdot (g^{bd}\, \boldsymbol{g}_d) = g_{ac}\, g^{bd}\, \delta^c{}_d = g_{ac}\, g^{bc}. \tag{7}$$

Let $[g_{ac}]$ denote the matrix representation of the coordinates g_{ac} of the covariant metric tensor with respect to the basis $\boldsymbol{g}^a \otimes \boldsymbol{g}^b$, alternative definitions are obvious, then we get the relation

$$[g_{ac}]^{-1} = [g^{ac}]. \tag{8}$$

Similar relations hold for the co- and contravariant representations of the base vectors and metric tensor in the reference placement. For the representations in Cartesian coordinates we arrive at the simple expressions $G_{AB} = G^{AB} = \delta_{AB}$ for the Lagrangian and $g_{ab} = g^{ab} = \delta_{ab}$ for the Eulerian metric tensors.

A fundamental kinematical quantity is $\boldsymbol{F} = \mathrm{Grad}\varphi_t(\boldsymbol{X})$, given by

$$\boldsymbol{F} = F^a{}_A\, \boldsymbol{g}_a \otimes \boldsymbol{G}^A \quad \text{with} \qquad F^a{}_A = \frac{\partial x^a}{\partial X^A}. \tag{9}$$

The deformation gradient $\boldsymbol{F}$ maps infinitesimal line elements $\mathrm{d}\boldsymbol{X}$ of the reference configuration to the line element $\mathrm{d}\boldsymbol{x}$ of the current configuration, i.e.,

$$\mathrm{d}\boldsymbol{x} = \boldsymbol{F}\, \mathrm{d}\boldsymbol{X}. \tag{10}$$

Let $\mathrm{d}\boldsymbol{x} = \boldsymbol{F}\,\mathrm{d}\boldsymbol{X}$ and $\mathrm{d}\boldsymbol{y} = \boldsymbol{F}\,\mathrm{d}\boldsymbol{Y}$ denote two independent infinitesimal line elements, then the infinitesimal vectorial area element $\mathrm{d}\boldsymbol{a}$ spanned by $\mathrm{d}\boldsymbol{x}$ and $\mathrm{d}\boldsymbol{y}$ is given by the cross product, i.e., $\mathrm{d}\boldsymbol{a} = \mathrm{d}\boldsymbol{x} \times \mathrm{d}\boldsymbol{y}$; in detail:

$$\mathrm{d}\boldsymbol{a} = (\boldsymbol{F}\,\mathrm{d}\boldsymbol{X}) \times (\boldsymbol{F}\,\mathrm{d}\boldsymbol{Y}) = \mathrm{Cof}[\boldsymbol{F}]\,(\mathrm{d}\boldsymbol{X} \times \mathrm{d}\boldsymbol{Y}) = \mathrm{Cof}[\boldsymbol{F}]\;\mathrm{d}\boldsymbol{A}. \tag{11}$$

Obviously, the cofactor of the deformation gradient Cof $\boldsymbol{F}$ maps infinitesimal vectorial area elements $\mathrm{d}\boldsymbol{A} = \boldsymbol{N}\,\mathrm{d}A$ of the reference configuration onto infinitesimal area elements $\mathrm{d}\boldsymbol{a} = \boldsymbol{n}\,\mathrm{d}a$ of the current configuration. $\boldsymbol{N}$ and $\boldsymbol{n}$ characterize the unit normal vectors of the associated area elements. Calculation rules concerning the cofactor are, e.g., given in Schröder and Neff (2003), more advanced rules are summarized in de Boer (1993)[1]. Evaluating (11) allows to identify the components of the cofactor of $\boldsymbol{F}$ as

$$\mathrm{Cof}\;\boldsymbol{F} = \begin{bmatrix} F_{22}F_{33} - F_{32}F_{23} & F_{31}F_{23} - F_{21}F_{33} & F_{21}F_{32} - F_{31}F_{22} \\ F_{32}F_{13} - F_{12}F_{33} & F_{11}F_{33} - F_{31}F_{13} & F_{31}F_{12} - F_{11}F_{32} \\ F_{12}F_{23} - F_{22}F_{13} & F_{21}F_{13} - F_{11}F_{23} & F_{11}F_{22} - F_{21}F_{12} \end{bmatrix}.$$

[1]de Boer introduced the tensor cross product $\boldsymbol{F} \overset{\times}{\times} \boldsymbol{F} = 2\,\mathrm{Cof}\;\boldsymbol{F}$ instead of the cofactor.

If the deformation gradient is invertible the following relation holds:

$$\operatorname{Cof} \boldsymbol{F} = \det[\boldsymbol{F}]\boldsymbol{F}^{-T}.$$

The inverse deformation gradient $\boldsymbol{F}^{-1} = \operatorname{grad} \varphi_t^{-1}(\boldsymbol{x})$ appears as

$$\boldsymbol{F}^{-1} = \{F^{-1}\}^A{}_a\, \boldsymbol{G}_A \otimes \boldsymbol{g}^a \quad \text{with} \qquad \{F^{-1}\}^A{}_a = \frac{\partial X^A}{\partial x^a}. \tag{12}$$

The transformation of the infinitesimal volume elements of the reference placement $\mathrm{d}V$ onto the one of the actual placement $\mathrm{d}v$ is given by the scalar triple product

$$\mathrm{d}v = [\,\mathrm{d}\boldsymbol{x},\ \mathrm{d}\boldsymbol{y},\ \mathrm{d}\boldsymbol{z}] = (\,\mathrm{d}\boldsymbol{x} \times \mathrm{d}\boldsymbol{y}) \cdot \mathrm{d}\boldsymbol{z} = \mathrm{d}\boldsymbol{a} \cdot \mathrm{d}\boldsymbol{z}, \tag{13}$$

with the linear independent line elements $\mathrm{d}\boldsymbol{x}$, $\mathrm{d}\boldsymbol{y}$, and $\mathrm{d}\boldsymbol{z}$. Applying (11), and using $\mathrm{d}\boldsymbol{z} = \boldsymbol{F}\,\mathrm{d}\boldsymbol{Z}$, we obtain

$$\mathrm{d}v = \mathrm{d}\boldsymbol{a} \cdot \mathrm{d}\boldsymbol{z} = (\operatorname{Cof}[\boldsymbol{F}]\ \mathrm{d}\boldsymbol{A}) \cdot (\boldsymbol{F}\,\mathrm{d}\boldsymbol{Z}) = \det[\boldsymbol{F}]\ \mathrm{d}V, \tag{14}$$

with

$$\mathrm{d}V = \mathrm{d}\boldsymbol{A} \cdot \mathrm{d}\boldsymbol{Z} = (\,\mathrm{d}\boldsymbol{X} \times \mathrm{d}\boldsymbol{Y}) \cdot \mathrm{d}\boldsymbol{Z} = [\,\mathrm{d}\boldsymbol{X},\ \mathrm{d}\boldsymbol{Y},\ \mathrm{d}\boldsymbol{Z}]. \tag{15}$$

Using the abbreviation $J = \det \boldsymbol{F}$ for the determinant of the deformation gradient, we write

$$\mathrm{d}v = J\ \mathrm{d}V \quad \text{with} \qquad J > 0. \tag{16}$$

Note that the arguments $\{\boldsymbol{F}, \operatorname{Cof} \boldsymbol{F}, \det \boldsymbol{F}\}$ govern the transformations of the infinitesimal line, vectorial area, and volume elements from the reference onto the actual placement, as depicted in Fig. 1.

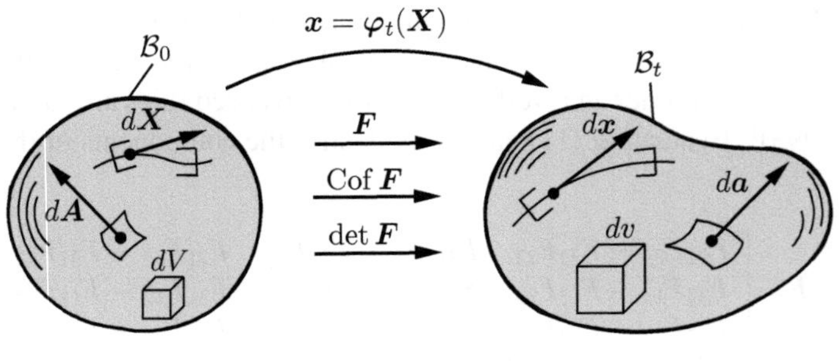

Fig. 1 Mappings of the infinitesimal line, area, and volume elements, $\boldsymbol{F} : \mathrm{d}\boldsymbol{X} \mapsto \mathrm{d}\boldsymbol{x}$, $\operatorname{Cof} \boldsymbol{F} : \mathrm{d}\boldsymbol{A} \mapsto \mathrm{d}\boldsymbol{a}$, and $\det \boldsymbol{F} : \mathrm{d}V \mapsto \mathrm{d}v$

An important deformation measure for the construction of free energy functions is the right Cauchy–Green tensor:

$$C := F^T F = C_{AB}\, G^A \otimes G^B \quad \text{with} \quad C_{AB} = F^a{}_A\, g_{ab}\, F^b{}_B. \tag{17}$$

In hyperelasticity, we postulate the existence of a so-called Helmholtz free energy function (free energy) ψ, defined per unit mass, and accordingly a specific strain energy density W, defined per unit volume, satisfying $\rho_0 \psi(F) = W(F)$, where ρ_0 denotes the density in the reference placement. Furthermore, let $P = P_a{}^A g^a \otimes G_A$ denote the first Piola–Kirchhoff stress tensor. In the following we consider hyperelastic materials, which means that the internal dissipation $\mathcal{D}_{int}$ is zero for every admissible process. Therefore, the Clausius–Duhem inequality reduces to the equality

$$\mathcal{D}_{int} = P : \dot{F} - \rho_0 \dot{\psi} = \left(P - \frac{\partial W}{\partial F} \right) : \dot{F} = 0, \tag{18}$$

which has to be fulfilled for all possible thermodynamic processes. Here $\dot{F}$ denotes the material time derivative of the deformation gradient. Thus we conclude

$$P = \frac{\partial W}{\partial F} =: \partial_F W. \tag{19}$$

Applying the chain rule and assuming $W = W(C)$, we obtain the expression

$$S = 2\, \frac{\partial W}{\partial C} =: 2\, \partial_C W, \tag{20}$$

for the symmetric second Piola–Kirchhoff stress tensor $S = S^{AB} G_A \otimes G_B$.

Material Frame Indifference. The principle of material frame indifference requires the invariance of the constitutive equation under superimposed rigid body rotations $Q \in \mathcal{SO}(3)$ onto the current configuration, i.e.,

$$Q : x \in \mathcal{B}_t \mapsto Qx =: x^+. \tag{21}$$

The special orthogonal group is defined by

$$\mathcal{SO}(3) = \{ Q \mid QQ^T = \mathbf{I} \quad \text{and} \quad \det Q = 1 \}. \tag{22}$$

This superimposed rotation is associated with the deformation gradient

$$F^+ = \operatorname{Grad} x^+ = QF. \tag{23}$$

The principle of material frame indifference requires the invariance of the free energy[2] with respect to the superimposed rotations, i.e.,

$$W(\boldsymbol{F}) = W(\boldsymbol{F}^{+}) \quad \forall\, \boldsymbol{Q} \in \mathcal{SO}(3). \tag{24}$$

Inserting the right Cauchy–Green tensor $\boldsymbol{C}$ instead of $\boldsymbol{F}$ leads to the well-known reduced constitutive equations, which fulfill the principle of material objectivity (24) a priori, see, e.g., Truesdell and Noll (1965), i.e.,

$$W(\boldsymbol{F}^T\boldsymbol{F}) = W(\boldsymbol{F}^{+T}\boldsymbol{F}^{+}) \quad \forall\, \boldsymbol{Q} \in \mathcal{SO}(3). \tag{25}$$

This is obviously true, because

$$\boldsymbol{C}^{+} = \boldsymbol{F}^{+T}\boldsymbol{F}^{+} = (\boldsymbol{Q}\boldsymbol{F})^T(\boldsymbol{Q}\boldsymbol{F}) = \boldsymbol{F}^T\boldsymbol{F} = \boldsymbol{C}, \tag{26}$$

thus $W(\boldsymbol{C}) = W(\boldsymbol{C}^{+})$ holds for all $\boldsymbol{Q} \in \mathcal{SO}(3)$.

Principle of Material Symmetry. For the formulation of anisotropic hyperelastic energies, further restrictions, due to the present material symmetry, have to be taken into account. Therefore, we consider the invariance of the strain energy function with respect to superimposed rotations onto the reference placement, associated to the symmetry group $\mathcal{G}$ of the material, i.e.,

$$W(\boldsymbol{F}) = W(\boldsymbol{F}\boldsymbol{Q}^T) \quad \forall\, \boldsymbol{Q} \in \mathcal{G} \subset \mathcal{O}(3). \tag{27}$$

The material symmetry group $\mathcal{G}$ is a subset of the full orthogonal group

$$\mathcal{O}(3) = \{\boldsymbol{Q} \mid \boldsymbol{Q}\boldsymbol{Q}^T = \mathbf{I} \quad \text{and} \quad \det \boldsymbol{Q} = \pm 1\}. \tag{28}$$

The reduced constitutive equations must fulfill the invariance condition

$$W(\boldsymbol{C}) = W(\boldsymbol{Q}\boldsymbol{C}\boldsymbol{Q}^T) \quad \forall\, \boldsymbol{Q} \in \mathcal{G}. \tag{29}$$

An isotropic material is characterized by $\mathcal{G} \equiv \mathcal{O}(3)$, i.e., the free energy is invariant under all transformations of the full orthogonal group,

$$W_{iso}(\boldsymbol{C}) = W_{iso}(\boldsymbol{Q}\boldsymbol{C}\boldsymbol{Q}^T) \quad \forall \quad \boldsymbol{Q} \in \mathcal{O}(3)\,. \tag{30}$$

Moreover, (30) is the definition of a scalar-valued isotropic tensor function. As a consequence we can express $W_{iso}(\boldsymbol{C})$ in terms of the principal invariants

$$I_1 = \operatorname{tr} \boldsymbol{C}, \quad I_2 = \operatorname{tr}[\operatorname{Cof}\, \boldsymbol{C}], \quad I_3 = \det \boldsymbol{C} \tag{31}$$

[2]Alternative invariance requirements in terms of first/second Piola–Kirchhoff stress tensors are $\boldsymbol{Q}\boldsymbol{P}(\boldsymbol{F}) = \boldsymbol{P}(\boldsymbol{F}^{+}) \forall \boldsymbol{Q} \in \mathcal{SO}(3)$ and $\boldsymbol{S}(\boldsymbol{F}) = \boldsymbol{S}(\boldsymbol{F}^{+}) \forall \boldsymbol{Q} \in \mathcal{SO}(3)$, respectively.

or in terms of the main invariants

$$J_1 = \mathrm{tr}\, \boldsymbol{C}, \quad J_2 = \mathrm{tr}[\boldsymbol{C}^2], \quad J_3 = \mathrm{tr}[\boldsymbol{C}^3] \tag{32}$$

of the right Cauchy–Green tensor. In order to derive the relations between the principal and main invariants we use the Cayley–Hamilton theorem, which states that every square matrix satisfies its own characteristic equation. For the right Cauchy–Green, which is a second-order tensor, we obtain

$$-\boldsymbol{C}^3 + I_1\, \boldsymbol{C}^2 - I_2\, \boldsymbol{C} + I_3 \mathbf{I} = \mathbf{0}. \tag{33}$$

After some algebraic manipulations we obtain the following expressions:

$$I_2 = \frac{1}{2}\left([\mathrm{tr}\, \boldsymbol{C}]^2 - \mathrm{tr}[\boldsymbol{C}^2]\right) = \frac{1}{2}(J_1^2 - J_2) \tag{34}$$

and

$$I_3 = \frac{1}{3}\left(\mathrm{tr}[\boldsymbol{C}^3] - I_1\, \mathrm{tr}[\boldsymbol{C}^2] + I_2\, \mathrm{tr}\, \boldsymbol{C}\right) = \frac{1}{3}\left(J_3 - \frac{3}{2} J_1 J_2 + \frac{1}{2} J_1^3\right). \tag{35}$$

For the construction of anisotropic free energy functions, Neumann's Principle, Neumann (1885), plays an important role; it states:

> The symmetry group of a considered material must be included in the symmetry group of any tensor function of the constitutive laws of the material.

This principle leads to several restrictions of the specific form of the free energy function and is therefore important for the representation of the associated tensor functions. For the construction of anisotropic response functions we use the concept of structural tensors and apply representation theorems of tensor functions. The concept of structural tensors was first introduced in an attractive way with important applications by Boehler (1978, 1979), see also Boehler (1987), although some similar ideas might have been touched on earlier. This procedure results from the isotropicization theorem of anisotropic tensor functions, see, e.g., Zheng (1994):

> An anisotropic constitutive law of a material point, which possesses a physical symmetry group, can be formally expressed as an isotropic function by introducing a set of structural tensors, provided that the set characterizes the underlying symmetry group of the material.

The structural tensors act as additional variables in the constitutive laws. For the representation of the energy function as isotropic tensor functions we need the whole set of scalar-valued invariants in terms of a deformation measure and the structural tensors. Therefore, one of the main goals in the invariant theory is to find a set of basic invariants for a given set of tensor arguments, from which all other invariants can be generated. This is possible, due to *Hilbert's Theorem*, which states:

> For any finite number of tensor arguments of any order, which is invariant with respect to an anisotropy group, there exists a finite number of invariants.

In order to make this contribution self-explanatory, we end this section with the definitions of the integrity and functional basis.

Integrity basis: The set of invariants for a given set of tensor arguments relative to a fixed symmetry group $\mathcal{G}$ is called an integrity basis, if an arbitrary polynomial invariant of the same arguments can be expressed as a polynomial in the basic invariants. If no element of the considered set can be expressed as a polynomial in the remaining invariants of the integrity basis, it is called irreducible.

Functional basis: The set of invariants is called a functional basis, if an arbitrary invariant in terms of the underlying arguments can be expressed as a function (not necessarily polynomial) in terms of the elements of the basis. It is said to be irreducible, if none of the elements of the basic set can be expressed as a function of the remaining invariants.

Pioneering works in this field go back to Wang (1969a, b, 1970a, b, 1971) and Smith (1970, 1971). In this context we also refer to Spencer (1971, 1965) and Smith et al. (1963). A unified approach summarizing the developments in this field can be found in the review of Zheng (1994).

2.2 Transverse Isotropic Hyperelasticity

Let the vector $\boldsymbol{a}$ of unit length be the preferred direction of a transversely isotropic material and $\boldsymbol{Q}(\alpha, \boldsymbol{a})$ all rotations about the $\boldsymbol{a}$-axis, then the associated material symmetry group is defined by

$$\mathcal{G}_{ti} := \{\pm \mathbf{I};\ \boldsymbol{Q}(\alpha, \boldsymbol{a}) \mid 0 \leq \alpha < 2\,\pi\}. \tag{36}$$

The structural tensor $\boldsymbol{M}$ whose invariance group preserves the material symmetry group $\mathcal{G}_{ti}$ is given by the rank-one tensor

$$\boldsymbol{M} = \boldsymbol{a} \otimes \boldsymbol{a}. \tag{37}$$

The strain energy density $W_{ti} = W_{ti}(\boldsymbol{C}, \boldsymbol{M})$ is an anisotropic scalar-valued function with respect to transformations of $\boldsymbol{C}$ alone, i.e.,

$$W_{ti}(\boldsymbol{C}, \boldsymbol{M}) = W(\boldsymbol{Q}\boldsymbol{C}\boldsymbol{Q}^T, \boldsymbol{M}) \quad \forall\, \boldsymbol{Q} \in \mathcal{G}_{ti}. \tag{38}$$

Obviously, the structural tensor $\boldsymbol{M}$ characterizes the material symmetry group of the considered material, i.e.,

$$\boldsymbol{M} = \boldsymbol{Q}\boldsymbol{M}\boldsymbol{Q}^T \quad \forall\, \boldsymbol{Q} \in \mathcal{G}_{ti}. \tag{39}$$

Therefore, W_{ti} is an isotropic function with respect to transformations of whole argument list $\{\boldsymbol{C}, \boldsymbol{M}\}$:

$$W_{ti}(\boldsymbol{C}, \boldsymbol{M}) = W(\boldsymbol{QCQ}^T, \boldsymbol{QMQ}^T) \quad \forall\, \boldsymbol{Q} \in \mathcal{O}(3). \tag{40}$$

Thus, W_{ti} can be formulated in terms of the associated principal/main and mixed invariants of the arguments $\{\boldsymbol{C}, \boldsymbol{M}\}$. Exploiting the fact that the powers of the structural tensor are the structural tensor itself, the mixed invariants of the two symmetric tensors $\boldsymbol{C}$ and $\boldsymbol{M}$ are

$$J_4 = \mathrm{tr}[\boldsymbol{CM}] \quad \text{and} \quad J_5^* = \mathrm{tr}[\boldsymbol{C}^2\boldsymbol{M}]. \tag{41}$$

However, instead of using J_5^*, which is a mixed invariant quadratic in the right Cauchy–Green tensor, it is convenient, in view to the proof of the polyconvexity to use the alternative quadratic expression

$$J_5 = \mathrm{tr}[\mathrm{Cof}[\boldsymbol{C}]\boldsymbol{M}]. \tag{42}$$

A detailed discussion of this point is given in Schröder and Neff (2001, 2003), Schröder et al. (2008), Schröder (2010), Ebbing (2010), and Ball (2006). For the strain energy function, we assume the general form

$$W_{ti} = \sum_k W_k^{ti}(I_1, I_2, I_3, J_4, J_5) + c. \tag{43}$$

In order to enforce $W(\mathbf{I}, \boldsymbol{M}) = 0$ we introduce the constant $c \in \mathbb{R}$.

2.3 Generalized Convexity Conditions

The mathematical treatment of boundary value problems is mainly based on the direct methods of variations. A sufficient condition for the existence of minimizers is the sequential-weak-lower-semicontinuity (s.w.l.s.) of the stored energy

$$\int_{\mathcal{B}} W(\boldsymbol{F})\, dV \tag{44}$$

in combination with the coercivity of the function $W(\boldsymbol{F})$. For the notion of s.w.l.s. we refer to advanced textbooks on variational methods. From a physical point of view, it is desirable that infinite stresses correspond with extreme strains, in this context see the remarks in Ciarlet (1988) and Antman (1995). Let λ_{min} and λ_{max} be the smallest and largest eigenvalues of $\boldsymbol{C}$ in the interval $]0, \infty[$. In order to capture this behavior for large strains we require that

$$W(\boldsymbol{F}) \to +\infty \quad \text{for} \quad \det \boldsymbol{F} \to 0^+ \tag{45}$$

and

$$W(\boldsymbol{F}) \to +\infty \quad \text{for} \quad (\| \boldsymbol{F} \| + \| \operatorname{Cof} \boldsymbol{F} \| + \det \boldsymbol{F}) \to +\infty. \tag{46}$$

The latter relation is replaced by the sharper coercivity inequality

$$\hat{W}(\boldsymbol{F}) \geq c_1 \left[\| \boldsymbol{F} \|^p + \| \operatorname{Cof} \boldsymbol{F} \|^q + (\det \boldsymbol{F})^r \right] + c_0$$

with the constants $c_1 > 0, p > 0, q > 0, r > 0$ and c_0. The latter condition must be fulfilled for all possible deformation gradients. To obtain useful existence results, the exponent parameters must satisfy

$$p \geq 2, \ q \geq p/(p-1), \ r > 1. \tag{47}$$

It can be shown that quasiconvexity and therefore polyconvexity of the stored energy imply that the corresponding acoustic tensor is elliptic (rank-one convex). This implies that the associated Euler equations of the functional are elliptic for all possible deformations. For finite-valued, continuous functions we summarize the important implications

$$\text{convexity} \to \text{polyconvexity} \to \text{quasiconvexity} \to \text{rank-one convexity}.$$

That means, a function which is convex is also polyconvex. A polyconvex function is always quasiconvex and a quasiconvex function is always rank-one convex. The converse implications are not true; for more informations, we refer to Dacorogna (1989), Šilhavý (1997) and the references therein. The relations between the generalized convexity conditions and the existence of minimizers are illustrated in Fig. 2.

The concept of quasiconvexity was introduced by Morrey (1952, 1966). It is formulated as an integral inequality over an arbitrary domain subjected to affine Dirichlet boundary conditions. If an additional growth condition, which bounds the

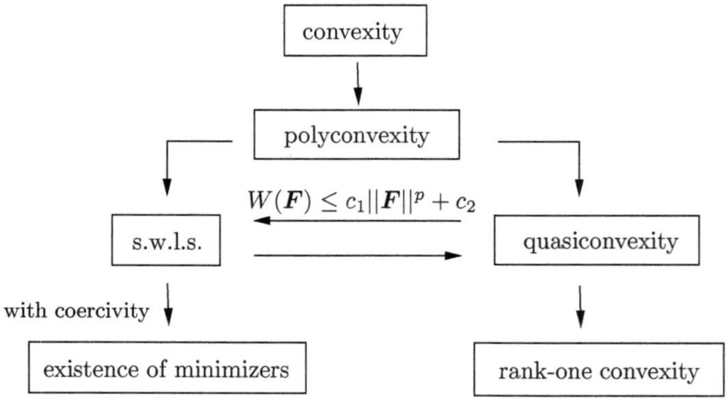

Fig. 2 Implications of the generalized convexity conditions

rate of growth of the strain energy function from above, is fulfilled, the s.w.l.s. condition is ensured.

Definition of Quasiconvexity. Morrey (1952): an elastic stored energy density is quasiconvex whenever for all $\mathcal{B} \subset \mathbb{R}^3$, all constant deformation gradients $\overline{\boldsymbol{F}} \in \mathbb{R}^{3\times 3}$, and all superposed fluctuation fields $\boldsymbol{w} \in C_0^\infty(\mathcal{B})$ (that means with $\boldsymbol{w} = \boldsymbol{0}$ on $\partial\mathcal{B}$), the following integral inequality is valid

$$\int_{\mathcal{B}} W(\overline{\boldsymbol{F}} + \mathrm{Grad}\boldsymbol{w})\, dV \geq \int_{\mathcal{B}} W(\overline{\boldsymbol{F}})\, dV = W(\overline{\boldsymbol{F}}) \times \mathrm{Vol}(\mathcal{B}).$$

□

This integral inequality must hold for arbitrary fluctuations $\boldsymbol{w}$. However, if a quasiconvex functional additionally satisfies the growth condition

$$\hat{W}(\boldsymbol{F}) \leq c_1 \parallel \boldsymbol{F} \parallel^p + c_2 \quad \text{with} \quad c_1 > 0,\ c_2 > 0,\ p > 1 \tag{48}$$

it is sequentially weakly lower semicontinuous. This condition is rather complicated to prove for explicit functions. Thus, an important concept is the notion of polyconvexity introduced by Ball (1977a, b), see also Marsden and Hughes (1983) and Ciarlet (1988). To prove polyconvexity of a specific function, we have to show that the function is convex with respect to $\{\boldsymbol{F}, \mathrm{Cof}\ \boldsymbol{F}, \det \boldsymbol{F}\} \in \mathbb{R}^{19}$.

Definition of Polyconvexity. The mapping $\boldsymbol{F} \mapsto W(\boldsymbol{F})$ is polyconvex if and only if there exists a function $P : \mathbb{R}^{3\times 3} \times \mathbb{R}^{3\times 3} \times \mathbb{R} \mapsto \mathbb{R}$ such that

$$W(\boldsymbol{F}) = P(\boldsymbol{F}, \mathrm{Cof}\ \boldsymbol{F}, \det \boldsymbol{F})$$

and the function $(\boldsymbol{F}, \mathrm{Cof}\ \boldsymbol{F}, \det \boldsymbol{F}) \in \mathbb{R}^{19} \mapsto P(\boldsymbol{F}, \mathrm{Cof}\ \boldsymbol{F}, \det \boldsymbol{F}) \in \mathbb{R}$ is convex for all points $\boldsymbol{X} \in \mathbb{R}^3$. □

In the following we drop the $\boldsymbol{X}$-dependence of the individual functions if there is no danger of confusion. A consequence of the definition of polyconvexity for a more restrictive class of energy densities is that a function

$$W(\boldsymbol{F}) = W_1(\boldsymbol{F}) + W_2(\mathrm{Cof}\ \boldsymbol{F}) + W_3(\det \boldsymbol{F}) \tag{49}$$

is polyconvex if W_1 and W_2 are convex in $\boldsymbol{F}$ and $\mathrm{Cof}\ \boldsymbol{F}$, respectively, and $W_3 : \det \boldsymbol{F} \in \mathbb{R}^+ \mapsto \mathbb{R}$ is also convex in the associated variable.

3 Least-Squares Method

In this chapter the necessary tools for mixed least-squares finite element formulations are presented. Besides the general least-squares approach, we provide also a one-dimensional introductory example where we discuss all steps for the solution of a boundary value problem in detail.

3.1 General Least-Squares Approach

An advantage of the least-squares method is the flexibility to design suited functionals directly approximating the unknown field variables of interest. Hence, the first step lies always in the construction of a functional containing the governing equations. In general there are different possibilities, e.g., different norms, in order to define a least-squares functional, see Bochev and Gunzburger (2009). In this contribution we use for the construction a squared $L^2(\mathcal{B})$-norm, i.e.,

$$||\bullet||^2_{L^2(\mathcal{B})} = \int_{\mathcal{B}} |\bullet|^2 \, \mathrm{d}V \, . \tag{50}$$

In order to define the minimization problem, we apply the squared $L^2(\mathcal{B})$-norm directly to a first-order system of i differential equations written in residual forms $\mathcal{R}_i = \mathbf{0}$, as

$$\mathcal{F}(\boldsymbol{a}, \boldsymbol{b}, \boldsymbol{c}, ...) = \sum_i \frac{1}{2} ||\omega_i \mathcal{R}_i||^2_{L^2(\mathcal{B})} = \sum_i \int_{\mathcal{B}} \frac{1}{2} \omega_i^2 \mathcal{R}_i \cdot \mathcal{R}_i \, \mathrm{d}V \rightarrow \min, \tag{51}$$

with the weights ω_i and the solution quantities $\boldsymbol{a}, \boldsymbol{b},$ and $\boldsymbol{c}$. The minimization condition is that the first variations with respect to the unknowns $\boldsymbol{a}, \boldsymbol{b}, \boldsymbol{c}$ vanish, i.e.,

$$\begin{aligned}
\delta_a \mathcal{F} &= \delta \boldsymbol{a} \cdot \frac{\partial \mathcal{F}}{\partial \boldsymbol{a}} = \sum_i \int_{\mathcal{B}} \omega_i^2 \delta_a \mathcal{R}_i \cdot \mathcal{R}_i \, dV = 0 \\
\delta_b \mathcal{F} &= \delta \boldsymbol{b} \cdot \frac{\partial \mathcal{F}}{\partial \boldsymbol{b}} = \sum_i \int_{\mathcal{B}} \omega_i^2 \delta_b \mathcal{R}_i \cdot \mathcal{R}_i \, dV = 0 \\
\delta_c \mathcal{F} &= \delta \boldsymbol{c} \cdot \frac{\partial \mathcal{F}}{\partial \boldsymbol{c}} = \sum_i \int_{\mathcal{B}} \omega_i^2 \delta_c \mathcal{R}_i \cdot \mathcal{R}_i \, dV = 0 \\
\dots \quad & \dots \qquad\qquad \dots \qquad\qquad\qquad\qquad = 0.
\end{aligned} \tag{52}$$

The discretization of the latter expression involves a linear system of algebraic equations, which directly yields the minimizer of the least-squares functional. In case of nonlinearities, iterative procedures as, for example, the standard Newton or the Gauss–Newton method can be used in order to obtain the final solution. In the next subsection, we provide an illustrative one-dimensional example for a mixed problem exploiting all steps in detail for the treatment of a boundary value problem with the LSFEM.

3.2 Introductory Example

As a simple application of the mixed least-squares finite element method, we consider a one-dimensional example with a second-order differential equation

$$u'' - f = 0 \tag{53}$$

on a domain $\mathcal{B}$. For the suitable interpolation of u we have to choose $\mathcal{C}^1$ continuous functions. To circumvent this and enabling the use of $\mathcal{C}^0$ continuous interpolation functions, we transform the differential equation of second order into a system of differential equations of first order. Therefore, we introduce a new independent variable $\varepsilon = u'$, leading to the first-order system in residual form

$$\varepsilon' - f = 0 \;\text{ and }\; \varepsilon - u' = 0 \tag{54}$$

with the boundary conditions

$$u = g \quad \text{on } \partial\mathcal{B}_u \subseteq \partial\mathcal{B} \quad \text{and} \quad \varepsilon = h \quad \text{on } \partial\mathcal{B}_\varepsilon \subseteq \partial\mathcal{B} \tag{55}$$

and the decomposition

$$\partial\mathcal{B} = \partial\mathcal{B}_u \cup \partial\mathcal{B}_\varepsilon \;\wedge\; \partial\mathcal{B}_u \cap \partial\mathcal{B}_\varepsilon = \emptyset. \tag{56}$$

Applying a squared $L^2(\mathcal{B})$-norm leads to the least-squares functional

$$\begin{aligned}\mathcal{F} &= \frac{1}{2}||\varepsilon' - f||^2_{L^2(\mathcal{B})} + \frac{1}{2}||\varepsilon - u'||^2_{L^2(\mathcal{B})}\\ &= \frac{1}{2}\int_{\mathcal{B}} (\varepsilon' - f)^2\,\mathrm{d}x + \frac{1}{2}\int_{\mathcal{B}} (\varepsilon - u')^2\,\mathrm{d}x \quad\rightarrow\quad \min,\end{aligned} \tag{57}$$

which has to be minimized. Thus, the first variations $\delta_u\mathcal{F}$ and $\delta_\varepsilon\mathcal{F}$ with respect to the solution variables $\{u, \varepsilon\}$ have to be zero:

$$\begin{aligned}\delta_u\mathcal{F} &= -\int_{\mathcal{B}} \delta u'(\varepsilon - u')\,\mathrm{d}x = 0,\\ \delta_\varepsilon\mathcal{F} &= \int_{\mathcal{B}} \delta\varepsilon'(\varepsilon' - f)\,\mathrm{d}x + \int_{\mathcal{B}} \delta\varepsilon(\varepsilon - u')\,\mathrm{d}x = 0,\end{aligned} \tag{58}$$

with the conditions

$$\delta u = 0 \text{ on } \partial\mathcal{B}_u, \qquad \delta\varepsilon = 0 \text{ on } \partial\mathcal{B}_\varepsilon. \tag{59}$$

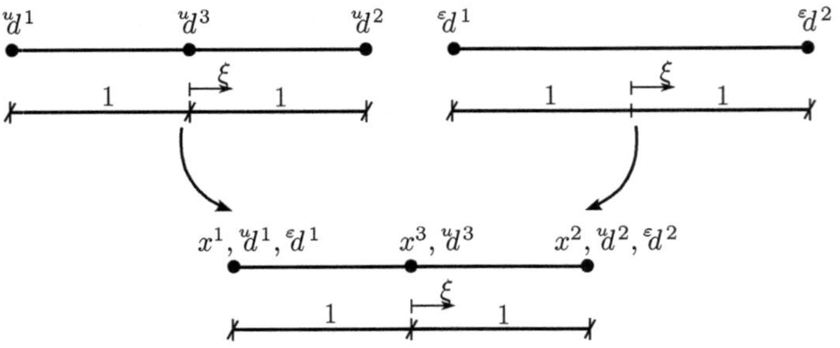

Fig. 3 Mixed reference element (Ω^e) in 1D

For the discretization we use mixed finite elements consisting of a combination of a quadratic three-noded element (for u) and a linear two-noded element (for ε) with one degree of freedom at each node. This leads to a reference element in the parameter space (ξ) with three nodes. Here the first two nodes have two degrees of freedom and the midnode has just one degree of freedom, see Fig. 3. The vectors of the nodal degrees of freedom for u, ε, and the geometry x are given as

$$\overset{u}{\boldsymbol{d}} = [\overset{u}{d}{}^1,\, \overset{u}{d}{}^2,\, \overset{u}{d}{}^3]^T, \quad \overset{\varepsilon}{\boldsymbol{d}} = [\overset{\varepsilon}{d}{}^1,\, \overset{\varepsilon}{d}{}^2]^T \quad \text{and} \quad \boldsymbol{x} = [x^1, x^2, x^3]^T. \tag{60}$$

For the approximation of the unknown field u and the geometry x we use quadratic polynomial interpolation on the reference element $\Omega^e = [-1, 1]$ with the shape functions summarized in a vector $\boldsymbol{N}_u$,

$$\boldsymbol{N}_u(\xi) = \left[N_u^1,\, N_u^2,\, N_u^3\right] = \left[\frac{1}{2}(\xi^2 - \xi),\, \frac{1}{2}(\xi^2 + \xi),\, 1 - \xi^2\right] \tag{61}$$

and the respective vector $\boldsymbol{B}_u = \partial_x \boldsymbol{N}_u$ containing the derivatives of the shape functions with respect to x

$$\boldsymbol{B}_u(\xi) = \left[B_u^1,\, B_u^2,\, B_u^3\right] = \frac{2}{l^e}\left[\xi - \frac{1}{2},\, \xi + \frac{1}{2},\, -2\xi\right]. \tag{62}$$

Here, the derivatives of the shape functions are derived by the chain rule

$$\frac{\partial N}{\partial x} = \frac{\partial N}{\partial \xi}\frac{\partial \xi}{\partial x} = \frac{\partial N}{\partial \xi} J_{11}^{-1} \tag{63}$$

with the transformation J_{11}. With $\boldsymbol{x} = (0, l^e, l^e/2)^T$ we obtain

$$J_{11} = \frac{\partial x}{\partial \xi} = \sum_{I=1}^{3} N_{u,\xi}^I x^I = \left(\xi - \frac{1}{2}\right) 0 + \left(\xi + \frac{1}{2}\right) l^e - (2\xi)\frac{l^e}{2} = \frac{l^e}{2}. \tag{64}$$

With the element nodes $I = 1, 2, 3$ the interpolation of u, u' and the variations δu, $\delta u'$ is

$$
\begin{aligned}
u(\xi) &= \boldsymbol{N}_u(\xi)\overset{u}{\boldsymbol{d}}, \;\; \delta u = \boldsymbol{N}_u(\xi)\delta\overset{u}{\boldsymbol{d}}, \\
u'(\xi) &= \boldsymbol{B}_u(\xi)\overset{u}{\boldsymbol{d}}, \;\; \delta u' = \boldsymbol{B}_u(\xi)\delta\overset{u}{\boldsymbol{d}}
\end{aligned}
\tag{65}
$$

and the geometry approximation

$$
x = \boldsymbol{N}_u(\xi)\boldsymbol{x}. \tag{66}
$$

For the approximation of ε we choose polynomial interpolation of order one on the reference element domain $\Omega^e = [-1, 1]$. The choice of a linear approach for ε is reasonable based on the quadratic interpolation of u and the resulting interpolation order of u' and its relation to ε. Here, the free choice of the interpolation orders for the unknowns is possible due to the fact that least-squares mixed finite elements are not restricted to the LBB condition, unlike, e.g., mixed Galerkin elements compare for instance Jiang (1998), Bochev and Gunzburger (2009), and Braess (1997). The vector of the shape functions of first-order $\boldsymbol{N}_\varepsilon$ appears as

$$
\boldsymbol{N}_\varepsilon = \left[N_\varepsilon^1, N_\varepsilon^2\right] = \left[\frac{1}{2}(1-\xi), \frac{1}{2}(1+\xi)\right]. \tag{67}
$$

The vector of the derivatives of the shape functions $\boldsymbol{B}_\varepsilon = \partial_x \boldsymbol{N}_\varepsilon$ has the form

$$
\boldsymbol{B}_\varepsilon = \left[B_\varepsilon^1, B_\varepsilon^2\right] = \frac{2}{l^e}\left[-\frac{1}{2}, \frac{1}{2}\right]. \tag{68}
$$

We obtain, with the interpolation of ε, ε' and their variations $\delta\varepsilon$ and $\delta\varepsilon'$, the matrix expressions

$$
\begin{aligned}
\varepsilon &= \sum_{J=1}^{2} N_\varepsilon^J \overset{\varepsilon}{d}{}^J = \boldsymbol{N}_\varepsilon \overset{\varepsilon}{\boldsymbol{d}}, \quad \delta\varepsilon = \sum_{J=1}^{2} N_\varepsilon^J \delta\overset{\varepsilon}{d}{}^J = \boldsymbol{N}_\varepsilon \delta\overset{\varepsilon}{\boldsymbol{d}}, \\
\varepsilon' &= \sum_{J=1}^{2} B_\varepsilon^J \overset{\varepsilon}{d}{}^J = \boldsymbol{B}_\varepsilon \overset{\varepsilon}{\boldsymbol{d}}, \quad \delta\varepsilon' = \sum_{J=1}^{2} B_\varepsilon^J \delta\overset{\varepsilon}{d}{}^J = \boldsymbol{B}_\varepsilon \delta\overset{\varepsilon}{\boldsymbol{d}}.
\end{aligned}
\tag{69}
$$

The domain is discretized with n_{ele} finite elements, i.e., $\mathcal{B} = \mathcal{B}^h = \bigcup_{e=1}^{n_{ele}} \mathcal{B}^e$. We obtain the first variation as $\delta\mathcal{F} = \sum_e \delta\mathcal{F}^e$ with the contribution $\delta\mathcal{F}^e$ of a typical element

$$
\delta\mathcal{F}^e = \delta\overset{u}{\boldsymbol{d}}{}^T \underbrace{\int_{\mathcal{B}^e} -\boldsymbol{B}_u^T \boldsymbol{N}_\varepsilon \,\mathrm{d}x}_{\boldsymbol{K}_{u\varepsilon}^e} \overset{\varepsilon}{\boldsymbol{d}} + \delta\overset{u}{\boldsymbol{d}}{}^T \underbrace{\int_{\mathcal{B}^e} \boldsymbol{B}_u^T \boldsymbol{B}_u \,\mathrm{d}x}_{\boldsymbol{K}_{uu}^e} \overset{u}{\boldsymbol{d}}
$$

$$+ \delta \overset{\varepsilon}{\boldsymbol{d}}^T \underbrace{\int_{\mathcal{B}^e} \boldsymbol{B}_{\varepsilon}^T \boldsymbol{B}_{\varepsilon} + \boldsymbol{N}_{\varepsilon}^T \boldsymbol{N}_{\varepsilon} \, \mathrm{d}x}_{\boldsymbol{K}^e_{\varepsilon\varepsilon}} \overset{\varepsilon}{\boldsymbol{d}}$$

$$+ \delta \overset{\varepsilon}{\boldsymbol{d}}^T \underbrace{\int_{\mathcal{B}^e} -\boldsymbol{N}_{\varepsilon}^T \boldsymbol{B}_u \, \mathrm{d}x}_{\boldsymbol{K}^e_{\varepsilon u}} \overset{u}{\boldsymbol{d}} + \delta \overset{\varepsilon}{\boldsymbol{d}}^T \underbrace{\int_{\mathcal{B}^e} -\boldsymbol{B}_{\varepsilon}^T f \, \mathrm{d}x}_{\boldsymbol{r}^e_{\varepsilon}} = 0, \tag{70}$$

where $\boldsymbol{K}^e_{uu}, \boldsymbol{K}^e_{u\varepsilon}, \boldsymbol{K}^e_{\varepsilon u}, \boldsymbol{K}^e_{\varepsilon\varepsilon}$ denote the submatrices of the element stiffness matrix $\boldsymbol{K}^e$ and $\boldsymbol{r}^e_{\varepsilon}$ denotes a part of the element right-hand-side $\boldsymbol{r}^e = \left[0, \boldsymbol{r}^e_{\varepsilon}\right]^T$. This leads to the system of equations for one element

$$\underbrace{\begin{bmatrix} \boldsymbol{K}^e_{uu} & \boldsymbol{K}^e_{u\varepsilon} \\ \boldsymbol{K}^e_{\varepsilon u} & \boldsymbol{K}^e_{\varepsilon\varepsilon} \end{bmatrix}}_{\boldsymbol{K}^e} \underbrace{\begin{bmatrix} \overset{u}{\boldsymbol{d}}{}^e \\ \overset{\varepsilon}{\boldsymbol{d}}{}^e \end{bmatrix}}_{\boldsymbol{d}^e} = - \underbrace{\begin{bmatrix} 0 \\ \boldsymbol{r}^e_{\varepsilon} \end{bmatrix}}_{\boldsymbol{r}^e}, \tag{71}$$

with the element vector of degrees of freedom $\boldsymbol{d}^e = [\overset{u}{\boldsymbol{d}}{}^e, \overset{\varepsilon}{\boldsymbol{d}}{}^e]^T$. The transformation of the integral from the physical space (here $\mathrm{d}x$) to the reference domain $\Omega^e = [-1, 1]$ (here $\mathrm{d}\xi$) is based on

$$\int_{\mathcal{B}^e} \mathrm{d}x = \int_{-1}^{1} J_{11} \, \mathrm{d}\xi, \tag{72}$$

with the Jacobian J_{11} given in (64). The evaluation of the integral expressions can be done analytically or, as mostly used in the field of finite element formulations by numerical integration schemes as, for instance, Gaussian quadrature, see, e.g., Wriggers (2001). Transformation and evaluation of the integral lead to the linear system of equations

$$\underbrace{\begin{bmatrix} \frac{7}{3l^e} & \frac{1}{3l^e} & -\frac{8}{3l^e} & \frac{5}{6} & \frac{1}{6} \\ \frac{1}{3l^e} & \frac{7}{3l^e} & -\frac{8}{3l^e} & -\frac{1}{6} & -\frac{5}{6} \\ -\frac{8}{3l^e} & -\frac{8}{3l^e} & \frac{16}{3l^e} & -\frac{2}{3} & \frac{2}{3} \\ \frac{5}{6} & -\frac{1}{6} & -\frac{2}{3} & (\frac{l^e}{3} + \frac{1}{l^e}) & (\frac{l^e}{6} - \frac{1}{l^e}) \\ \frac{1}{6} & -\frac{5}{6} & \frac{2}{3} & (\frac{l^e}{6} - \frac{1}{l^e}) & (\frac{l^e}{3} + \frac{1}{l^e}) \end{bmatrix}}_{\boldsymbol{K}^e} \underbrace{\begin{bmatrix} \overset{u}{d}{}^1 \\ \overset{u}{d}{}^2 \\ \overset{u}{d}{}^3 \\ \overset{\varepsilon}{d}{}^1 \\ \overset{\varepsilon}{d}{}^2 \end{bmatrix}}_{\boldsymbol{d}^e} = \underbrace{\begin{bmatrix} 0 \\ 0 \\ 0 \\ -f \\ f \end{bmatrix}}_{-\boldsymbol{r}^e} . \tag{73}$$

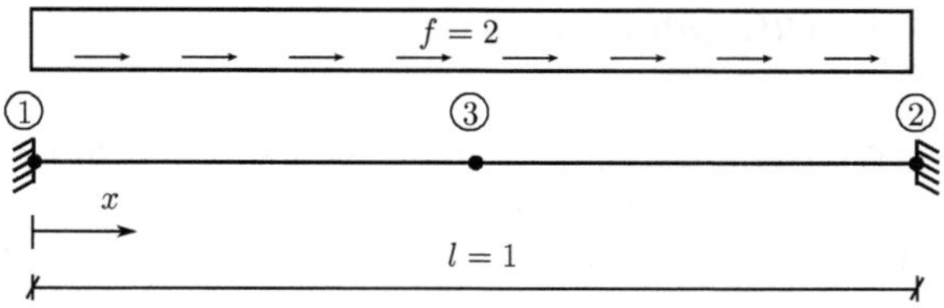

Fig. 4 Setup boundary value problem in 1D

As essential boundary conditions, we fix u at the left- and the right-hand side ($u(0) = u(l) = 0$) and choose $f = 2$, see also Fig. 4. Here, due to the fact that we only consider one element, no assembly procedure is neccesary. Applying the essential boundary conditions on the final system of equations (deleting the first two columns and rows) leads to the reduced system

$$\begin{bmatrix} \frac{16}{3} & -\frac{2}{3} & \frac{2}{3} \\ -\frac{2}{3} & \frac{4}{3} & -\frac{5}{6} \\ \frac{2}{3} & -\frac{5}{6} & \frac{4}{3} \end{bmatrix} \begin{bmatrix} {}^{u}d^{3} \\ {}^{\varepsilon}d^{1} \\ {}^{\varepsilon}d^{2} \end{bmatrix} = \begin{bmatrix} 0 \\ -2 \\ 2 \end{bmatrix}. \tag{74}$$

The vector of unknowns can be computed as $[{}^{u}d^{3}, {}^{\varepsilon}d^{1}, {}^{\varepsilon}d^{2}]^{T} = [-1/4, -1, 1]^{T}$. To check the result, we investigate the analytical solution of the problem. Therefore, we integrate the given differential equation $u'' = f$ twice with the integration constants c_1 and c_2 and obtain

$$\begin{aligned} u' &= \int f dx &&= fx + c_1, \\ u &= \int (fx + c_1) dx &&= \frac{1}{2} fx^2 + c_1 x + c_2. \end{aligned} \tag{75}$$

Insertion of the boundary conditions $u(0) = u(l) = 0$ yields $c_1 = -l$ and $c_2 = 0$, thus we obtain

$$u' = fx - l = 2x - 1 \;\text{ and }\; u = \frac{1}{2} fx^2 - lx = x^2 - x. \tag{76}$$

With this in hand we can compute the unknown degrees of freedom for comparison with the latter computed result and can see, that our finite element solution is in accordance with the analytical one,

$$u\left(\frac{1}{2}\right) = -\frac{1}{4} \rightarrow {}^{u}d^{3}, \quad u'(0) = -1 \rightarrow {}^{\varepsilon}d^{1}, \quad u'(l = 1) = 1 \rightarrow {}^{\varepsilon}d^{2}. \tag{77}$$

3.3 Interpolation Spaces

The mixed least-squares finite element formulations presented in this contribution are based on displacement–stress functionals. Hence, the solution variables are the displacements ($\boldsymbol{u}$) and the stresses ($\boldsymbol{\sigma}, \boldsymbol{P}$). For the interpolation of these unknowns appropriate approximation spaces have to be chosen. For the displacements $W^{1,p}(\mathcal{B})$ is an appropriate choice due to its restrictions that the unknown function as well as their derivative has to fulfill the $L^p(\mathcal{B})$-norm

$$||\bullet||_{L^p(\mathcal{B})} = \sqrt[p]{\int_{\mathcal{B}} |\bullet|^p \, \mathrm{d}V} \,. \tag{78}$$

This leads to the definition of the Sobolev space

$$W^{1,p}(\mathcal{B}) = \{\boldsymbol{u} \in L^p(\mathcal{B}) : \nabla \boldsymbol{u} \in L^p(\mathcal{B})\},$$

with $||\boldsymbol{u}||_{L^p(\mathcal{B})} < \infty$ and $||\nabla \boldsymbol{u}||_{L^p(\mathcal{B})} < \infty$. For the interpolation of the stresses the space $W^q(\text{div}, \mathcal{B})$ is a suitable choice, compare, e.g., Müller et al. (2014). The restriction here is that the function as well as its divergence has to fulfill the $L^q(\mathcal{B})$-norm. With this in hand we obtain the Sobolev space

$$W^q(\text{div}, \mathcal{B}) = \{\boldsymbol{P} \in L^q(\mathcal{B})^2 : \text{div}\, \boldsymbol{P} \in L^q(\mathcal{B})\}.$$

Dependent on the formulation, p and q have to be chosen suitable. In the case of linear elastic problems, $p = 2$ and $q = 2$ can be chosen leading to the Sobolev spaces $W^{1,2}(\mathcal{B}) = H^1(\mathcal{B})$ and $W^2(\text{div}, \mathcal{B}) = H(\text{div}, \mathcal{B})$. In the following subsection, we provide interpolation functions, which guarantee a conforming discretization of the above-mentioned Sobolev spaces.

3.4 Interpolation Functions

For the approximation of quantities in the framework of the FEM appropriate interpolation functions have to be chosen. The choice of the functions is dependent on the interpolation space, see Sect. 3.3. In this chapter, we want to present the interpolation functions used for our least-squares mixed finite element formulations. Here, we differentiate between standard interpolation polynomials of Lagrangian type, which ensure conforming discretizations of $W^{1,p}(\mathcal{B})$ and vector-valued Raviart–Thomas interpolation functions which ensure conforming discretizations of $W^q(\text{div}, \mathcal{B})$.

Standard interpolation polynomials. For the interpolation of quantities where the function $\boldsymbol{u}(\boldsymbol{x})$ as well as the derivative $\boldsymbol{u}'(\boldsymbol{x})$ has to satisfy the $L^p(\mathcal{B})$-norm

$$||\boldsymbol{u}||_{L^p(\mathcal{B})} < \infty \quad \text{and} \quad ||\boldsymbol{u}'||_{L^p(\mathcal{B})} < \infty, \tag{79}$$

we choose standard interpolation polynomials of Lagrangian type. In the following we consider the construction of these polynomials in two dimensions for a triangular finite element domain in the parameter space $\boldsymbol{\xi} = (\xi, \eta)^T$ in detail. Therefore, we start with a general polynomial

$$N(\xi, \eta) = a_1 + a_2\xi + a_3\eta + a_4\xi^2 + a_5\xi\eta + a_6\eta^2 \dots . \tag{80}$$

The related monomials can be identified, for instance, using the Pascal's triangle, see, e.g., Zhu et al. (2005). In order to construct the interpolation functions N^I we solve for each interpolation site I (associated to a node) a system of equations enforcing that the interpolation polynomial has to be one at the respective node coordinates and zero at all other nodes

$$N^I(\xi_J, \eta_J) = \begin{cases} 1, & \text{for } I = J \\ 0, & \text{for } I \neq J, \end{cases} \tag{81}$$

with the nodal coordinates $(\xi_J, \eta_J)^T$. For instance, for the first node we obtain the system of equations

$$\begin{aligned} & a_1^1 + a_2^1\xi_1 + a_3^1\eta_1 + a_4^1\xi_1^2 + a_5^1\xi_1\eta_1 + a_6^1\eta_1^2 \dots = 1 \\ \wedge\ & a_1^1 + a_2^1\xi_2 + a_3^1\eta_2 + a_4^1\xi_2^2 + a_5^1\xi_2\eta_2 + a_6^1\eta_2^2 \dots = 0 \\ \wedge\ & a_1^1 + a_2^1\xi_3 + a_3^1\eta_3 + a_4^1\xi_3^2 + a_5^1\xi_3\eta_3 + a_6^1\eta_3^2 \dots = 0 \\ \wedge\ & \dots \end{aligned} \tag{82}$$

from which we compute the coefficients a_i^1. By solving the system of equations with respect to a changed right-hand-side vector (the position of the "one" is changing) we obtain the seeked coefficients a_i^I. Inserting them into the general form of the interpolation polynomial (80) yields the function for each interpolation site.

Vector-valued Raviart–Thomas Interpolation functions. For the interpolation of quantities where the function $\boldsymbol{\sigma}(\boldsymbol{x})$ as well as the divergence div $\boldsymbol{\sigma}(\boldsymbol{x})$ has to satisfy the $L^q(\mathcal{B})$-norm (78)

$$||\boldsymbol{\sigma}||_{L^q(\mathcal{B})} < \infty \quad \text{and} \quad ||\operatorname{div}\boldsymbol{\sigma}||_{L^q(\mathcal{B})} < \infty, \tag{83}$$

we choose vector-valued Raviart–Thomas interpolation functions Ψ_m^J where m denotes the interpolation order and J the associated interpolation site. We differentiate between outer and inner interpolation sites J^{out} and J^{in}. The total number of interpolation sites is then given as $|J|_C = |J^{out}|_C + |J^{in}|_C = m^2 + 4m + 3$ with $|J^{out}|_C = 3(m+1)$ and $|J^{in}|_C = m(m+1)$. Here, $|A|_C$ denotes the cardinality of the set A, i.e., the number of elements in the set A.

The construction is shown again on a two-dimensional triangular finite element domain in the parameter space $\boldsymbol{\xi} = (\xi, \eta)^T$. The outer interpolation sites are related to the respective element edges e^L (with $|e^L|_C = 3$) and their associated normals $\boldsymbol{n}^L$

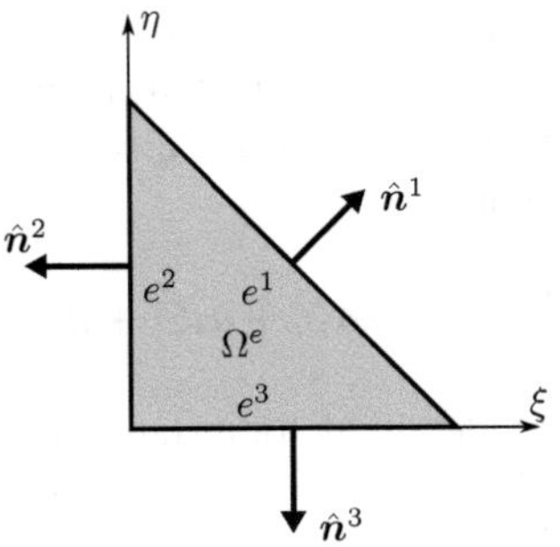

Fig. 5 Numbering of edges e^L and their associated normals $\boldsymbol{n}^L$

Table 1 Raviart–Thomas setups ($m = 0, 1, 2$) in two dimensions

RT_m	Pol. order of Ψ_m^J	$\|J^{out}\|_C$ per e^L	$\|J^{out}\|_C$	$\|J^{in}\|_C$	$\|J\|_C$
RT_0	quasi-linear	1	3	0	3
RT_1	quasi-quadratic	2	6	2	8
RT_2	quasi-cubic	3	9	6	15

(with $|\boldsymbol{n}^L|_C = 3$) and the inner ones to the triangular domain denoted by Ω^e, see also Fig. 5. The interpolation order as well as the number of interpolation sites for a two-dimensional triangular element up to order $m = 2$ is given in Table 1. The term "quasi" in Table 1 means that the interpolation function has a full polynomial of order m but also some terms of the next interpolation level. In the framework of this contribution, we restrict ourselves to meshes with non-curved edges. An enhancement to curved boundaries has been done in Bertrand et al. (2014). The general form of the vectorial basis functions of order m in the parameter space $\boldsymbol{\xi} = (\xi, \eta)^T$ is given by

$$\hat{\boldsymbol{v}}_m(\xi, \eta) = \hat{\boldsymbol{p}}_m(\xi, \eta) + \hat{p}_m(\xi, \eta) \begin{pmatrix} \xi \\ \eta \end{pmatrix}, \tag{84}$$

where $\hat{p}_m, \hat{\boldsymbol{p}}_m$ are general scalar-valued and vectorial functions of order m with J unknown coefficients $(a, b, c, \ldots)$, see exemplary equation (93). For the construction we have to evaluate the expressions

$$\begin{aligned} M_{out}^{L,K} &= \int_{e^L} (\hat{\boldsymbol{v}}_m \cdot \hat{\boldsymbol{n}}^L)\, \hat{q}_m^{L,K}\, ds \\ \text{and (for } m \geq 1) \quad M_{in}^I &= \int_{\Omega^e} \hat{\boldsymbol{v}}_m \cdot \hat{\boldsymbol{q}}_{m-1}^I\, da \end{aligned} \tag{85}$$

for each interpolation site J, that means for each n^L, $\hat{q}_m^{L,K}$ and $\hat{\boldsymbol{q}}_{m-1}^I$. Here $\hat{q}_m^{L,K}$, $\hat{\boldsymbol{q}}_{m-1}^I$ are explicit, scalar-valued, and vectorial functions of order m and $m-1$ corresponding to the interpolation site J, which have to be chosen to be linear independent. The

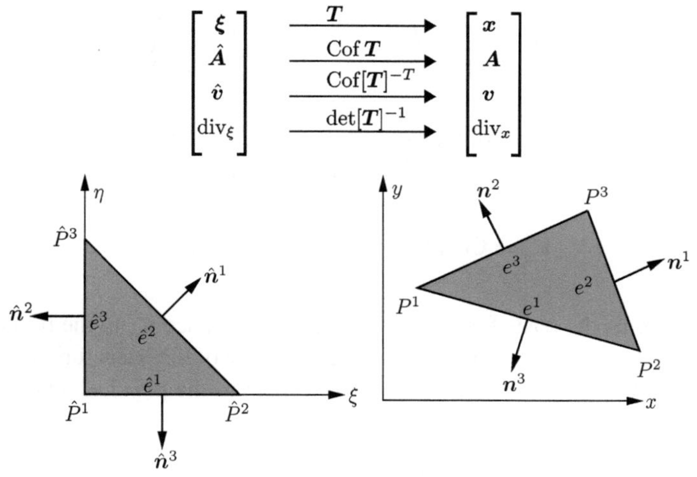

Fig. 6 Piola transformation

linear independency has to be guaranteed for the inner interpolation sites and the individual edges, independently. With the conditions

$$M^{L,K}_{out}(\hat{\boldsymbol{n}}^O, \hat{q}^{OP}_m) = \begin{cases} 1, & \text{for } L = O \wedge K = P \\ 0, & \text{for } L \neq O \vee K \neq P \end{cases} \tag{86}$$

and

$$M^I_{in}(\hat{\boldsymbol{q}}^P_{m-1}) = \begin{cases} 1, & \text{for } I = P \\ 0, & \text{for } I \neq P \end{cases} \tag{87}$$

we obtain for each interpolation site J a system of equations in which solutions are the unknown coefficients. These yields, entering in (84), the (linear independent, compare Ciarlet 1991) vectorial basis functions $\hat{\boldsymbol{v}}^J_m(\xi, \eta)$ for each interpolation site J of the reference triangle. For the transformation of the basis function from the parameter space $(\hat{P}^i, \hat{\boldsymbol{n}}^i, \mathrm{d}\hat{\boldsymbol{A}}, \hat{\boldsymbol{v}}^J_m \ldots)$ to the physical space $(P^i, \boldsymbol{n}^i, \mathrm{d}\boldsymbol{A}, \boldsymbol{v}^J_m \ldots)$, see also Fig. 6, we have to fulfill the requirement that the flux over the element edges is equal in both configurations. Let $(\mathrm{d}\hat{\boldsymbol{A}}, \mathrm{d}\boldsymbol{A})$ denote the vectorial area element of the parameter and physical space, respectively, then we demand

$$\hat{\boldsymbol{v}}^J_m \cdot \mathrm{d}\hat{\boldsymbol{A}} = \boldsymbol{v}^J_m \cdot \mathrm{d}\boldsymbol{A}. \tag{88}$$

Now we insert the mapping of the vectorial area element (11) and assume a linear geometry transformation $\boldsymbol{\varphi}_t$, see Fig. 1. This leads to a constant transformation matrix $\boldsymbol{T}$ associated to the unit triangle in the parameter space $\boldsymbol{\xi}(\xi, \eta)$ and the triangular element in the physical space $\boldsymbol{x}(x_1, x_2)$:

$$\boldsymbol{T} = \frac{\partial \boldsymbol{x}}{\partial \boldsymbol{\xi}} = \begin{pmatrix} -x_1^1 + x_1^2 & -x_1^1 + x_1^3 \\ -x_2^1 + x_2^2 & -x_2^1 + x_2^3 \end{pmatrix} \tag{89}$$

with the known coordinates of the vertices $P^I = (x_1^I, x_2^I)$ in the physical space. With this in hand and $\text{Cof}\,\boldsymbol{T} = \det[\boldsymbol{T}]\boldsymbol{T}^{-T}$ we obtain from (88)

$$\hat{\boldsymbol{v}}_m^J = \text{Cof}[\boldsymbol{T}]^T \boldsymbol{v}_m^J \rightsquigarrow \boldsymbol{v}_m^J = \frac{1}{\det \boldsymbol{T}} \boldsymbol{T} \hat{\boldsymbol{v}}_m^J, \tag{90}$$

which transforms the basis function of the parameter space $\hat{\boldsymbol{v}}_m^J$ to the basis function of the physical space $\boldsymbol{v}_m^J$. Furthermore, the divergence of the basis function has to be transformed. Applying the divergence with respect to the physical space (div_x) on both sides of (90) yields

$$\text{div}_x \boldsymbol{v}_m^J = \text{div}_x \left[\frac{1}{\det \boldsymbol{T}} \boldsymbol{T} \hat{\boldsymbol{v}}_m^J \right] = \frac{1}{\det \boldsymbol{T}} \text{div}_\xi \hat{\boldsymbol{v}}_m^J, \tag{91}$$

using the relations

$$\boldsymbol{T} \text{div}_x \hat{\boldsymbol{v}}_m^J = \text{div}_\xi \hat{\boldsymbol{v}}_m^J \quad \text{and} \quad \text{div}_x \left[\frac{1}{\det \boldsymbol{T}} \boldsymbol{T} \right] = 0, \tag{92}$$

because $\boldsymbol{T}$ is a constant matrix. Finally, to obtain the vector-valued Raviart–Thomas interpolation functions Ψ_m^J, we have to apply a normalization condition on $\boldsymbol{v}_m^J$ in order to get suitable functions for Ψ_m^J.

> The sum of all Raviart–Thomas shape functions Ψ_m^J belonging to one edge multiplied with the associated normal of this edge has to be equal to one.

In the following we will provide the construction of the interpolation functions for $m = 1$.

Basis functions for order $m = 1$. The general form of the vectorial basis function (84) for the order $m = 1$ is given as

$$\hat{\boldsymbol{v}}_1(\xi, \eta) = \underbrace{\begin{pmatrix} a_1 + a_2\xi + a_3\eta \\ b_1 + b_2\xi + b_3\eta \end{pmatrix}}_{\hat{\boldsymbol{p}}_1} + \underbrace{(c_1\xi + c_2\eta)}_{\hat{p}_1} \begin{pmatrix} \xi \\ \eta \end{pmatrix}. \tag{93}$$

Since $m \geq 1$, both parts of (85) have to be evaluated for the interpolation sites $J = 1 \dots 8$ leading to eight equations

$$M_{out}^{L,K} \text{ with } L = 1 \dots 3, K = 1 \dots 2 \quad \text{and} \quad M_{in}^I \text{ with } I = 1 \dots 2. \tag{94}$$

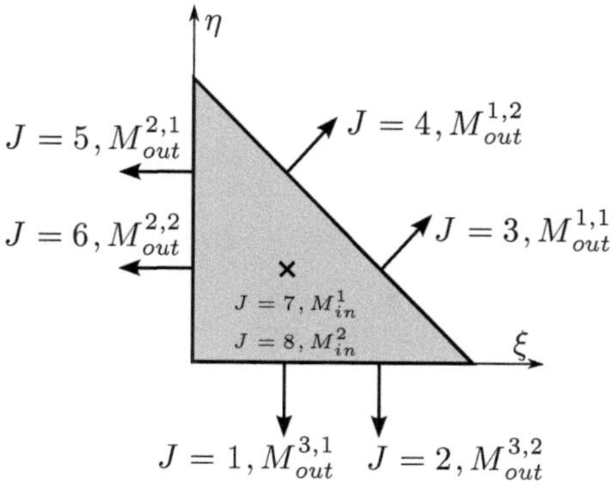

Fig. 7 Numbering of interpolation sites J for RT_1

Exemplarily, we will consider the evaluation for the sites $J = 3$ ($M^{1,1}_{out}$) and $J = 8$ (M^2_{in}) in more detail, see Fig. 7.

$\underline{J = 3, M^{1,1}_{out}}$: We obtain, evaluating the first part of (85) with $\hat{\boldsymbol{n}}^1 = \sqrt{\frac{1}{2}}\begin{pmatrix}1\\1\end{pmatrix}$, $\hat{q}^{1,1}_1 = \xi$ and the correlation for the coordinates $\xi + \eta = 1$

$$
\begin{aligned}
M^{1,1}_{out} &= \int_{e^1} (\hat{\boldsymbol{v}}_1 \cdot \hat{\boldsymbol{n}}^1)\, \hat{q}^{1,1}_1 \, ds \\
&= \int_{e^1} \begin{pmatrix} a_1 + a_2\xi + a_3\eta + c_1\xi^2 + c_2\xi\eta \\ b_1 + b_2\xi + b_3\eta + c_1\xi\eta + c_2\eta^2 \end{pmatrix} \cdot \frac{1}{\sqrt{2}} \begin{pmatrix}1\\1\end{pmatrix} \xi \, ds \\
&= \frac{a_1}{2} + \frac{a_2}{3} + \frac{a_3}{6} + \frac{b_1}{2} + \frac{b_2}{3} + \frac{b_3}{6} + \frac{c_1}{3} + \frac{c_2}{6}.
\end{aligned}
\tag{95}
$$

$\underline{J = 8, M^2_{in}}$: For this interpolation site we choose $\hat{\boldsymbol{q}}^2_0 = (0, 1)^T$ and obtain evaluating the second part of (85)

$$
\begin{aligned}
M^2_{in} &= \int_D \hat{\boldsymbol{v}}_m \cdot \hat{\boldsymbol{q}}^2_0 \, da \\
&= \int_0^1 \int_0^{1-\eta} \begin{pmatrix} a_1 + a_2\xi + a_3\eta + c_1\xi^2 + c_2\xi\eta \\ b_1 + b_2\xi + b_3\eta + c_1\xi\eta + c_2\eta^2 \end{pmatrix} \begin{pmatrix}0\\1\end{pmatrix} d\xi d\eta \\
&= \frac{b_1}{2} + \frac{b_2}{6} + \frac{b_3}{6} + \frac{c_1}{24} + \frac{c_2}{12}.
\end{aligned}
\tag{96}
$$

The computations at each interpolation site, as exemplary done for $J = 3$ (95) and $J = 8$ (96), yield under consideration of (86) and (87) a system of equations, which has to be solved for each interpolation site J. The right-hand-side vector has a nonvanishing entry at the Jth entry, which is equal to one. Solving the system of equations yields the parameters a_i, b_i for $i = 1, 2, 3$ and c_1, c_2 of the vectorial basis functions (93). We obtain the eight vectorial basis functions $\hat{\boldsymbol{v}}_1^J$ for the considered RT_1-case as

$$\hat{\boldsymbol{v}}_1^1 = \begin{pmatrix} -8\eta\xi - 8\xi^2 + 6\xi \\ -8\eta^2 - 8\eta\xi + 12\eta + 6\xi - 4 \end{pmatrix}$$
$$\hat{\boldsymbol{v}}_1^2 = \begin{pmatrix} 8\xi^2 - 4\xi \\ 8\eta\xi - 2\eta - 6\xi + 2 \end{pmatrix}$$
$$\hat{\boldsymbol{v}}_1^3 = \begin{pmatrix} 8\xi^2 - 4\xi \\ 8\eta\xi - 2\eta \end{pmatrix}$$
$$\hat{\boldsymbol{v}}_1^4 = \begin{pmatrix} 8\eta\xi - 2\xi \\ 8\eta^2 - 4\eta \end{pmatrix}$$
$$\hat{\boldsymbol{v}}_1^5 = \begin{pmatrix} 8\eta\xi - 6\eta - 2\xi + 2 \\ 8\eta^2 - 4\eta \end{pmatrix} \tag{97}$$
$$\hat{\boldsymbol{v}}_1^6 = \begin{pmatrix} -8\eta\xi + 6\eta - 8\xi^2 + 12\xi - 4 \\ -8\eta^2 - 8\eta\xi + 6\eta \end{pmatrix}$$
$$\hat{\boldsymbol{v}}_1^7 = \begin{pmatrix} -8\eta\xi - 16\xi^2 + 16\xi \\ -8\eta^2 - 16\eta\xi + 8\eta \end{pmatrix}$$
$$\hat{\boldsymbol{v}}_1^8 = \begin{pmatrix} -16\eta\xi - 8\xi^2 + 8\xi \\ -16\eta^2 - 8\eta\xi + 16\eta \end{pmatrix}.$$

Now the obtained vector-valued basis functions (as well as their divergences) have to be transformed to the physical space by (90) and (91). The evaluation of the normalization condition for RT_1 yields

$$\Psi_1^J = \frac{l}{2}\,\boldsymbol{v}_1^J \quad \text{and} \quad \text{div}\ \Psi_1^J = \frac{l}{2}\ \text{div}\ \boldsymbol{v}_1^J \tag{98}$$

where l denotes the associated length of the edge of the interpolation site under consideration. For further details the reader is referred to, e.g., Raviart and Thomas (1977).

4 LSFEM–Linear Elasticity

In this section, we provide least-squares formulations for the theory of linear elasticity. Although this theory is only valid for problems undergoing small deformations and displacements, it is widely used in engineering. Beside compressible problems, we consider the case of quasi-incompressible materials, which could lead to unsatisfying solutions, as, e.g., stress oscillations or locking effects, when used in combination with common approaches as the standard Galerkin method. In detail, we introduce the governing differential equations for the geometrically linear setup and derive the construction of a two-field least-squares functional related to stresses and displacements. Following the ideas of the variational least-squares approach from Sect. 3.1, we provide the necessary first and second variations of the functional. The discretization of the variational problem involves a linear system of algebraic equations, which directly yields the solution. For a convenient extension to nonlinear material behavior, we retain the formalism of linearization when applying a standard Newton method (even for the linear problem) for finding the root of the first variation in the considered problems.

4.1 Linear Elasticity

In order to define linear elastic material behavior, we introduce the free energy function

$$\psi(\boldsymbol{\varepsilon}) = \frac{1}{2}\lambda(\operatorname{tr}\boldsymbol{\varepsilon})^2 + \mu \operatorname{tr}\boldsymbol{\varepsilon}^2, \tag{99}$$

with the symmetric displacement gradient $\boldsymbol{\varepsilon} = \nabla^s \boldsymbol{u} = 1/2(\nabla \boldsymbol{u} + (\nabla \boldsymbol{u})^T)$ and the Lamé material parameters λ and μ. The linear stress–strain relation yields

$$\boldsymbol{\sigma} = \frac{\partial \psi(\boldsymbol{\varepsilon})}{\partial \boldsymbol{\varepsilon}} = \lambda \operatorname{tr}[\boldsymbol{\varepsilon}]\mathbf{I} + 2\mu\boldsymbol{\varepsilon}, \tag{100}$$

with the Cauchy stress tensor $\boldsymbol{\sigma}$. The second derivative of the free energy with respect to the strains $\boldsymbol{\varepsilon}$ yields the fourth-order elasticity tensor

$$\mathbb{C} = \frac{\partial^2 \psi(\boldsymbol{\varepsilon})}{\partial \boldsymbol{\varepsilon} \partial \boldsymbol{\varepsilon}} = \lambda \mathbf{I} \otimes \mathbf{I} + 2\mu \mathbb{I}. \tag{101}$$

Here, the index representations of the identity tensors of second order ($\mathbf{I}$) and fourth order ($\mathbb{I}$) are given as

$$\mathrm{I}_{ij} = \delta_{ij} \quad \text{and} \quad \mathbb{I}_{ijkl} = \delta_{ik}\delta_{jl}, \tag{102}$$

where δ_{ij} denotes the Kronecker Delta. The inverse of the elasticity tensor (101) yields the so-called compliance tensor

$$\mathbb{C}^{-1} = \frac{1}{2\mu}\mathbb{I} - \frac{\lambda}{2\mu(2\mu + 3\lambda)}\mathbf{I} \otimes \mathbf{I}. \tag{103}$$

With this in hand we use a set of partial differential equations of first order, namely the balance of momentum, the constitutive relation, and the stress symmetry condition all written in residual form as

$$
\begin{aligned}
\mathcal{R}_1 &:= \quad \operatorname{div} \boldsymbol{\sigma} + \boldsymbol{f} = \mathbf{0} \rightarrow \text{balance of momentum},\\
\mathcal{R}_2 &:= \boldsymbol{\sigma} - \mathbb{C} : \nabla^s \boldsymbol{u} = \mathbf{0} \rightarrow \text{constitutive relation},\\
\mathcal{R}_3 &:= \quad \boldsymbol{\sigma} - \boldsymbol{\sigma}^T = \mathbf{0} \rightarrow \text{stress symmetry}\,.
\end{aligned} \tag{104}
$$

From the mathematical point of view the third residual is redundant and could be neglected, which has been proven by Cai and Starke (2004). However, from the practial point of view it seems to be advantageous to control the lack of symmetry of the stress tensor (associated to the balance of moment of momentum) directly, see Schwarz et al. (2014). It is known that in the case of quasi-incompressible elasticity, characterized by Poisson's ratio $\nu \rightarrow 0.5$ and therefore $\lambda \rightarrow \infty$, problems can arise for numerical methods. In order to circumvent these problems, a complementary formulation based on the compliance tensor (103) could be utilized. By tensor multiplication of the second and third residuals in (104) with $\mathbb{C}^{-1}$ we obtain the final formulation for the case of linear elastostatics

$$
\begin{aligned}
\mathcal{R}_1 &:= \quad \operatorname{div} \boldsymbol{\sigma} + \boldsymbol{f} = \mathbf{0},\\
\mathcal{R}_2 &:= \mathbb{C}^{-1} : \boldsymbol{\sigma} - \nabla^s \boldsymbol{u} = \mathbf{0},\\
\mathcal{R}_3 &:= \mathbb{C}^{-1} : (\boldsymbol{\sigma} - \boldsymbol{\sigma}^T) = \mathbf{0}.
\end{aligned} \tag{105}
$$

It should be noted that the multiplication of $(105)_2$ and $(105)_3$ with the compliance tensor $\mathbb{C}^{-1} \in \mathbb{R}^{3x3x3x3}$ is **not** enforcing the symmetry condition sym$[\mathbb{C}^{-1} : \boldsymbol{\sigma} - \nabla^s \boldsymbol{u}]$ or sym$[\mathbb{C}^{-1} : (\boldsymbol{\sigma} - \boldsymbol{\sigma}^T)]$. Following (51) and using the residuals defined in (105), we obtain the two-field least-squares functional

$$
\begin{aligned}
\mathcal{F}(\boldsymbol{\sigma}, \boldsymbol{u}) = &\frac{1}{2} \int_{\mathcal{B}} \omega_1^2 \left((\operatorname{div} \boldsymbol{\sigma} + \boldsymbol{f}) \cdot (\operatorname{div} \boldsymbol{\sigma} + \boldsymbol{f}) \right) \mathrm{d}V\\
&+ \frac{1}{2} \int_{\mathcal{B}} \omega_2^2 \left((\mathbb{C}^{-1} : \boldsymbol{\sigma} - \nabla^s \boldsymbol{u}) : (\mathbb{C}^{-1} : \boldsymbol{\sigma} - \nabla^s \boldsymbol{u}) \right) \mathrm{d}V\\
&+ \frac{1}{2} \int_{\mathcal{B}} \omega_3^2 \left((\mathbb{C}^{-1} : (\boldsymbol{\sigma} - \boldsymbol{\sigma}^T)) : (\mathbb{C}^{-1} : (\boldsymbol{\sigma} - \boldsymbol{\sigma}^T) \right) \mathrm{d}V.
\end{aligned} \tag{106}
$$

A prove for ellipticity and continuity for a least-squares functional in the $H(\text{div}) \times H^1$-norm, which includes optimal error estimates and offers ideal multigrid solutions can be found in Cai and Starke (2004). The minimization of this functional presupposes the first variation with respect to stresses and displacements, which has to be zero

$$
\begin{aligned}
\delta_u \mathcal{F} &= \sum_{i=1}^{3} \int_{\mathcal{B}} \omega_i^2 \delta_u \mathcal{R}_i \cdot \mathcal{R}_i \, dV = 0,\\
\delta_\sigma \mathcal{F} &= \sum_{i=1}^{3} \int_{\mathcal{B}} \omega_i^2 \delta_\sigma \mathcal{R}_i \cdot \mathcal{R}_i \, dV = 0,
\end{aligned} \tag{107}
$$

with

$$\delta \boldsymbol{u} = 0 \text{ on } \partial\mathcal{B}_u \quad \text{and} \quad \delta\boldsymbol{\sigma} = 0 \text{ on } \partial\mathcal{B}_\sigma. \tag{108}$$

The linearizations of (107) for the application of a Newton scheme are

$$\begin{aligned}
\Delta_u\delta_u\mathcal{F} &= \sum_{i=1}^{3}\int_{\mathcal{B}} \omega_i^2(\Delta_u\delta_u\mathcal{R}_i\cdot\mathcal{R}_i + \delta_u\mathcal{R}_i\cdot\Delta_u\mathcal{R}_i)\,dV,\\
\Delta_\sigma\delta_u\mathcal{F} &= \sum_{i=1}^{3}\int_{\mathcal{B}} \omega_i^2(\Delta_\sigma\delta_u\mathcal{R}_i\cdot\mathcal{R}_i + \delta_u\mathcal{R}_i\cdot\Delta_\sigma\mathcal{R}_i)\,dV,\\
\Delta_u\delta_\sigma\mathcal{F} &= \sum_{i=1}^{3}\int_{\mathcal{B}} \omega_i^2(\Delta_u\delta_\sigma\mathcal{R}_i\cdot\mathcal{R}_i + \delta_\sigma\mathcal{R}_i\cdot\Delta_u\mathcal{R}_i)\,dV,\\
\Delta_\sigma\delta_\sigma\mathcal{F} &= \sum_{i=1}^{3}\int_{\mathcal{B}} \omega_i^2(\Delta_\sigma\delta_\sigma\mathcal{R}_i\cdot\mathcal{R}_i + \delta_\sigma\mathcal{R}_i\cdot\Delta_\sigma\mathcal{R}_i)\,dV.
\end{aligned} \tag{109}$$

Due to the fact that we consider a linear problem, all terms with "mixed" variations as, e.g., $\Delta_u\delta_\sigma\mathcal{R}_i$ are equal to zero. The non-vanishing terms for the first variation are

$$\begin{aligned}
&\delta_\sigma\mathcal{R}_1 = \operatorname{div}\delta\boldsymbol{\sigma}, \quad \delta_u\mathcal{R}_2 = -\nabla^s\delta\boldsymbol{u},\\
&\delta_\sigma\mathcal{R}_2 = \mathbb{C}^{-1} : \delta\boldsymbol{\sigma},\ \delta_\sigma\mathcal{R}_3 = \mathbb{C}^{-1} : (\delta\boldsymbol{\sigma} - \delta\boldsymbol{\sigma}^T).
\end{aligned} \tag{110}$$

Analogously, we obtain the linearized residuals $\Delta_u\mathcal{R}_i$ and $\Delta_\sigma\mathcal{R}_i$. At this point, all quantities for the implementation of the least-squares mixed finite element RT_mP_k are provided. In the next section some numerical examples for the compressible and quasi-incompressible cases are presented, where we investigate in detail the resulting approximation quality of stresses and displacements. Here, all given quantities as, e.g., dimensions and material parameters are denoted without any units and have to be chosen in a consistent manner (as, e.g., in SI units). The element implementations and computations have been done using the *AceGen* and *AceFEM* packages (version 6.503), see Korelc (1997, 2002), of *Mathematica* (version 10.1), see Wolfram Research (2015). Furthermore, for the vizualizations *Paraview* (version 4.3.1), see Ahrens et al. (2005), has been used. Basis for the contour plots of the stresses are the quantities evaluated at the corner nodes of each triangle. Inside an element the values are interpolated linearly, between elements the plot is discontinous.

4.2 Cantilever Beam, Linear Elastic

First, we want to consider the performance independent of the redundant residual $\mathcal{R}_3$, i.e., $\omega_3 = 0$, $\omega_1 = \omega_2 = 1$. In detail we want to investigate the convergence of the vertical displacement at the upper right node of a cantilever beam using different interpolation orders for the displacements ($k = 1, 2, 3, 4$) and the stresses

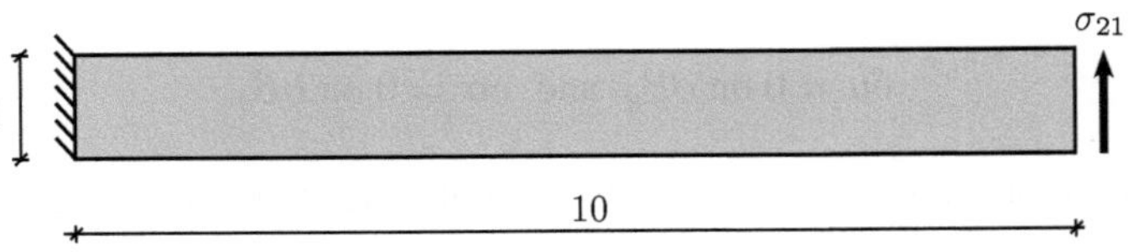

Fig. 8 Geometry of the cantilever beam

Fig. 9 Displacement convergence for u_2 of the upper right node (10, 1) over number of equations (neq) of the final system of equations for RT_0P_k

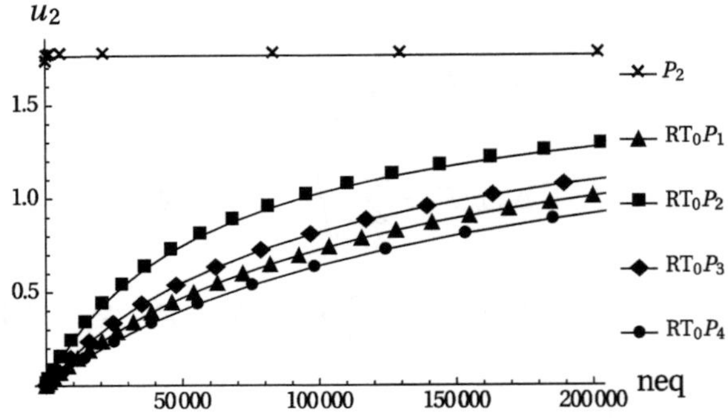

($m = 0, 1, 2$) leading to elements RT_mP_k. As already mentioned the polynomial orders could be chosen arbitrarily since the method is not restricted to the LBB condition. Since we consider here a compressible material (Young's modulus $E = 200$, Poisson's ratio $\nu = 0.35$), we use a standard Galerkin element (triangle, purely displacement based, quadratic interpolation→ P_2) as a reference solution. The final outcome of this investigation should be a statement, which interpolation combination is recommendable. As a boundary value problem we choose a cantilever beam under plain strain conditions, see Fig. 8. The left side of the beam has a fixed displacement boundary ($u_1 = u_2 = 0$). The upper and lower edges are assumed to be stress-free ($\boldsymbol{\sigma n} = (0, 0)^T$). The system is loaded by a shear stress on the right side ($\boldsymbol{\sigma n} = (0, 0.1)^T$).

It can be seen (in Figs. 9, 10, and 11), that the choice of $m = 0$ for the stress interpolation does not give good results for the standard formulation (with $\omega_3 = 0$), even if the interpolation of the displacement is of quartic order. Furthermore, the choice of $k = 1$ for the displacements is not recommendable. The best interpolation combinations for the displacements tested here are additonally depicted (and compared) in Fig. 12.

Furthermore, we want to investigate the influence of the third (redundant) residual $\mathcal{R}_3$ on the performance. Therefore, we consider the same boundary value problem, see Fig. 8, and vary the weighting factor $\omega_3 = 0, 1, 5, 10$, whereas $\omega_1 = \omega_2 = 1$. As an element type we choose RT_1P_2. The results are depicted in Fig. 13. Here, the crucial impact of a suitable choice of ω_3 can be seen. The best performance is reached using $\omega_3 = 5$ or $\omega_3 = 10$.

These findings can be confirmed considering the convergence of the squared $L^2(\mathcal{B})$-norms of the individual residuals ($||\mathcal{R}_i||^2_{L^2(\mathcal{B})}$) plotted over logarithmic scales,

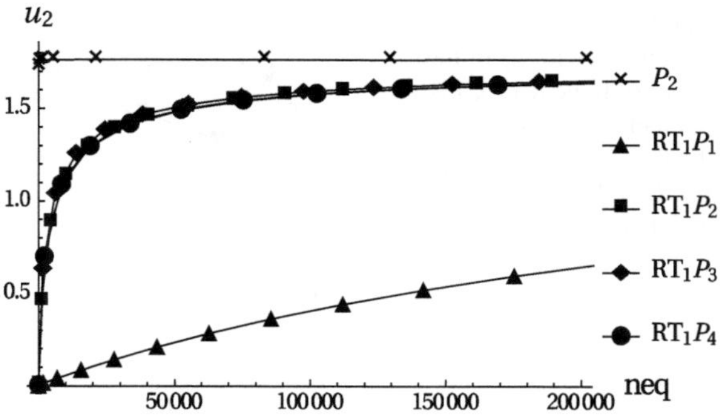

Fig. 10 Displacement convergence for u_2 of the upper right node (10, 1) over number of equations (neq) of the final system of equations for RT_1P_k

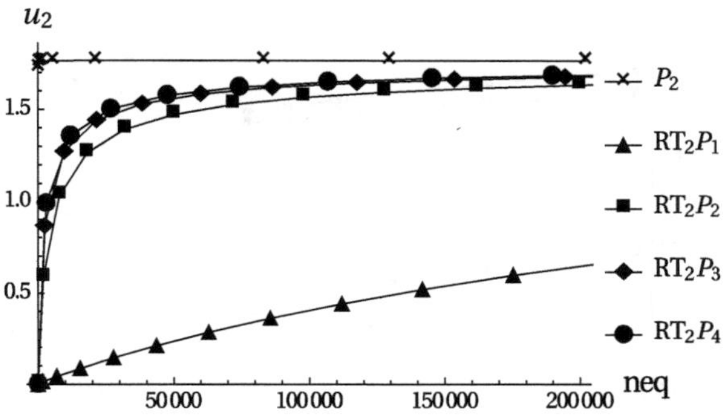

Fig. 11 Displacement convergence for u_2 of the upper right node (10, 1) over number of equations (neq) of the final system of equations for RT_2P_k

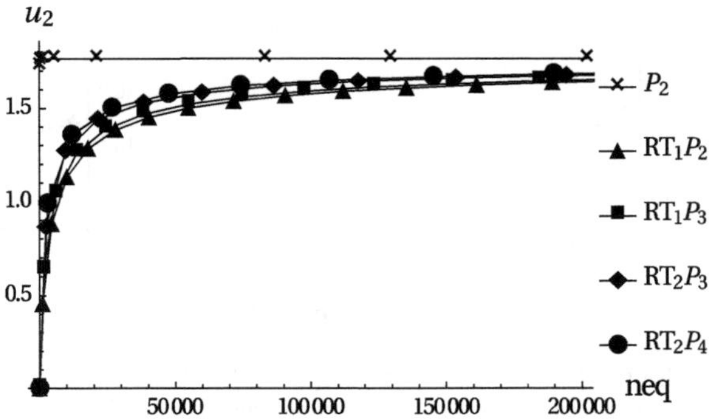

Fig. 12 Displacement convergence for u_2 of the upper right node (10, 1) over number of equations (neq) of the final system of equations for best tested combinations RT_mP_k

see Fig. 14. In the plot of the functional $\mathcal{F}$, see Fig. 15, it can be seen clearly, that the convergence order (the slope of the provided data) is steeper for $\omega_3 = 5$ and $\omega_3 = 10$ than for $\omega_3 = 0$ and $\omega_3 = 1$.

4.3 Cook's Membrane, Quasi-incompressible Elastic

As already seen in the last example, the provided formulation can give reliable results with respect to the displacement. In the following example, we want to concentrate

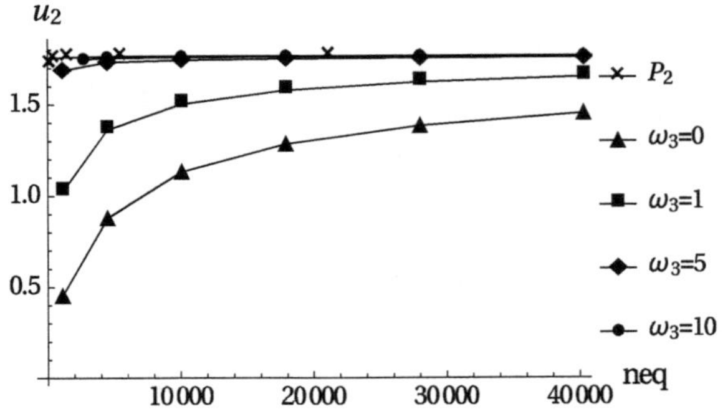

Fig. 13 Displacement convergence for u_2 of the upper right node (10, 1) over number of equations (neq) of the final system of equations for RT_1P_2 using different weights $\omega_3 = 0, 1, 5, 10$

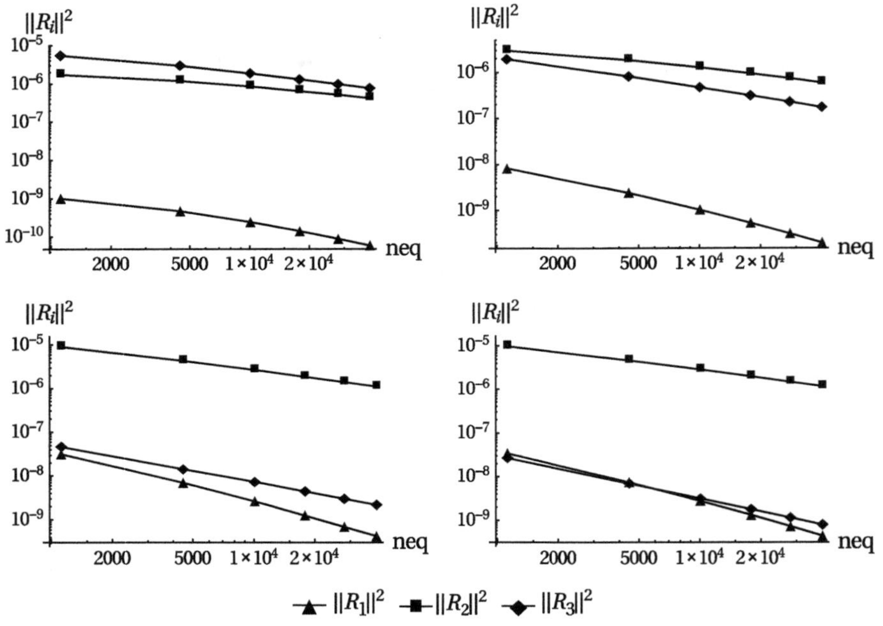

Fig. 14 Convergence for $||\mathcal{R}_i||^2_{L^2(\mathcal{B})}$ over number of equations (neq) of the final system of equations for $\omega_3 = 0$ (*upper left*), $\omega_3 = 1$ (*upper right*), $\omega_3 = 5$ (*lower left*), and $\omega_3 = 10$ (*lower right*)

on the stress approximation of the mixed least-squares finite element. Therefore, we consider the well-known Cook's Membrane problem for a quasi-incompressible material. This benchmark is well suited since it exhibits a stress singularity in the upper left corner. Furthermore, it is known that lower order standard methods produce for quasi-incompressible elastic material behavior an oscillating stress field. For comparison we again use the standard Galerkin quadratic triangular element (P_2) and furthermore a mixed Galerkin element of type T_2P_0 (triangular element, quadratic displacement, constant pressure). As a further result of our computation we will show the fulfillment of the functional of the least-squares formulation as a contour plot over

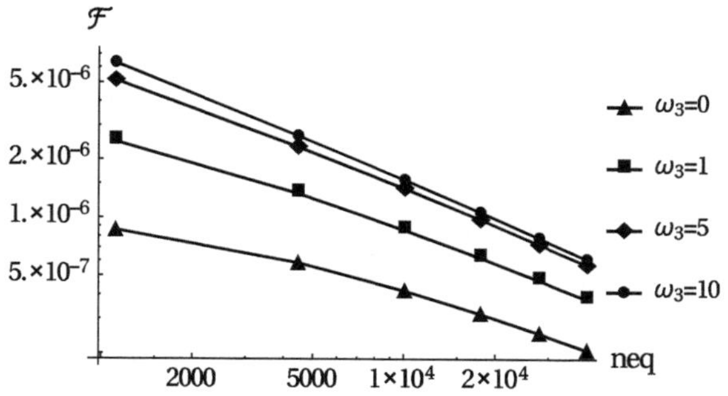

Fig. 15 Convergence for $\mathcal{F}$ over number of equations (neq) of the final system of equations for different weights ω_i

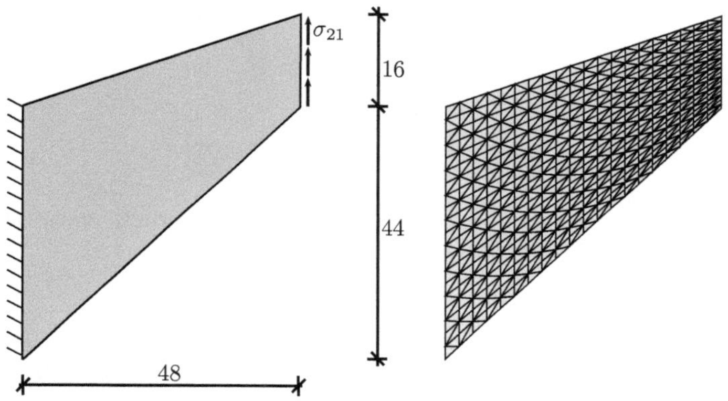

Fig. 16 Geometry and mesh (20 × 20 elements) of the Cook's membrane

the domain. In the geometry with dimensions of 48 × 60 the left side is assumed to be a displacement boundary and clamped. The other edges are stress boundaries with the right edge $\boldsymbol{\sigma n} = (0, 1)$ and stress-free boundary conditions at the upper and lower edges. In order to obtain a quasi-incompressible material behavior we apply Poisson's ratio as $\nu = 0.499$. The Young's modulus is chosen as $E = 200$. The conversion to the Lamé parameters yields $\lambda = 33288.86$ and $\mu = 66.71$. Here, the weights ω_i are chosen as $\omega_1 = \omega_2 = \omega_3 = 1$. The geometry and the utilized structured mesh with 800 elements are shown in Fig. 16.

In Figs. 17 and 18 the results for the stress component σ_{11} are plotted as an "out of the plane" value. Using this graphical representation, it can be seen clearly that the standard Galerkin formulation (P_2) shows oscillating stresses over the domain, whereas the mixed Galerkin formulation (T_2P_0) as well as the least-squares mixed finite element (RT_1P_2) provide a smooth stress approximation. Furthermore, quantitatively, the results are of the same size. Due to this graphical representation of the stress, the discontinuous stresses for the Galerkin elements and in contrast to that the continuous stress interpolation (for the normal components) for the least-squares element become viewable. In Fig. 18 (right) the distribution of the functional (error) over the domain is shown, which could be used, e.g., for an adaptive mesh refinement. It can be seen clearly that the functional satisfies the condition $\mathcal{F} \to 0$ in an

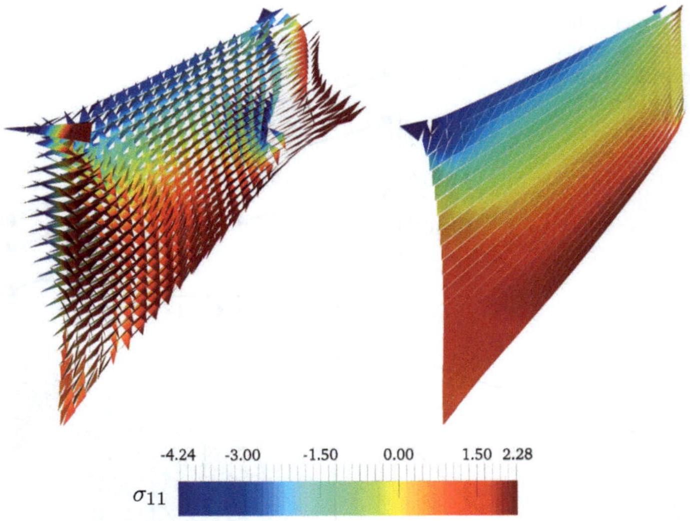

Fig. 17 Plot of the σ_{11} stresses for a quasi-incompressible elastic material for the standard (*left*, P_2) and the mixed (*right*, T_2P_0) Galerkin elements

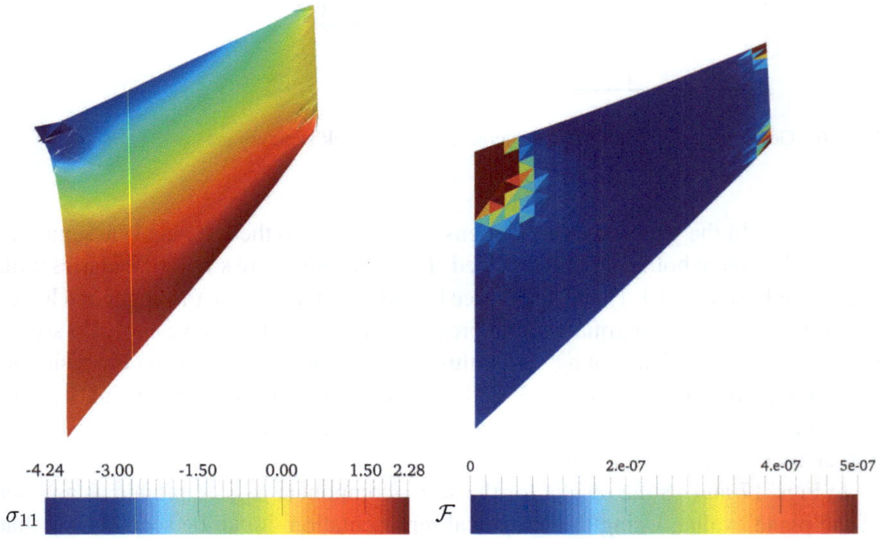

Fig. 18 Plot of the σ_{11} stresses for a quasi-incompressible elastic material and of the functional value for the mixed least-squares finite element (RT_1P_2)

appropriate way over the whole domain. In the area of the singularity, higher values show up.

5 LSFEM–Hyperelasticity

Materials which behave purely elastic also in the case of large strain are called hyperelastic. In the framework of this contribution we will consider a free energy function of neo-Hookean type in order to describe the stress response of the material. For the development of a hyperelastic least-squares formulation, we follow the general rules described in Sect. 3.1. As a starting point for the construction of a least-squares functional for hyperelasticity, we define the residuals

$$\begin{aligned} \mathcal{R}_1 &= \qquad \operatorname{Div} \boldsymbol{P} + \boldsymbol{f} = \boldsymbol{0} \rightarrow \text{balance of momentum}, \\ \mathcal{R}_2 &= \quad \boldsymbol{P} - \rho_0 \partial_F \psi(\boldsymbol{C}) = \boldsymbol{0} \rightarrow \text{constitutive relation}, \\ \mathcal{R}_3 &= \boldsymbol{F}^{-1}\boldsymbol{P} - (\boldsymbol{F}^{-1}\boldsymbol{P})^T = \boldsymbol{0} \rightarrow \text{stress symmetry} \end{aligned} \tag{111}$$

where $\boldsymbol{f}$ denotes the body force. Following (51), we obtain a general least-squares functional for hyperelasticity with the solution quantities displacements and first Piola–Kirchhoff stress tensor as

$$\begin{aligned} \mathcal{F}(\boldsymbol{P}, \boldsymbol{u}) = \frac{1}{2} \int_{\mathcal{B}} \omega_1^2 (\operatorname{Div} \boldsymbol{P} + \boldsymbol{f}) \cdot (\operatorname{Div} \boldsymbol{P} + \boldsymbol{f})\, dV \\ + \frac{1}{2} \int_{\mathcal{B}} \omega_2^2 (\boldsymbol{P} - \rho_0 \partial_F \psi(\boldsymbol{C})) : (\boldsymbol{P} - \rho_0 \partial_F \psi(\boldsymbol{C}))\, dV \\ + \frac{1}{2} \int_{\mathcal{B}} \omega_3^2 (\boldsymbol{F}^{-1}\boldsymbol{P} - (\boldsymbol{F}^{-1}\boldsymbol{P})^T) : (\boldsymbol{F}^{-1}\boldsymbol{P} - (\boldsymbol{F}^{-1}\boldsymbol{P})^T)\, dV, \end{aligned} \tag{112}$$

which has to be minimized. Therefore, we need the first variations to be zero leading to

$$\begin{aligned} \delta_u \mathcal{F} &= \sum_{i=1}^{3} \int_{\mathcal{B}} \omega_i^2 \delta_u \mathcal{R}_i \cdot \mathcal{R}_i \, dV = 0, \\ \delta_P \mathcal{F} &= \sum_{i=1}^{3} \int_{\mathcal{B}} \omega_i^2 \delta_P \mathcal{R}_i \cdot \mathcal{R}_i \, dV = 0, \end{aligned} \tag{113}$$

with

$$\delta \boldsymbol{u} = 0 \text{ on } \partial \mathcal{B}_u \text{ and } \delta \boldsymbol{P} = 0 \text{ on } \partial \mathcal{B}_P. \tag{114}$$

The linearization for the application of a Newton scheme yields

$$
\begin{aligned}
\Delta_u \delta_u \mathcal{F} &= \sum_{i=1}^{3} \int_{\mathcal{B}} \omega_i^2 (\Delta_u \delta_u \mathcal{R}_i \cdot \mathcal{R}_i + \delta_u \mathcal{R}_i \cdot \Delta_u \mathcal{R}_i)\, dV, \\
\Delta_P \delta_u \mathcal{F} &= \sum_{i=1}^{3} \int_{\mathcal{B}} \omega_i^2 (\Delta_P \delta_u \mathcal{R}_i \cdot \mathcal{R}_i + \delta_u \mathcal{R}_i \cdot \Delta_P \mathcal{R}_i)\, dV, \\
\Delta_u \delta_P \mathcal{F} &= \sum_{i=1}^{3} \int_{\mathcal{B}} \omega_i^2 (\Delta_u \delta_P \mathcal{R}_i \cdot \mathcal{R}_i + \delta_P \mathcal{R}_i \cdot \Delta_u \mathcal{R}_i)\, dV, \\
\Delta_P \delta_P \mathcal{F} &= \sum_{i=1}^{3} \int_{\mathcal{B}} \omega_i^2 (\Delta_P \delta_P \mathcal{R}_i \cdot \mathcal{R}_i + \delta_P \mathcal{R}_i \cdot \Delta_P \mathcal{R}_i)\, dV.
\end{aligned}
\tag{115}
$$

Here, the nontrivial variations are given by

$$
\begin{aligned}
\delta_P \mathcal{R}_1 &= \mathrm{Div}\, \delta \boldsymbol{P}, \\
\delta_u \mathcal{R}_2 &= -\rho_0 \partial^2_{FF} \psi(\boldsymbol{C}) \delta \boldsymbol{F}, \quad \delta_P \mathcal{R}_2 = \delta \boldsymbol{P}, \\
\delta_u \mathcal{R}_3 &= \delta \boldsymbol{F}^{-1} \boldsymbol{P} - (\delta \boldsymbol{F}^{-1} \boldsymbol{P})^T, \\
\delta_P \mathcal{R}_3 &= \boldsymbol{F}^{-1} \delta \boldsymbol{P} - (\boldsymbol{F}^{-1} \delta \boldsymbol{P})^T
\end{aligned}
\tag{116}
$$

and the associated linear increments appear as

$$
\begin{aligned}
\Delta_P \mathcal{R}_1 &= \mathrm{Div}\, \Delta \boldsymbol{P}, \\
\Delta_u \mathcal{R}_2 &= -\rho_0 \partial^2_{FF} \psi(\boldsymbol{C}) \Delta \boldsymbol{F}, \quad \Delta_P \mathcal{R}_2 = \Delta \boldsymbol{P}, \\
\Delta_u \mathcal{R}_3 &= \Delta \boldsymbol{F}^{-1} \boldsymbol{P} - (\Delta \boldsymbol{F}^{-1} \boldsymbol{P})^T, \\
\Delta_P \mathcal{R}_3 &= \boldsymbol{F}^{-1} \Delta \boldsymbol{P} - (\boldsymbol{F}^{-1} \Delta \boldsymbol{P})^T.
\end{aligned}
\tag{117}
$$

The nonvanishing mixed terms for the second variation are

$$
\begin{aligned}
\Delta_u \delta_u \mathcal{R}_2 &= -\partial_F (\partial^2_{FF} \psi(\boldsymbol{C}) \delta \boldsymbol{F}) \Delta \boldsymbol{F}, \\
\Delta_u \delta_u \mathcal{R}_3 &= \Delta \delta \boldsymbol{F}^{-1} \boldsymbol{P} - (\Delta \delta \boldsymbol{F}^{-1} \boldsymbol{P})^T, \\
\Delta_P \delta_u \mathcal{R}_3 &= \delta \boldsymbol{F}^{-1} \Delta \boldsymbol{P} - (\delta \boldsymbol{F}^{-1} \Delta \boldsymbol{P})^T, \\
\Delta_u \delta_P \mathcal{R}_3 &= \Delta \boldsymbol{F}^{-1} \delta \boldsymbol{P} - (\Delta \boldsymbol{F}^{-1} \delta \boldsymbol{P})^T.
\end{aligned}
\tag{118}
$$

5.1 Isotropic Hyperelasticity

For the description of the stress response of the materials, we use the derivative of a free energy function based on the invariants of the right Cauchy–Green deformation tensor $\boldsymbol{C}$, see (31). For the finite element formulation in Sects. 5.2 and 5.3, we consider an isotropic free energy function of neo-Hookean type given as

$$\psi_{NH}^{iso} = \frac{\lambda}{4}(I_3 - 1) - \left(\frac{\lambda}{2} + \mu\right)\ln(\sqrt{I_3}) + \frac{\mu}{2}(I_1 - 3)\,, \tag{119}$$

see Schwarz (2009). The first Piola–Kirchhoff stress tensor is given by the derivative of the free energy with respect to the deformation gradient $\boldsymbol{F}$, see also (19). We obtain for (119) the first Piola–Kirchhoff stress tensor as

$$\boldsymbol{P} = \frac{\lambda}{2}(I_3 - 1)\,\boldsymbol{F}^{-T} + \mu\left(\boldsymbol{F} - \boldsymbol{F}^{-T}\right). \tag{120}$$

5.2 3D Plate, Hyperelastic

We have a look on the boundary value problem of a three-dimensional plate. Due to symmetry conditions we reduce it to one fourth of the domain under consideration of symmetry boundary conditions. The reduced geometry is clamped at two sides, see Fig. 19.

At the symmetry sides the tangential entries of the local traction vector and the normal displacements are set to zero. The upper and lower faces are stress-free with $\boldsymbol{PN} = (0, 0, 0)^T$. As a load we apply a body force $\boldsymbol{f} = (0, 0, -2)^T$. The material parameters are chosen as $E = 200$ and $\nu = 0.35$ and the weighting factors are $\omega_1 = 1, \omega_2 = \omega_3 = 1/\mu$. Figure 20 shows a convergence study for the vertical displacement at the point $(10, 10, 1)$. Here, we compare the performance of a least-squares element (RT_0P_2) with the a ten-noded standard Galerkin finite element (P_2). It can be seen that the provided least-squares element performs well for the given boundary value problem.

Furthermore, a plot of the P_{11} stresses for both elements is provided in Fig. 21. Unless the low interpolation order for the stresses, the result for the least-squares element is in line with the solution computed by the Galerkin element.

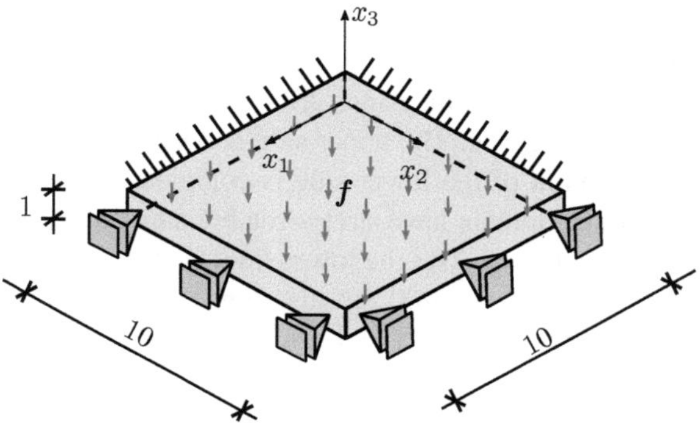

Fig. 19 Geometry of the clamped plate

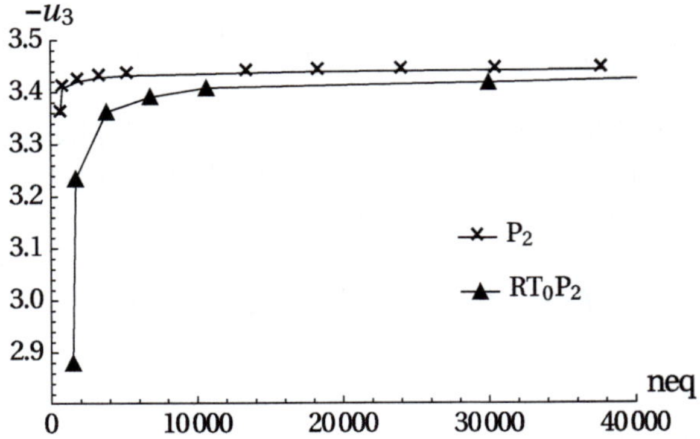

Fig. 20 Displacement convergence for u_2 of node (10, 10, 1) over the number of equations (neq) of the final system of equations for RT_0P_2 and P_2

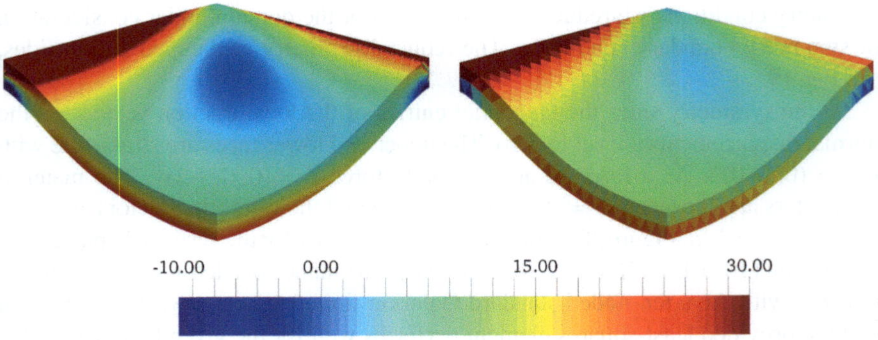

Fig. 21 Stress component P_{11} on the deformed shape of the clamped plate under body force for P_2 *(left)* and RT_0P_2 *(right)*

5.3 Compression Test, Quasi-incompressible, Hyperelastic

As a further example we want to consider a two-dimensional compression test. The material is chosen to be nearly incompressible ($E = 240.566$, $\nu = 0.498$). The domain with the dimensions 20×10 has a displacement bound in vertical direction for the lower side. The upper side is bounded in horizontal direction. Additionally, the midpoint of the lower side is completely fixed $\boldsymbol{u} = (0, 0)^T$. The left and right side of the domain have stress-free boundaries ($\boldsymbol{PN} = (0, 0)^T$), as well as the shear component $P_{12} = 0$ at the lower side. On the upper side the P_{22} stresses are given as $P_{22} = 0$ or $P_{22} = -400$, see also Fig. 22. The weighting factors are chosen as in the previous example as $\omega_1 = 1$, $\omega_2 = \omega_3 = 1/\mu$.

We consider a least-squares element of the type RT_2P_3 with $neq = 47{,}999$. Therefore, we choose a structured (and symmetric) mesh with 40×40 triangular elements to avoid possible influences of mesh anisotropy, as shown in the deformed configuration in Fig. 23. For comparison we use a mixed Galerkin element (T_2P_0, triangular

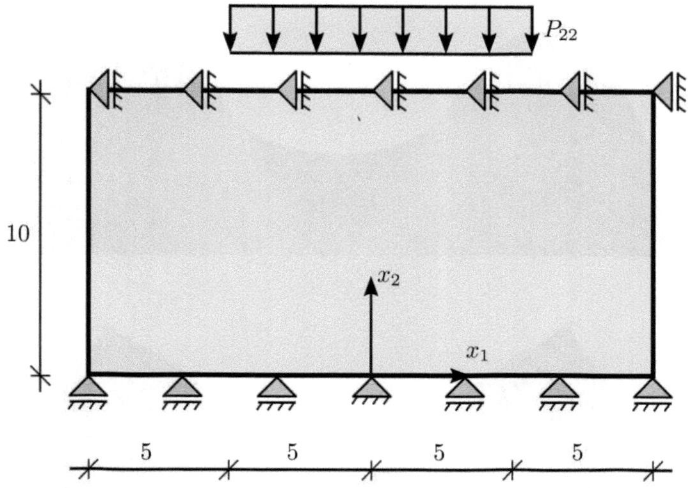

Fig. 22 Geometry of the compression test

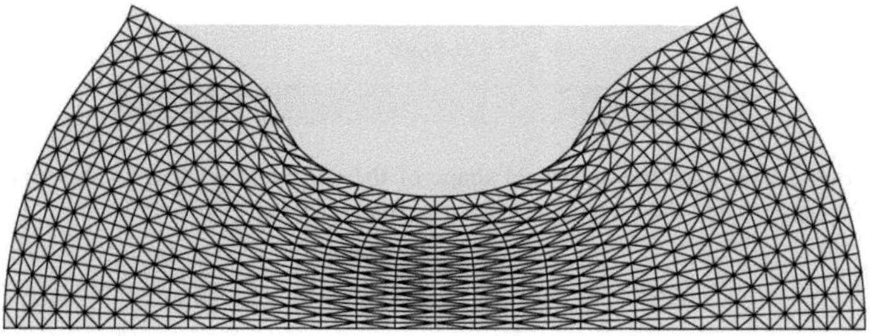

Fig. 23 Mesh for RT_2P_3 of the compression test for the deformed configuration

element, quadratic displacement, constant pressure) with $neq = 48{,}620$. Both elements lead to the convergent solution of $u_2 \approx 5.67$ for the vertical displacement of the upper midpoint with the coordinates (0, 10). Furthermore, we check the fulfillment of the incompressibility constraint. Therefore, the distribution of $\det \boldsymbol{F}$ is plotted over the domain for both elements, see Fig. 24. Here, both elements obtain similar results and show a nearly volume preserving behavior.

5.4 *Transverse Isotropic Hyperelasticity*

As the isotropic basis of the material behavior the free energy given in equation (119) is used for the proposed formulation. Adding a transverse isotropic part in terms of the mixed invariant J_4, see (37) and (41), we obtain the transverse isotropic free energy

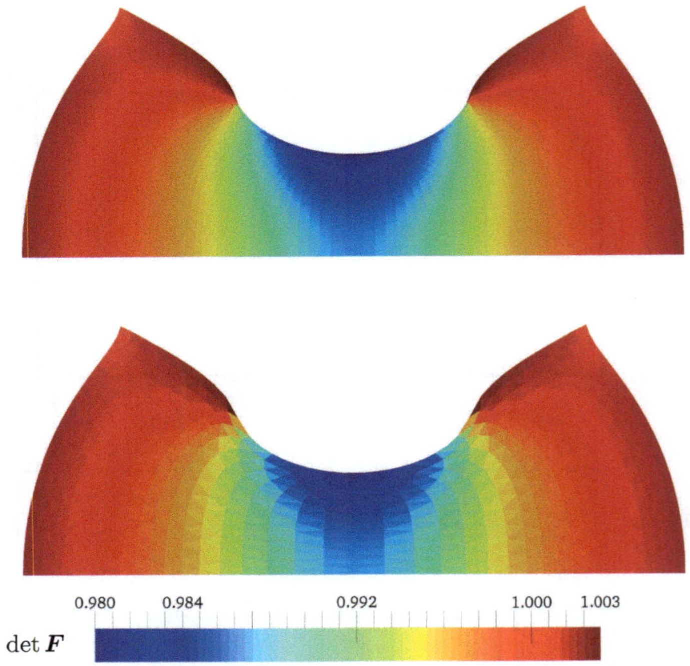

Fig. 24 Plot of $\det \boldsymbol{F}$ over the deformed shape of the compression test for the mixed Galerkin element T_2P_0 (*upper*) and the mixed least-squares element RT_2P_3 (*lower*)

$$\psi_{NH}^{ti} = \frac{\lambda}{4}(I_3 - 1) - \left(\frac{\lambda}{2} + \mu\right)\ln(\sqrt{I_3}) + \frac{\mu}{2}(I_1 - 3) + \alpha_1 \langle J_4 - 1 \rangle^{\alpha_2} \tag{121}$$

with the definition of the Macauly brackets

$$\langle \beta \rangle := \frac{1}{2}(\beta + |\beta|) \tag{122}$$

and the requirement of the parameters $\alpha_1 \geq 0$ and $\alpha_2 > 1$, see Balzani et al. (2006). The derivative of the free energy (121) with respect to $\boldsymbol{F}$, $\partial_{\boldsymbol{F}} \psi_{NH}^{ti}$, yields

$$\boldsymbol{P} = \frac{\lambda}{2}(I_3 - 1)\boldsymbol{F}^{-T} + \mu\left(\boldsymbol{F} - \boldsymbol{F}^{-T}\right) + 2\boldsymbol{F}\alpha_2\alpha_1\langle J_4 - 1 \rangle^{\alpha_2 - 1}\boldsymbol{M}. \tag{123}$$

5.5 Cantilever Beam, Transverse Isotropic, Hyperelastic

In this numerical example, we want to investigate the influence of a transversely isotropic material behavior on the vertical displacement at the point with the coordinates (10, 1) of the cantilever beam, see also Fig. 25.

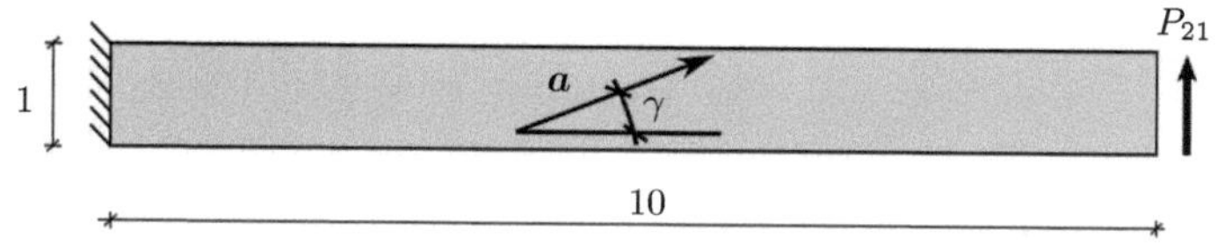

Fig. 25 Geometry of the cantilever beam

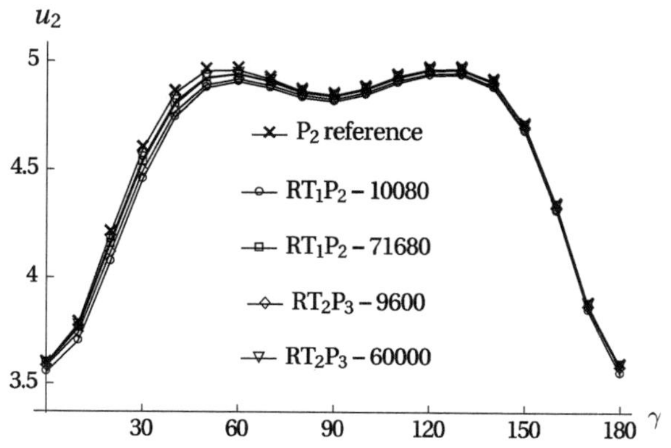

Fig. 26 Displacement of the upper right node of the cantilever beam over different angles γ denoting the preferred direction

Therefore, we consider different values for the angle of the preferred direction $0° \leq \gamma \leq 180°$ leading to $\boldsymbol{a} = (\cos[\gamma], \sin[\gamma])^T$. The domain of consideration has the dimensions 10×1. The left side of the beam has a fixed displacement boundary ($u_1 = u_2 = 0$). The local traction vectors of the upper and lower edges are assumed to be stress-free leading to the essential boundary conditions ($\boldsymbol{PN} = (0, 0)^T$). The system is loaded by a shear stress on the right side ($\boldsymbol{PN} = (0, 0.1)^T$).

Since we consider here a compressible material we use a quadratic standard Galerkin element (P_2) as a reference solution. For the least-squares mixed finite element formulation, we consider the element types RT_1P_2 as well as RT_2P_3. Due to the outcome of the example in Sect. 4.2, we choose the weights $\omega_1 = 1$, $\omega_2 = 1/\mu$ and $\omega_3 = 10/\mu$. The Young's modulus is set to $E = 200$ and the Poisson's ratio to $\nu = 0.35$. The material parameters with respect to the transverse isotropy are chosen as $\alpha_1 = 10{,}000$ and $\alpha_2 = 4$.

Figure 26 shows that both element types (RT_1P_2, RT_2P_3) almost reach the reference solution for different number of equations of the final system (stated in the legend).

Acknowledgments The authors would like to thank the German Research Foundation (DFG) for financial support: research grant SCHR 570/14-1. Furthermore, the authors are grateful to Gerhard Starke and Benjamin Müller with whom they have shared inspiring work on least-squares finite element methods. Special thanks go to Nils Viebahn for many hours of support especially for the implementation and visualization and to Maximilian Igelbüscher, Carina Nisters, and Serdar Serdas for fruitfull discussions.

References

Ahrens, J., Geveci, B., & Law, C. (2005). *ParaView: An end-user tool for large data visualization, visualization handbook*. Champaign: Elsevier. Version 10.1 edition.

Antman, S. S. (1995). *Nonlinear problems of elasticity*. New York: Springer.

Babuška, I. (1973). The finite element method with lagrangian multipliers. *Numerische Mathematik, 20*(3), 179–192.

Ball, J. M. (1977a). Convexity conditions and existence theorems in non-linear elasticity. *Archive of Rational Mechanics and Analysis, 63*, 337–403.

Ball, J. M. (1977b). Constitutive inequalities and existence theorems in nonlinear elastostatics. In R. J. Knops (Ed.), *Herriot Watt symposion: Nonlinear analysis and mechanics* (Vol. 1, pp. 187–238). London: Pitman.

Balzani, D. (2006) Polyconvex anisotropic energies and modeling of damage applied to arterial walls. *Ph.D. thesis*. University Duisburg-Essen, Verlag Glückauf Essen.

Balzani, D., Neff, P., Schröder, J., & Holzapfel, G. A. (2006). A polyconvex framework for soft biological tissues. Adjustment to experimental data. *International Journal for Numerical Methods in Engineering, 43*(20), 6052–6070.

Bathe, K. J. (1995). *Finite Element Procedures*. Prentice: Prentice Hall.

Bathe, K.-J. (2001). The inf-sup condition and its evaluation for mixed finite element methods. *Computers and Structures, 79*(2), 243–252.

Bertrand, F., Münzenmaier, S., & Starke, G. (2014). First-order system least squares on curved boundaries. *SIAM Journal on Scientific Computing, 52*, 880–894.

Bochev, P., & Gunzburger, M. (2009). *Least-squares finite element methods*. New York: Springer.

Boehler, J. P. (1978). Lois de comportement anisotrope des milieux continus. *Journal de Mécanique, 17*(2), 153–190.

Boehler, J. P. (1979). A simple derivation of representations for non-polynomial constitutive equations in some cases of anisotropy. *Zeitschrift für angewandte Mathematik und Mechanik, 59*, 157–167.

Boehler, J. P. (1987). Introduction to the invariant formulation of anisotropic constitutive equations. In J. P. Boehler (Ed.), *Applications of tensor functions in solid mechanics* (Vol. 292, pp. 13–30). CISM courses and lectures. Vienna: Springer.

Braess, D. (1997). *Finite Elemente* (2nd ed.). Berlin: Springer.

Brezzi, F. (1974). On the existence, uniqueness and approximation of saddle-point problems arising from lagrangian multipliers. *Revue française d'automatique, informatique, recherche opérationnelle. Analyse numérique, 8*(2), 129–151.

Cai, Z., & Starke, G. (2003). First-order system least squares for the stress-displacement formulation: Linear elasticity. *SIAM Journal on Numerical Analysis, 41*, 715–730.

Cai, Z., & Starke, G. (2004). Least-squares methods for linear elasticity. *SIAM Journal on Numerical Analysis, 42*, 826–842.

Cai, Z., & Westphal, C. R. (2009). An adaptive mixed least-squares finite element method for viscoelastic fluids of Oldroyd type. *Journal of Non-Newtonian Fluid Mechanics, 159*, 72–80.

Cai, Z., Manteuffel, T. A., & McCormick, S. F. (1995). First-order system least squares for velocity-vorticity-pressure form of the Stokes equations, with application to linear elasticity. *Electronic Transactions on Numerical Analysis, 3*, 150–159.

Cai, Z., Manteuffel, T. A., & McCormick, S. F. (1997). First-order system least squares for the Stokes equation, with application to linear elasticity. *SIAM Journal on Numerical Analysis, 34*, 1727–1741.

Cai, Z., Manteuffel, T. A., McCormick, S. F., & Parter, S. V. (1998). First-order system least squares (FOSLS) for planar linear elasticity: Pure traction problem. *SIAM Journal on Numerical Analysis, 35*, 320–335.

Cai, Z., Lee, C.-O., Manteuffel, T. A., & McCormick, S. F. (2000a). First-order system least squares for linear elasticity: Numerical results. *SIAM Journal on Scientific Computing, 21*, 1706–1727.

Cai, Z., Lee, C.-O., Manteuffel, T. A., & McCormick, S. F. (2000b). First-order system least squares for the Stokes and linear elasticity equations: Further results. *SIAM Journal on Scientific Computing*, *21*, 1728–1739.

Ciarlet, P. G. (1988). *Mathematical elasticity: Three-dimensional elasticity* (Vol. I). Amsterdam: Elsevier Science Ltd.

Ciarlet, P. G. (1991). *Handbook of numerical analysis, Vol II: Finite element methods, Part 1*. Amsterdam: Elsevier Science Ltd.

Dacorogna, B. (1989). *Direct methods in the calculus of variations* (1st ed., Vol. 78). Applied mathematical sciences. Berlin: Springer.

de Boer, R. (1993). *Vektor- und Tensorrechnung für Ingenieure*. Berlin: Springer.

Deang, J. M., & Gunzburger, M. D. (1998). Issues related to least-squares finite element methods for the Stokes equations. *SIAM Journal on Scientific Computing*, *20*(3), 878–906.

Eason, E. D. (1976). A review of least-squares methods for solving partial differential equations. *International Journal for Numerical Methods in Engineering*, *10*, 1021–1046.

Ebbing, V. (2010). Design of polyconvex energy functions for all anisotropy classes. *Ph.D. thesis*, Institut für Mechanik, Abteilung Bauwissenschaften, Fakultät für Ingenieurwissenschaften, Universität Duisburg-Essen.

Ern, A., & Guermond, J.-L. (2013). *Theory and practice of finite elements* (Vol. 159). New York: Springer Science & Business Media.

Jiang, B.-N. (1998). *The least-squares finite element method*. Berlin: Springer.

Jiang, B.-N., & Wu, J. (2002). The least-squares finite element method in elasticity. Part I: Plane stress or strain with drilling degrees of freedom. *International Journal for Numerical Methods in Engineering*, *53*, 621–636.

Korelc, J. (1997). Automatic generation of finite-element code by simultaneous optimization of expressions. *Theoretical Computer Science*, *187*, 231–248.

Korelc, J. (2002). Multi-language and multi-environment generation of nonlinear finite element codes. *Engineering with Computers*, *18*, 312–327.

Ladyzhenskaya, O. A. (1969). *The mathematical theory of viscous incompressible flow* (Vol. 76). New York: Gordon and Breach.

Lynn, P. P., & Arya, S. K. (1973). Use of the least squares criterion in the finite element formulation. *International Journal for Numerical Methods in Engineering*, *6*, 75–88.

Manteuffel, T. A., McCormick, S. F., Schmidt, J. G., & Westphal, C. R. (2006). First-order system least squares (FOSLS) for geometrically nonlinear elasticity. *SIAM Journal on Numerical Analysis*, *44*, 2057–2081.

Marsden, J. E., & Hughes, J. R. (1983). *Mathematical foundations of elasticity*. Prentice: Prentice-Hall.

Morrey, C. B. (1952). Quasi-convexity and the lower semicontinuity of multiple integrals. *Pacific Journal of Mathematics*, *2*, 25–53.

Morrey, C. B. (1966). *Multiple integrals in the calculus of variations*. Berlin: Springer.

Müller, B., Starke, G., Schwarz, A., & Schröder, J. (2014). A first-order system least squares method for hyperelasticity. *SIAM Journal on Scientific Computing*, *36*, 795–816.

Neumann, F. E. (1885). *Vorlesungen über die Theorie der Elastizität der festen Körper und des Lichtäthers*. Leipzig: Teubner.

Pontaza, J. P. (2003). Least-squares variational principles and the finite element method: theory, form, and model for solid and fluid mechanics. *Ph.D. thesis*. Texas A&M University.

Pontaza, J. P., & Reddy, J. N. (2003). Spectral/*hp* least-squares finite element formulation for the Navier-Stokes equation. *Journal of Computational Physics*, *190*, 523–549.

Raviart, P. A., & Thomas, J. M. (1977). A mixed finite element method for 2nd order elliptic problems. *Mathematical aspects of finite element methods*. Lecture notes in mathematics. New York: Springer.

Schröder, J. (2010). Anisotropic polyconvex energies. In J. Schröder & P. Neff (Eds.), *Poly-, quasi- and rank-one convexity in applied mechanics* (Vol. 516, pp. 53–106). CISM courses and lectures. Vienna: Springer.

Schröder, J., & Neff, P. (2001). On the construction of polyconvex anisotropic free energy functions. In Miehe, C. (Ed.) *Proceedings of the IUTAM Symposium Computational Mechanics of Solid Materials at Large Strains* (pp. 171–180). Kluwer Academic Publishers.

Schröder, J., & Neff, P. (2003). Invariant formulation of hyperelastic transverse isotropy based on polyconvex free energy functions. *International Journal of Solids and Structures*, *40*, 401–445.

Schröder, J., Neff, P., & Ebbing, V. (2008). Anisotropic polyconvex energies on the basis of crystallographic motivated structural tensors. *Journal of the Mechanics and Physics of Solids*, *56*(12), 3486–3506.

Schwarz, A. (2009). Least-squares mixed finite elements for solid mechanics. *Ph.D. thesis*, University Duisburg-Essen.

Schwarz, A., & Schröder, J. (2007). Least-squares mixed finite elements with applications to anisotropic elasticity and viscoplasticity. *Proceedings of Applied Mathematics and Mechanics*, *7*, 4040043–4040044.

Schwarz, A., Schröder, J., & Starke, G. (2009). Least-squares mixed finite elements for small strain elasto-viscoplasticity. *International Journal for Numerical Methods in Engineering*, *77*, 1351–1370.

Schwarz, A., Schröder, J., & Starke, G. (2010). A modified least-squares mixed finite element with improved momentum balance. *International Journal for Numerical Methods in Engineering*, *81*, 286–306.

Schwarz, A., Steeger, K., & Schröder, J. (2014). Weighted overconstrained least-squares mixed finite elements for static and dynamic problems in quasi-incompressible elasticity. *Computational Mechanics*, *54*(1), 603–612.

Schwarz, A., Steeger, K., & Schröder, J. (2015). Weighted overconstrained least-squares mixed finite elements for hyperelasticity. *Proceedings of Applied Mathematics and Mechanics*, *15*, 227–228.

Smith, G. F. (1970). On a fundamental error in two papers of C.-C. Wang, "On representations for isotropic functions, Parts I and II". *Archive for Rational Mechanics and Analysis*, *36*, 161–165.

Smith, G. F. (1971). On isotropic functions of symmetric tensors, skew-symmetric tensors and vectors. *International Journal of Engineering Science*, *9*, 899–916.

Smith, G. F., Smith, M. M., & Rivlin, R. S. (1963). Integrity bases for a symmetric tensor and a vector. The crystal classes. *Archive for Rational Mechanics and Analysis*, *12*, 93–133.

Spencer, A. J. M. (1965). Isotropic integrity bases for vectors and second-order tensors. *Archive for Rational Mechanics and Analysis*, *18*, 51–82.

Spencer, A. J. M. (1971). Theory of invariants. In A. C. Eringen (Ed.), *Continuum physics* (Vol. 1, pp. 239–353). New-York: Academic Press.

Tchonkova, M., & Sture, S. (1997). A mixed least squares method for solving problems in linear elasticity: Formulation and initial results. *Computational Mechanics*, *19*, 317–326.

Tchonkova, M., & Sture, S. (2002). A mixed least squares method for solving problems in linear elasticity: Theoretical study. *Computational Mechanics*, *29*, 332–339.

Truesdell, C., & Noll, W. (1965). The nonlinear field theories of mechanics. In S. Flügge (Ed.), *Handbuch der Physik III/3*. Berlin: Springer.

Šilhavý, M. (1997). *The mechanics and thermodynamics of continuous media*. Berlin: Springer.

Wang, C.-C. (1969a). On representations for isotropic functions. Part I. Isotropic functions of symmetric tensors and vectors. *Archive for Rational Mechanics and Analysis*, *33*, 249–267.

Wang, C.-C. (1969b). On representations for isotropic functions. Part II. Isotropic functions of skew-symmetric tensors, symmetric tensors, and vectors. *Archive for Rational Mechanics and Analysis*, *33*, 268–287.

Wang, C.-C. (1970a). A new representation theorem for isotropic functions: An answer to professor G. F. Smith's Criticism of my papers on representations for isotropic functions. Part 1. Scalar-valued isotropic functions. *Archive for Rational Mechanics and Analysis*, *36*, 166–197.

Wang, C.-C. (1970b). A new representation theorem for isotropic functions: An answer to professor G. F. Smith's criticism of my papers on representations for isotropic functions. Part 2. Vector-valued isotropic functions, symmetric tensor-valued isotropic functions, and skew-symmetric tensor-valued isotropic functions. *Archive for Rational Mechanics and Analysis*, *36*, 198–223.

Wang, C.-C. (1971). Corrigendum to my recent papers on "representations for isotropic functions". *Archive for Rational Mechanics and Analysis*, *43*, 392–395.

Wolfram Research Inc. (2015). *Mathematica*. Champaign: Wolfram Research, Inc. Version 10.1 edition.

Wriggers, P. (2001). *Nichtlineare Finite-Element-Methoden*. Berlin: Springer.

Zheng, Q.-S. (1994). Theory of representations for tensor functions - a unified invariant approach to constitutive equations. *Applied Mechanics Reviews*, *47*, 545–587.

Zhu, J. Z., Taylor, R. L., & Zienkiewicz, O. C. (2005). *The finite element method: its basis and fundamentals*. Oxford: Butterworth Heinemann.

Zienkiewicz, O. C., Owen, D. R., & Lee, K. N. (1974). Least square-finite element for elastostatic problems. Use of 'reduced' integration. *International Journal for Numerical Methods in Engineering*, *8*, 341–358.

Theoretical and Numerical Elastoplasticity

Batmanathan Dayanand Reddy

Abstract This chapter presents an overview of the theory of classical elastoplasticity and associated variational problems. The flow theory is presented as a normality relation for a convex yield function, or equivalently in terms of the dissipation function. The latter formulation provides the basis for the variational theory, for which results on well-posedness are presented. Predictor-corrector algorithms based on the time-discrete problem are reviewed. Aspects of the large-deformation theory, including algorithmic aspects, are also presented.

1 Introduction

The objective of this chapter is to present a reasonably self-contained overview of the theory of elastoplasticity, including aspects of the variational problems, discrete approximations, and associated solution algorithms. Both the small- and large-deformation theories are considered.

There are a number of works that deal in depth with the basic aspects of elastoplasticity presented in this chapter. For further background on classical plasticity the reader is referred to the texts by Gurtin et al. (2010) and Lubliner (1990). Comprehensive treatments of computational plasticity may be found for example in the works by de Souza Neto et al. (2008), Simo and Hughes (1998), while the monograph by Han and Reddy (2013) treats mathematical aspects as well as numerical analysis of problems of small-strain plasticity.

B.D. Reddy (✉)
Centre for Research in Computational and Applied Mechanics,
University of Cape Town, Rondebosch, South Africa
e-mail: daya.reddy@uct.ac.za

J. Schröder and P. Wriggers (eds.), *Advanced Finite Element Technologies*,
CISM International Centre for Mechanical Sciences 566,
DOI 10.1007/978-3-319-31925-4_7

1.1 *Elastic–Plastic Behaviour in One Dimension*

Consider the stress–strain behaviour of a bar in uniaxial stress as shown in Fig. 1. Starting at the origin, the path OA is purely elastic. At $\sigma = \sigma_0$, where σ_0 is the initial yield stress, irreversible plastic behaviour becomes possible along the curve AB. The stress σ_1 at the point B is now the new or current yield stress: a further increase in stress will result in continued plastic behaviour along the path BC, while a decrease in stress will result in an *elastic* response, along the curve BD. This elastic behaviour will continue until the stress reaches the value $\sigma = -\sigma_1$ in the compressive range, beyond which plastic behaviour occurs once again along the curve DE.

The total strain is made up additively of an elastic component ε^e and a plastic component ε^p, with the elastic part of the strain given by Hooke's law: that is,

$$\varepsilon = \varepsilon^e + \varepsilon^p, \quad \sigma = E\varepsilon^e = E(\varepsilon - \varepsilon^p),$$

with E being Young's modulus. Determination of the plastic strain requires information about the stress history and its evolution, and the plastic strain rate is as follows:

$$\dot{\varepsilon}^p = \begin{cases} 0 & \text{if } \sigma \in (-\sigma_1, \sigma_1) \text{ or } \sigma_1\dot{\sigma} < 0, \\ \lambda \operatorname{sgn} \sigma & \text{if } |\sigma| = \sigma_1 \text{ and } \sigma_1\dot{\sigma} > 0. \end{cases} \tag{1}$$

More concisely, we may write

$$\dot{\varepsilon}^p = \lambda \frac{d\varphi}{d\sigma}, \qquad \lambda \geq 0,\ \varphi(\sigma) \leq 0,\ \lambda\varphi(\sigma) = 0. \tag{2}$$

The initial elastic range is the set of stresses σ for which $\varphi_0(\sigma) = |\sigma| - \sigma_0 \leq 0$, while the current elastic range, as a result of the plastic deformation, is given by $\varphi(\sigma) = |\sigma| - \sigma_1 \leq 0$.

The flow relation (2) can be 'inverted' in the sense that the stress may be given in terms of the plastic strain rate. The key quantity that makes this possible is the

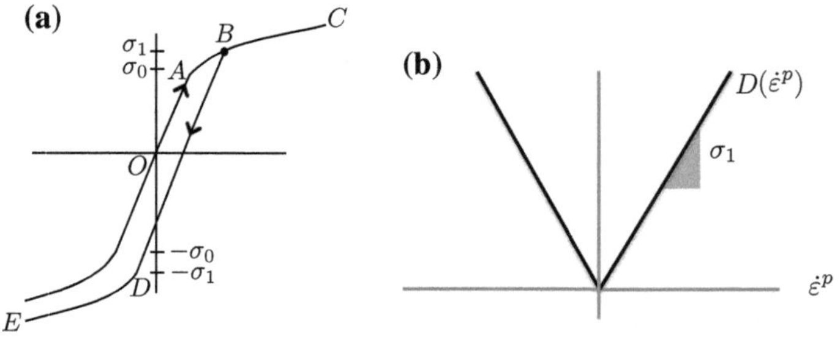

Fig. 1 **a** Behaviour in uniaxial tension and compression of an elastic-plastic material; **b** the dissipation function $D(\dot{\varepsilon}^p)$

dissipation function D. From (2), since $\lambda = |\dot{\varepsilon}^p|$ and $|\sigma| = \sigma_1$ it follows by inverting this relation that

$$\sigma = \sigma_1 \frac{\dot{\varepsilon}^p}{|\dot{\varepsilon}^p|} = \left. \frac{dD(q)}{dq} \right|_{q=\dot{\varepsilon}^p} \tag{3}$$

where the *dissipation function D* is defined by

$$D(\dot{\varepsilon}^p) = \sigma_1 |\dot{\varepsilon}^p|. \tag{4}$$

This function is *convex* and *positively homogeneous* (that is, $D(c\dot{\varepsilon}^p) = |c|D(\dot{\varepsilon}^p)$), and differentiable everywhere except at $\dot{\varepsilon}^p = 0$ (Fig. 1b). With the dissipation function at our disposal, it is possible to capture the entire flow relation in a single *inequality*, viz

$$D(q) - D(\dot{\varepsilon}^p) - \sigma(q - \dot{\varepsilon}^p) \geq 0 \quad \text{for all } q. \tag{5}$$

To see this, consider first the case of elastic behaviour, for which $\dot{\varepsilon}^p = 0$. Then (5) becomes

$$\sigma q \leq \sigma_1 |q| \qquad \text{or} \qquad |\sigma| \leq \sigma_1,$$

which is precisely the requirement that the stress lies in the elastic region. On the other hand, D is differentiable when $\dot{\varepsilon}^p \neq 0$, and in this case (5) is easily shown to be *equivalent* to the relation (3).

2 Three-Dimensional Elastoplastic Behaviour

We will assume isothermal conditions throughout for convenience. Furthermore, we develop a theory of rate-independent plasticity for quasistatic situations in which inertial terms may be neglected in the equation of motion. A further discussion of rate-independence can be found in Gurtin et al. (2010, p. 428). The appropriate extension to rate-dependent behaviour is the theory of viscoplasticity (see, for example, Gurtin et al. (2010), Lubliner (1990), Simo and Hughes (1998) for accounts of viscoplasticity).

Consider a body that occupies a domain $\Omega \subset \mathbb{R}^d$ $(d = 2, 3)$ with boundary Γ. The Cauchy stress tensor, denoted by $\boldsymbol{\sigma}$, is symmetric and satisfies the equation of equilibrium

$$-\operatorname{div} \boldsymbol{\sigma} = \boldsymbol{b}, \tag{6}$$

where $\boldsymbol{b}$ is the body force.

The linearized strain ε is given by

$$\varepsilon(\boldsymbol{u}) = \frac{1}{2}\left(\nabla \boldsymbol{u} + [\nabla \boldsymbol{u}]^T\right). \tag{7}$$

The boundary Γ has unit outward normal $\boldsymbol{n}$ and is partitioned into nonoverlapping subsets Γ_u and Γ_t such that $\Gamma_u \cup \Gamma_t = \Gamma$. Then possible boundary conditions would be

$$\boldsymbol{u} = \bar{\boldsymbol{u}} \text{ on } \Gamma_u, \quad \boldsymbol{\sigma n} = \bar{\boldsymbol{t}} \text{ on } \Gamma_t. \tag{8}$$

Here $\bar{\boldsymbol{u}}$ and $\bar{\boldsymbol{t}}$ are a prescribed displacement and surface traction, respectively.

Elastoplastic behaviour. The total strain ε is assumed additively decomposable into an elastic part ε^e and a plastic part ε^p: that is,

$$\varepsilon = \varepsilon^e + \varepsilon^p. \tag{9}$$

The elastic component of strain satisfies the elastic constitutive relation; assuming isotropy this is given by

$$\boldsymbol{\sigma} = \mathbb{C}\varepsilon^e = \lambda(\operatorname{tr}\varepsilon^e)\boldsymbol{I} + 2\mu\varepsilon^e. \tag{10}$$

Here $\mathbb{C}$ is the elasticity tensor and λ and μ are the Lamé parameters.

An assumption based on physical behaviour is that of no change in volume accompanying plastic deformation; thus we impose the condition $\operatorname{tr}\varepsilon^p = \varepsilon^p_{ii} = 0$. To model hardening behaviour we introduce the back-stress $\boldsymbol{\alpha}$, which is a symmetric second-order tensor and which accounts for kinematic hardening, and a scalar variable η which accounts for isotropic hardening. A typical choice for η is the accumulated plastic strain, so that

$$\dot{\eta} = |\dot{\varepsilon}^p|. \tag{11}$$

We define two force-like variables: a symmetric tensor $\boldsymbol{a}$ and a scalar g that are conjugate, respectively, to $\boldsymbol{\alpha}$ and η. We collect these two pairs of variables in arrays p and s and write $\mathsf{p} = (\boldsymbol{\alpha}, \eta)$, $\mathsf{s} = (\boldsymbol{a}, g)$, with inner product $\mathsf{s} \circ \mathsf{p} = \boldsymbol{a} : \boldsymbol{\alpha} + g\eta$. To develop the equations for plastic flow within a thermodynamic framework, we define the free energy ψ which is assumed to be an additive function of the elastic strain ε^e and the internal variables p: that is,

$$\psi = \psi(\varepsilon^e, \mathsf{p}) = \psi^e(\varepsilon^e) + \psi^p(\boldsymbol{\alpha}, \eta). \tag{12}$$

The free-energy imbalance takes the form $\dot{\psi} \le \boldsymbol{\sigma} : \dot{\varepsilon} + \mathsf{s} \circ \dot{\mathsf{p}}$. For linear elastic materials $\psi^e(\varepsilon^e) = \frac{1}{2}\varepsilon^e : \mathbb{C}\varepsilon^e$; from this relation and (10) the reduced dissipation inequality

$$\boldsymbol{\sigma} : \dot{\boldsymbol{\varepsilon}}^p + \mathsf{s} \circ \dot{\mathsf{p}} \geq 0 \tag{13}$$

follows, where the conjugate forces are defined by

$$\boldsymbol{a} = -\frac{\partial \psi^p}{\partial \boldsymbol{\alpha}}, \qquad g = -\frac{\partial \psi^p}{\partial \eta}. \tag{14}$$

For linear hardening behaviour, for example, the plastic part of ψ has the quadratic form

$$\psi^p(\boldsymbol{\alpha}, \eta) = \tfrac{1}{2}k_1|\boldsymbol{\alpha}|^2 + \tfrac{1}{2}k_2\eta^2 \tag{15}$$

where k_1 and k_2 are nonnegative scalars associated with kinematic and isotropic hardening, respectively. The conjugate forces are immediately obtained from (14) and are

$$\boldsymbol{a} = -k_1\boldsymbol{\alpha}, \qquad g = -k_2\eta. \tag{16}$$

The case of perfect plasticity is recovered by setting $k_1 \equiv 0$ and $k_2 \equiv 0$.

The elastic region and plastic flow The classical theory of plasticity constrains the stresses to lie, pointwise, in a region of admissible stresses $\mathcal{E}$. Plastic flow takes place only when the stresses lie on the boundary $\mathcal{S}$ of $\mathcal{E}$, with its exterior assumed to be not attainable. The structure of the region $\mathcal{E}$ and the form taken by the flow relations are determined by the principle of maximum plastic work, which has its origins in the work of von Mises, Taylor, and Bishop and Hill (see Lubliner (1990) for further details). According to the principle in its original form, for the case of perfect plasticity, given a state of stress $\boldsymbol{\sigma} \in \mathcal{E}$ and an associated plastic strain rate $\dot{\boldsymbol{\varepsilon}}^p$, the pair $(\boldsymbol{\sigma}, \dot{\boldsymbol{\varepsilon}}^p)$ is such as to maximize the plastic work among all admissible stresses: that is,

$$\boldsymbol{\sigma} : \dot{\boldsymbol{\varepsilon}}^p \geq \boldsymbol{\tau} : \dot{\boldsymbol{\varepsilon}}^p \quad \text{for all } \boldsymbol{\tau} \in \mathcal{E}. \tag{17}$$

We generalize to hardening plasticity by defining the generalized stress S and generalized plastic strain P by

$$\mathsf{S} := (\boldsymbol{\sigma}, \boldsymbol{\alpha}, g), \quad \mathsf{P} = (\boldsymbol{\varepsilon}^p, \boldsymbol{\alpha}, \eta) \tag{18}$$

and requiring that $\mathsf{S} \in \mathcal{E}$, a set of admissible generalized stresses. Purely elastic behaviour takes place when S lies in the interior of $\mathcal{E}$, while plastic loading may take place only when S lies on the yield surface $\mathcal{S}$. The postulate of maximum plastic work is now as follows: given a generalized stress $\mathsf{S} \in \mathcal{E}$ and an associated generalized plastic strain rate $\dot{\mathsf{P}}$, the pair $(\mathsf{S}, \dot{\mathsf{P}})$ satisfies

$$\mathsf{S} \circ \dot{\mathsf{P}} \geq \mathsf{T} \circ \dot{\mathsf{P}} \quad \text{for all } \mathsf{T} \in \mathcal{E}. \tag{19}$$

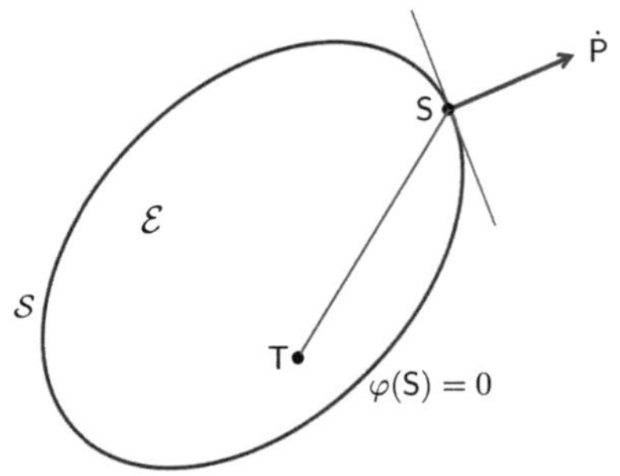

Fig. 2 The elastic region, yield function, and normality relation in generalized stress space

Here the inner product between conjugate generalized quantities is defined by $\mathsf{S} \circ \dot{\mathbf{P}} = \boldsymbol{\sigma} : \dot{\varepsilon} + \boldsymbol{a} : \dot{\alpha} + g\dot{\eta}$.

There are two major consequences of the maximum plastic work inequality. First, it can be shown that the generalized plastic strain rate $\dot{\mathsf{P}}$ associated with a generalized stress S on the yield surface $\mathcal{S}$ is normal to the tangent hyper-plane at the point S to $\mathcal{S}$. This result is generally referred to as the *normality law*. Second, it can be shown that the region $\mathcal{E}$ is *convex*. These notions are illustrated in Fig. 2. We can describe the surface $\mathcal{S}$ and the elastic region $\mathcal{E}$ with the use of a function φ, called the *yield function*:

$$\mathcal{S} := \{\mathsf{S} : \varphi(\mathsf{S}) = 0\} \quad \text{and} \quad \mathcal{E} := \{\mathsf{S} : \varphi(\mathsf{S}) < 0\}. \tag{20}$$

Then plastic flow takes place only when S lies on the yield surface so that $\varphi(\mathsf{S}) = 0$, while it is necessarily zero for any S such that $\varphi(\mathsf{S}) < 0$. The requirement that $\varphi = \dot{\varphi} = 0$ during plastic loading is known as the *consistency condition*.

If the yield surface is *smooth*, the normality relation becomes

$$\dot{\mathsf{P}} = \lambda\, \nabla\phi(\mathsf{S}), \tag{21}$$

where λ is a nonnegative scalar, called the *plastic multiplier*. Further, we have the *complementarity condition* $\lambda \geq 0, \quad \phi \leq 0, \quad \lambda\phi = 0$. An equivalent form of stating this set of relations is through the inequality (cf. Fig. 2)

$$\dot{\mathsf{P}} \circ (\mathsf{T} - \mathsf{S}) \leq 0 \quad \text{for all } \varphi(\mathsf{T}) \leq 0. \tag{22}$$

A flow law in which the yield function serves as a potential, in the sense that the generalized plastic strain rate lies in the normal to the yield surface, is called an associative flow law.[1]

[1]Non-associative laws are important in certain applications. Two examples are the Mohr–Coulomb and Drucker–Prager laws, which are used to model plastic behaviour in materials such as concrete, soil, and rock; see for example Lubliner (1990) for a summary account. The theory corresponding to non-associative flow laws is more complex and requires a distinct setting.

If the yield surface for the case of perfect plasticity is defined by the function $\varphi(\boldsymbol{\sigma}) = \Phi(\boldsymbol{\sigma}) - c_0 = 0$, where $c_0 > 0$ is a constant, then the extension to kinematic and isotropic hardening entails the introduction of terms that describe translation and expansion of the yield surface. That is, the yield function now becomes $\mathsf{S} = (\boldsymbol{\sigma}, \boldsymbol{a}, g)$,

$$\varphi(\mathsf{S}) = \Phi(\boldsymbol{\sigma} + \boldsymbol{a}) + g - c_0. \tag{23}$$

From (21) when expanded we obtain

$$\dot{\varepsilon}^p = \lambda \frac{\partial \Phi(\boldsymbol{\sigma} + \boldsymbol{a})}{\partial \boldsymbol{\sigma}}, \quad \dot{\boldsymbol{\alpha}} = \lambda \frac{\partial \Phi(\boldsymbol{\sigma} + \boldsymbol{a})}{\partial \boldsymbol{a}} = \dot{\varepsilon}^p, \quad \dot{\eta} = \lambda \frac{\partial \Phi(\mathsf{S})}{\partial g} = \lambda. \tag{24}$$

Examples of yield criteria We confine attention here to perfectly plastic behaviour, in which the hardening variables are absent. The assumption of plastic incompressibility permits a further simplification, in that it suffices to write φ as a function of the invariants of the stress deviator $\operatorname{dev} \boldsymbol{\sigma} := \boldsymbol{\sigma} - \frac{1}{3}(\operatorname{tr} \boldsymbol{\sigma})\boldsymbol{I}$.

The *Mises–Hill yield criterion* is based on the assumption that the threshold of elastic behaviour is determined by the elastic shear energy density, so that

$$\varphi(\boldsymbol{\sigma}) = |\operatorname{dev} \boldsymbol{\sigma}| - c_0. \tag{25}$$

The normality law (21) is thus

$$\dot{\varepsilon}^p = \lambda \nabla \varphi(\boldsymbol{\sigma}) = \lambda \frac{\operatorname{dev} \boldsymbol{\sigma}}{|\operatorname{dev} \boldsymbol{\sigma}|}. \tag{26}$$

The *Tresca yield criterion* is based on the assumption that the elastic threshold is reached when the maximum shear stress reaches a critical value. In terms of the principal stresses $\boldsymbol{\sigma}_i$, the maximum shear stress is given by $\frac{1}{2} \max_{i \neq j} |\sigma_i - \sigma_j|$ and the yield function is given by

$$\varphi(\boldsymbol{\sigma}) = \max_{i \neq j} |\sigma_i - \sigma_j| - \sigma_0 = 0. \tag{27}$$

The Tresca yield surface is not smooth so that the normality relation has to be suitably interpreted at corners or edges on this surface.

Hardening laws. A typical choice for the isotropic hardening parameter is that of the *equivalent plastic strain* $\eta(t)$, defined in (11). For example, for the situation of biaxial stresses the Mises–Hill yield surface at any time is an ellipse B that is similar to the initial yield surface A, as shown in Fig. 3. The most common form of kinematic hardening is associated with the names of Prager and Ziegler. The variable $\boldsymbol{\alpha}$ is taken to be the plastic strain tensor ε^p and the plastic part of the free energy $\psi^h(\boldsymbol{\alpha})$ is given by the quadratic function $\psi^h(\varepsilon^p) = \frac{1}{2}k_1|\varepsilon^p|^2$, in which $k_1 > 0$ is the hardening constant. The corresponding conjugate force $\boldsymbol{a}$ is found from (16) to be

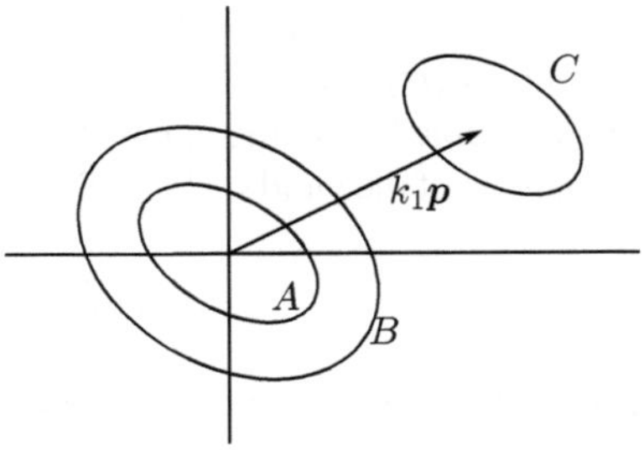

Fig. 3 Isotropic and kinematic hardening behaviour: *A* is the initial yield surface, *B* and *C* are subsequent yield surfaces after isotropic and kinematic hardening, respectively

$\boldsymbol{a} = -k_1\boldsymbol{\varepsilon}^p$. The yield function then translates by an amount $\boldsymbol{a}$ and is given by (Fig. 3)

$$\phi(\boldsymbol{\sigma}, \boldsymbol{a}) = \Phi(\boldsymbol{\sigma} + \boldsymbol{a}) - \sqrt{2/3}\,\sigma_0. \tag{28}$$

The flow relation in terms of the dissipation function The three-dimensional version of the flow relation has an analogue to the inverted one-dimensional form (5). We define the dissipation function D by

$$D(\dot{\mathsf{P}}) = \sup\{\mathsf{S} \circ \dot{\mathsf{P}} \mid \varphi(\mathsf{S}) \leq 0\}. \tag{29}$$

Then the flow relation (21) or (22) can be written *equivalently* as

$$D(\mathsf{Q}) - D(\dot{\mathsf{P}}) - \mathsf{S} \circ (\mathsf{Q} - \dot{\mathsf{P}}) \geq 0 \quad \text{for all } \mathsf{Q}. \tag{30}$$

To see this, consider for example the case of perfect plasticity with the Mises flow relation, for which the yield function is given by (25). Then the dissipation is, from (29), $D(\dot{\boldsymbol{\varepsilon}}^p) = c_0|\dot{\boldsymbol{\varepsilon}}^p|$. When $\dot{\boldsymbol{\varepsilon}}^p \neq \mathbf{0}$ the inequality (30) reduces to the *equation*

$$\boldsymbol{\sigma} = \left.\frac{\partial D}{\partial \boldsymbol{q}}\right|_{q=\dot{\varepsilon}^p} = c_0\frac{\dot{\boldsymbol{\varepsilon}}^p}{|\dot{\boldsymbol{\varepsilon}}^p|},$$

which could be obtained directly by inversion of (26). On the other hand, when $\dot{\boldsymbol{\varepsilon}}^p = \mathbf{0}$ then (30) reduces to

$$\boldsymbol{\sigma} : \boldsymbol{q} \leq D(\boldsymbol{q}) \quad \text{or} \quad |\mathrm{dev}\,\boldsymbol{\sigma}| \leq c_0;$$

this states that the stress lies in the elastic region. Thus the inequality (30) captures in a single expression the possibilities of elastic behaviour or plastic flow. For the case of kinematic and isotropic hardening with $\mathsf{P} = (\boldsymbol{\varepsilon}^p, \eta)$, it can be shown that the dissipation function is given by

$$D(\dot{\mathsf{P}}) = \begin{cases} c_0|\dot{\boldsymbol{\varepsilon}}^p| & |\dot{\boldsymbol{\varepsilon}}^p| \leq \dot{\eta}\,, \\ +\infty & \text{otherwise.} \end{cases} \tag{31}$$

3 The Primal Variational Problem

There are two alternative formulations of the flow relations that are of interest, as set out in the previous section. We describe as the *primal variational problem* the version that uses the flow law (30) in terms of the dissipation, while the *dual variational problem* is formulated using the flow law (21) or (22) that makes use of the yield function. We focus here on the primal version, which has as its basic unknown variables the displacement $\boldsymbol{u}$, plastic strain ε^p, and set of hardening variables $\mathsf{p} = (\boldsymbol{\alpha}, \eta)$. In what follows we shall equate $\boldsymbol{\alpha}$ with the plastic strain ε^p. We are then required to find $(\boldsymbol{u}, \varepsilon^p, \eta)$ that satisfy the equation of equilibrium (6), the elastic relation (10), and the flow relation in primal form (30). For convenience, we assume the homogeneous Dirichlet boundary condition $\boldsymbol{u} = \mathbf{0}$ on Γ.

The spaces V, Q, and M of displacements, plastic strains, and hardening variables are defined, respectively, by[2]

$$\begin{aligned} V &:= [H_0^1(\Omega)]^3, \\ Q &:= \left\{\boldsymbol{q} = (q_{ij})_{3\times 3} : \ q_{ji} = q_{ij}, \ q_{ij} \in L^2(\Omega), \ \operatorname{tr} \boldsymbol{q} = 0 \text{ a.e. in } \Omega\right\}, \\ M &:= L^2(\Omega). \end{aligned}$$

We set $W := V \times Q \times M$, which is a Hilbert space with the natural inner product $(\mathsf{w}, \mathsf{z})_W := (\boldsymbol{u}, \boldsymbol{v})_V + (\varepsilon^p, \boldsymbol{q})_Q + (\eta, \zeta)_M$ and the norm $\|\mathsf{z}\|_W := (\mathsf{z}, \mathsf{z})_W^{1/2}$, where $\mathsf{w} = (\boldsymbol{u}, \varepsilon^p, \eta)$ and $\mathsf{z} = (\boldsymbol{v}, \boldsymbol{q}, \zeta)$, and define the closed, convex subset

$$W_p = \{\mathsf{w} \in W : \ |\boldsymbol{q}| \leq \zeta \text{ a.e. in } \Omega\}. \tag{32}$$

We assume that the elasticity tensor $\mathbb{C}$ is pointwise stable, so that for isotropic materials the Lamé constants in (10) satisfy $\mu > 0$ and $3\lambda + 2\mu > 0$. We will pay particular attention to the special case of an elastoplastic material with linearly kinematic and isotropic hardening, together or separately, defined for example through (15) and (16). We assume that the hardening constants k_1 and k_2 satisfy

$$k_1, k_2 \in L^\infty(\Omega), \quad k_1 \geq 0, \quad k_2 \geq 0, \quad k_1 > 0 \text{ or } k_2 > 0. \tag{33}$$

The equilibrium equation. We take the scalar product of (6) with $\boldsymbol{v} - \dot{\boldsymbol{u}}$ for arbitrary $\boldsymbol{v} \in V$, integrate over Ω, and perform an integration by parts with the use of the elastic relation to obtain

$$\int_\Omega \mathbb{C}(\varepsilon(\boldsymbol{u}) - \varepsilon^p) : (\varepsilon(\boldsymbol{v}) - \varepsilon(\dot{\boldsymbol{u}}))\, dx = \int_\Omega \boldsymbol{f} \cdot (\boldsymbol{v} - \dot{\boldsymbol{u}})\, dx \quad \forall\, \boldsymbol{v} \in V. \tag{34a}$$

The flow relation. We integrate the relation (30) over Ω and seek $(\dot{\varepsilon}^p, \dot{\eta}) \in W_p$ that satisfies

[2]For details of function spaces see Chap. Functional Analysis, Boundary Value Problems and Finite Elements.

$$\int_\Omega D(\boldsymbol{q},\zeta)\,dx \geq \int_\Omega D(\dot{\varepsilon}^p,\dot{\eta})\,dx + \int_\Omega (\boldsymbol{\sigma}(\boldsymbol{u},\varepsilon^p) - k_1\varepsilon^p) : (\boldsymbol{q} - \dot{\varepsilon}^p)dx - \int_\Omega k_2\eta\,(\zeta - \dot{\eta})dx \qquad (34b)$$

for all $(\boldsymbol{q},\zeta) \in W_p$. The problem may be cast in the form of a variational inequality as follows: setting $\mathsf{w} = (\boldsymbol{u},\varepsilon^p,\eta)$ and $\mathsf{z} = (\boldsymbol{v},\boldsymbol{q},\zeta)$ as before, we define

$$a : W \times W \to \mathbb{R}, \quad a(\mathsf{w},\mathsf{z}) = \int_\Omega \Big[\boldsymbol{\sigma}(\boldsymbol{u},\varepsilon^p) : \varepsilon(\boldsymbol{v} - \boldsymbol{q}) + k_1\varepsilon^p : \boldsymbol{q} + k_2\eta\zeta\Big]\,dx, \qquad (35a)$$

$$j : W \times \mathbb{R}, \qquad j(\mathsf{z}) = \int_\Omega D(\boldsymbol{q},\zeta)\,dx, \qquad (35b)$$

$$\ell : W \to \mathbb{R}, \qquad \langle \ell, \mathsf{z}\rangle = \int_\Omega \boldsymbol{f}\cdot\boldsymbol{v}\,dx. \qquad (35c)$$

The bilinear form $a(\cdot,\cdot)$ is symmetric as a result of the symmetry properties of $\mathbb{C}$. From the properties of D, $j(\cdot)$ is convex, positively homogeneous, and nonnegative. We now add (34b) and (34a) to obtain the variational inequality

$$a(\mathsf{w}(t),\mathsf{z} - \dot{\mathsf{w}}(t)) + j(\mathsf{z}) - j(\dot{\mathsf{w}}(t)) \geq \langle \ell(t), \mathsf{z} - \dot{\mathsf{w}}(t)\rangle \quad \forall\, \mathsf{z} \in W. \qquad (36)$$

Note that the variational inequality is posed on the whole space W rather than W_p, observing that $j(\mathsf{z}) = \infty$ for $\mathsf{z} \notin W_p$ and bearing in mind the requirement $\dot{\mathsf{w}}(t) \in W_p$. The primal variational problem of elastoplasticity thus takes the following form: given $\ell \in H^1(0,T;W')$, $\ell(0) = 0$, find $\mathsf{w} = (\boldsymbol{u},\varepsilon^p,\eta) : [0,T] \to W$, $\mathsf{w}(0) = 0$, such that for almost all $t \in (0,T)$, $\dot{\mathsf{w}}(t) \in W_p$ and (36) is satisfied for all $\mathsf{z} \in W$.

It is readily shown that a solution to the classical problem solves the variational inequality (36). Conversely, it can be shown that if w is a smooth solution of (36) then w is also a solution to the classical problem.

Linearly kinematic hardening corresponds to the special case $k_2 = 0$. The variables in this case are the displacement $\boldsymbol{u}$ and the plastic strain ε^p, and the spaces V and Q are as previously defined. The solution space is now $W_{\text{kin}} := V \times Q$, with the inner product $(\mathsf{w},\mathsf{z})_W := (\boldsymbol{u},\boldsymbol{v})_V + (\varepsilon^p,\boldsymbol{q})_Q$ and the norm $\|\mathsf{z}\|_{W_{\text{kin}}} := (\mathsf{z},\mathsf{z})^{1/2}_{W_{\text{kin}}}$, where $\mathsf{w} = (\boldsymbol{u},\varepsilon^p)$ and $\mathsf{z} = (\boldsymbol{v},\boldsymbol{q})$. For this case the dissipation function (31) becomes $D(\boldsymbol{q}) = c_0|\boldsymbol{q}| \;\; \forall\, \boldsymbol{q} \in Q$.

The case of linearly isotropic hardening only is obtained by setting $k_1 = 0$ in (34b) and (35a).

The problem (34a) with (34b) has a *unique solution* $(\boldsymbol{u}(t),\varepsilon^p(t),\eta(t)) \in V \times Q \times M$ provided that either kinematic or isotropic hardening behaviour is present. The case of perfect plasticity requires a different approach altogether, in order to allow for discontinuities in the solution in the form of slip bands, for example.

4 Solution Algorithms

We focus in this section on time-discrete approximations, and refer to the texts mentioned in the introduction for details of finite element approximations. Time-discretization involves a uniform partitioning of the time interval $[0, T]$ according to $0 = t_0 < t_1 < \cdots < t_N = T$, where $t_n - t_{n-1} = k$, $k = T/N$. We write $\ell_n = \ell(t_n)$ and define Δw_n to be the backward difference $w_n - w_{n-1}$. Focusing for convenience on the problem with isotropic hardening, the time-discrete approximation of (34) is as follows: given all quantities at time t_n and the loading $\boldsymbol{f}_{n+1}$, find the displacement $\boldsymbol{u}_{n+1}$, plastic strain ε^p_{n+1} and hardening variable η_{n+1} that satisfy (see 34)

$$\int_\Omega \mathbb{C}(\varepsilon(\boldsymbol{u}_{n+1}) - \varepsilon^p_{n+1}) : \varepsilon(\boldsymbol{v})\, dx = \int_\Omega \boldsymbol{f} \cdot \boldsymbol{v}\, dx \quad \forall \boldsymbol{v} \in V, \tag{37a}$$

$$\begin{aligned}\int_\Omega D(\boldsymbol{q}, \zeta)\, dx \geq \int_\Omega D(\Delta\varepsilon^p, \Delta\eta)\, dx + \int_\Omega \left(\boldsymbol{\sigma}_{n+1} - k_1\varepsilon^p_{n+1}\right) : (\boldsymbol{q} - \Delta\varepsilon^p)\, dx \\ + \int_\Omega g_{n+1}(\zeta - \Delta\eta) dx,\end{aligned} \tag{37b}$$

where $\boldsymbol{\sigma}_{n+1} = \mathbb{C}(\boldsymbol{u}_{n+1} - \varepsilon^p_{n+1})$ and $g_{n+1} = -k_2\eta_{n+1}$. This problem is equivalent to the *minimization problem*

$$\mathsf{w}_n = \operatorname{argmin} J(\mathsf{z}) := \frac{1}{2}a(\mathsf{z}, \mathsf{z}) + j(\Delta\mathsf{z}) - \ell(\mathsf{z}_h)$$

where the bilinear form $a(\cdot, \cdot)$, linear functional $\ell(\cdot)$, and functional $j(\cdot)$ are as defined in (35). We are interested in algorithms of predictor–corrector type for solving the problem (37). A now standard such algorithm is as follows:

Predictor step: Given $\boldsymbol{u}_n$, ε^p_n, η_n, solve

$$\int_\Omega \mathbb{C}(\varepsilon(\boldsymbol{u}^{(i)}_{n+1} - \varepsilon^p_n - \Delta\varepsilon^{p*(i)}) : \varepsilon(\boldsymbol{v})\, dx = \int_\Omega \boldsymbol{f}_{n+1} \cdot \boldsymbol{v}\, dx, \tag{38a}$$

$$D^{(i)}(\boldsymbol{q}) - D^{(i)}(\Delta\varepsilon^{p*(i)}) - \boldsymbol{\sigma}^{*(i)}_{n+1} : (\boldsymbol{q} - \Delta\varepsilon^{p*(i)}) - g^{*(i)}_{n+1}(\zeta - \Delta\eta^{*(i)}) \geq 0 \tag{38b}$$

for $\boldsymbol{u}^{(i)}_{n+1}$, $\Delta\varepsilon^{p*(i)}$ and $\Delta\eta^{*(i)}$, where $D^{(i)}$ is a *smooth* convex approximation of D which satisfies $D(\boldsymbol{q}) \leq D^{(i)}(\boldsymbol{q})$ and (see Fig. 4a)

$$D^{(i)}|_{\Delta\varepsilon^{p(i-1)}} = D|_{\Delta\varepsilon^{p(i-1)}}, \nabla D^{(i)}|_{\Delta\varepsilon^{p(i-1)}} = \nabla D|_{\Delta\varepsilon^{p(i-1)}}. \tag{39}$$

Corrector step: Given $\boldsymbol{u}^{(i)}_{n+1}$, solve for $\Delta\varepsilon^{p(i)}$ the flow relation

$$D(\boldsymbol{q}) - D(\Delta\varepsilon^{p(i)}) - \boldsymbol{\sigma}^{(i)}_{n+1} : (\boldsymbol{q} - \Delta\varepsilon^{p(i)}) - g^{(i)}_{n+1}(\zeta - \Delta\eta^{(i)}) \geq 0 \tag{40}$$

where $\boldsymbol{\sigma}^{(i)}_{n+1} = \mathbb{C}(\varepsilon(\boldsymbol{u}^{(i)}_{n+1} - \varepsilon^{p(i)}_{n+1})$, $g^{(i)}_{n+1} = -k\eta^{(i)}_{n+1}$.

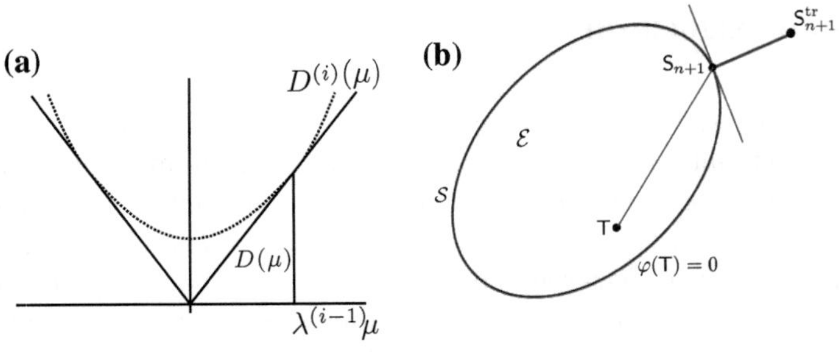

Fig. 4 **a** Approximation $D^{(i)}$ of D; **b** the closest-point projection in generalized stress space

The corrector step of the algorithm is equivalent to the displacement-driven *return mapping* algorithm (see for example Simo and Hughes (1998)), and may be interpreted as a *closest-point projection.* From (22) the corrector step (40) is equivalent to

$$\Delta\varepsilon^{p(i)} : (\boldsymbol{\tau} - \boldsymbol{\sigma}^{(i)}) + \Delta\eta^{(i)}(h - g) \leq 0 \quad \text{for } \varphi(\boldsymbol{\tau}, h) \leq 0. \tag{41}$$

Now the generalized stresses at time t_{n+1} may be written as $\boldsymbol{\sigma}_{n+1} = \boldsymbol{\sigma}^{\text{tr}}_{n+1} - \mathbb{C}\Delta\varepsilon^p$, $g_{n+1} = g_n - k_2\Delta\eta$ where $\boldsymbol{\sigma}^{\text{tr}}_{n+1} = \mathbb{C}[\varepsilon(\boldsymbol{u}_{n+1}) - \boldsymbol{p}_n]$ and $g^{\text{tr}}_{n+1} = -k_2\eta_n$ are the trial values at time t_{n+1} if no plastic flow were to take place in the time step $[t_n, t_{n+1}]$. Setting $\mathsf{S} = (\boldsymbol{\sigma}, g)$, $\mathsf{T} = (\boldsymbol{\tau}, h)$, and $\mathsf{S}^{\text{tr}} = (\boldsymbol{\sigma}^{\text{tr}}, g^{\text{tr}})$ the inequality (41) becomes, at time t_{n+1},

$$(\mathsf{T} - \mathsf{S}_{n+1}) : \mathsf{G}^{-1}(\mathsf{S}^{\text{tr}}_{n+1} - \mathsf{S}_{n+1}) \leq 0 \quad \text{for all } \mathsf{T} \in \mathcal{E} \quad \text{where} \quad \mathsf{G} = \begin{pmatrix} \mathbb{C} & 0 \\ 0 & k \end{pmatrix}.$$

In other words, the actual stress may be found as the orthogonal projection of the trial generalized stress S^{tr} onto the elastic region, in the inner product generated by G^{-1}. This is illustrated in Fig. 4b.

The problem may be formulated as the constrained minimization problem with the constraint $\mathsf{T} \in \mathcal{E}$ or $\varphi(\mathsf{T}) \leq 0$ imposed through a Lagrange multiplier λ: that is,

$$\mathsf{S}_{n+1} = \operatorname{argmin} (\mathsf{T} - \mathsf{S}^{\text{tr}}_{n+1}) : \mathsf{G}^{-1}(\mathsf{T} - \mathsf{S}^{\text{tr}}_{n+1}) + \lambda\varphi(\mathsf{T}). \tag{42}$$

For an account of the structure and implementation of the algorithm, see Simo and Hughes (1998), Chap. 3.

Examples of predictors *(a) Elastic predictor.* We set $\varepsilon^{p*(i)} = \varepsilon^{p(i-1)}$ so there is no need to define an approximate dissipation function $D^{(i)}$.
(b) Consistent tangent predictor. We define $D^{(i)}$ as the second order Taylor expansion of D about $\varepsilon^{p(i-1)}$. We put

$$D^{(i)}(\boldsymbol{q}) = D(\varepsilon^{p(i-1)}) + \nabla D(\varepsilon^{p(i-1)}) : (\boldsymbol{q} - \varepsilon^{p(i-1)}) \\ + \tfrac{1}{2}(\boldsymbol{q} - \varepsilon^{p(i-1)}) : [\boldsymbol{H}(\varepsilon^{p(i-1)}) + \rho \boldsymbol{I}](\boldsymbol{q} - \varepsilon^{p(i-1)}) \tag{43}$$

where $\boldsymbol{H} := \nabla^2 D(\varepsilon^{p(i-1)})$ and $\rho \geq 0$. With $\rho = 0$ and in a fully discrete setting, after the internal variables have been eliminated, (38a) and (38b) yield the set of displacement equations $\boldsymbol{K}_{\tan}\boldsymbol{d} = \boldsymbol{R}^i$, where $\boldsymbol{K}$ is the consistent tangent matrix and $\boldsymbol{R}^i$ the residual.

Convergence of the algorithm The predictor–corrector algorithm can be shown to converge, for sufficiently large hardening (see Djoko et al. (2007)). For example, assuming linear isotropic hardening with a hardening coefficient k_2 as in (15), it can be shown that the predictor–corrector algorithm converges provided that $r \sim (\lambda + 2\mu)/(2\mu(1 + k_2)) < 1/3$, and that

$$\|\mathsf{w} - \mathsf{w}^i\| \leq \left(\frac{2r}{1-r}\right)^i \|\mathsf{w}^1 - \mathsf{w}^0\|,$$

where w^i denotes the ith iterate in the algorithm.

5 Elastoplasticity at Large Deformations

We present here a brief account of the extension of parts of the theory in the earlier sections to large-deformation elastoplasticity. For details of the relevant concepts from continuum mechanics and elastoplasticity see, for example, de Souza Neto et al. (2008), Gurtin et al. (2010), Simo and Hughes (1998).

We identify a body with a bounded domain $\Omega \subset \mathbb{R}^d$ ($d = 2, 3$) with boundary Γ. Points in Ω are denoted by $\boldsymbol{X}$. For a given time interval $[0, T]$ a motion of the body is described by a function $\varphi : \Omega \times [0, T] \rightarrow \mathbb{R}^d$, so that the position of a material point initially at $\boldsymbol{X}$ is given by

$$\boldsymbol{X} \in \Omega \rightarrow \boldsymbol{x} = \varphi(\boldsymbol{X}, t) = \boldsymbol{u}(\boldsymbol{X}, t) + \boldsymbol{X}. \tag{44}$$

Here $\boldsymbol{u}$ is the displacement vector. The motion is assumed orientation-preserving, invertible, and such as to exclude interpenetration of matter: thus, the deformation gradient $\boldsymbol{F}(\boldsymbol{X}) = \operatorname{Grad}\varphi(\boldsymbol{X}, t)$ satisfies $J = \det \boldsymbol{F} > 0$ in Ω.

The right Cauchy–Green tensor $\boldsymbol{C}$ and Green-Lagrange strain tensor $\boldsymbol{E}$ are defined by

$$\boldsymbol{C} = \boldsymbol{F}^T\boldsymbol{F}, \tag{45a}$$

$$\boldsymbol{E} = \frac{1}{2}(\boldsymbol{C} - \boldsymbol{I}) = \tfrac{1}{2}\big(\operatorname{Grad}\boldsymbol{u} + [\operatorname{Grad}\boldsymbol{u}]^T + [\operatorname{Grad}\boldsymbol{u}]^T\operatorname{Grad}\boldsymbol{u}\big). \tag{45b}$$

Here Grad denotes the gradient operator in the reference configuration: that is, in component form $(\text{Grad}\,\boldsymbol{u})_{ij} = \partial u_i/\partial X_j$. The velocity $\boldsymbol{v}$ is defined by $\boldsymbol{v} = \bar{\boldsymbol{v}}(\boldsymbol{X}, t) = \dot{\varphi}$. Using the invertibility of the motion it may be written alternatively as $\boldsymbol{v} = \widehat{\boldsymbol{v}}(\boldsymbol{x}, t)$. The spatial gradient is denoted by grad, so that $(\text{grad}\,\boldsymbol{v})_{ij} = \partial v_i/\partial x_j$ for $\boldsymbol{v} = \boldsymbol{v}(\boldsymbol{x}, t)$. The velocity gradient $\boldsymbol{L}$ is a spatial field related to the deformation gradient by $\boldsymbol{L} = \text{grad}\,\boldsymbol{v} = \dot{\boldsymbol{F}}\boldsymbol{F}^{-1}$.

For an elastic–plastic body, the standard Kröner multiplicative decomposition is assumed: that is,

$$\boldsymbol{F} = \boldsymbol{F}^e\boldsymbol{F}^p, \tag{46}$$

in which $\boldsymbol{F}^e$ is the elastic part of the deformation, accounting for stretch and rotation of the lattice, while $\boldsymbol{F}^p$ represents the irreversible plastic distortion, resulting from formation and motion of dislocations. Consistent with the requirement $J > 0$, it is assumed that $\det \boldsymbol{F}^e > 0$ and $\det \boldsymbol{F}^p > 0$. Plastic deformation is further assumed to isochoric in nature, so that

$$\det \boldsymbol{F}^p = 1. \tag{47}$$

The elastic and plastic parts $\boldsymbol{F}^e$ and $\boldsymbol{F}^p$ of the deformation gradient are not gradients of a vector field, unlike the situation for $\boldsymbol{F}$. Corresponding to the Cauchy-Green and strain tensors in (45), elastic analogues may be defined according to

$$\boldsymbol{C}^e = \boldsymbol{F}^{eT}\boldsymbol{F}^e, \qquad \boldsymbol{E}^e = \tfrac{1}{2}(\boldsymbol{C}^e - \boldsymbol{I}). \tag{48}$$

The velocity gradient $\boldsymbol{L}$ may be decomposed additively into elastic and plastic parts; using the relation $\boldsymbol{L} = \dot{\boldsymbol{F}}\boldsymbol{F}^{-1}$ together with (46), we define

$$\boldsymbol{L}^e = \dot{\boldsymbol{F}}^e(\boldsymbol{F}^e)^{-1}, \quad \boldsymbol{L}^p = \dot{\boldsymbol{F}}^p[\boldsymbol{F}^p]^{-1}, \quad \boldsymbol{L} = \boldsymbol{L}^e + \boldsymbol{F}^e\boldsymbol{L}^p[\boldsymbol{F}^e]^{-1}. \tag{49}$$

Quasistatic behaviour is assumed throughout this work, so that inertial terms may be neglected. The equation of equilibrium in the reference configuration is

$$-\,\text{Div}\,\boldsymbol{P} = \boldsymbol{b}_0, \tag{50}$$

where $\boldsymbol{b}_0$ is the body force per unit reference volume and $\boldsymbol{P}$ is the first Piola–Kirchhoff stress, related to the Cauchy stress $\boldsymbol{\sigma}$ by $\boldsymbol{P} = J\boldsymbol{\sigma}\boldsymbol{F}^{-T}$.

Denoting by ψ the specific free energy of the body and by ρ_0 the mass density per unit reference volume, the free-energy imbalance, which follows from the second law of thermodynamics, takes the form

$$\rho_0\dot{\psi} - \boldsymbol{P} : \dot{\boldsymbol{F}} \le 0. \tag{51}$$

Two further useful stress measures, the elastic second Piola–Kirchhoff stress $\boldsymbol{S}^e$, and the Mandel stress $\boldsymbol{M}$, are defined by

$$\boldsymbol{S}^e = J[\boldsymbol{F}^e]^{-1}\boldsymbol{\sigma}[\boldsymbol{F}^e]^{-T}, \quad \boldsymbol{M} = J\boldsymbol{F}^{eT}\boldsymbol{\sigma}[\boldsymbol{F}^e]^{-T} = \boldsymbol{C}^e\boldsymbol{S}^e. \tag{52}$$

Then with the identity $\dot{\boldsymbol{E}}^e = [\boldsymbol{F}^e]^T\boldsymbol{D}^e\boldsymbol{F}^e$, the free-energy imbalance (51) can be recast in the form

$$\rho_0\dot{\psi} - \boldsymbol{S}^e : \dot{\boldsymbol{E}}^e - \boldsymbol{M} : \boldsymbol{L}^p \leq 0. \tag{53}$$

The free energy is assumed to be additively decomposable into an elastic part $\psi^e(\boldsymbol{F}^e)$ which captures the elastic response, and a plastic part ψ^p which captures features of the plastic behaviour such as hardening. Furthermore, the principle of material frame indifference requires the dependence to be on $\boldsymbol{C}^e = [\boldsymbol{F}^e]^T\boldsymbol{F}^e$ or equivalently $\boldsymbol{E}^e$. We restrict attention to isotropic hardening, which is captured by a scalar variable η, so that

$$\psi = \hat{\psi}^e(\boldsymbol{E}^e) + \hat{\psi}^p(\eta). \tag{54}$$

Application of the Coleman–Noll procedure leads, respectively, to the elastic relation, definition of the conjugate force g, and the reduced dissipation inequality:

$$\boldsymbol{S}^e = \rho_0\frac{\partial\hat{\psi}(\boldsymbol{E}^e)}{\partial\boldsymbol{E}^e}, \quad g := -\rho_0\frac{\partial\psi}{\partial\eta}, \quad \boldsymbol{M} : \boldsymbol{L}^p + g\dot{\eta} \geq 0. \tag{55}$$

The flow relation. We introduce conjugate pairs $\mathsf{S} := (\boldsymbol{M}, g)$ and $\mathsf{L}^\mathsf{p} := (\boldsymbol{L}^p, \dot{\eta})$, and the region $\mathcal{E}$ of admissible generalized stresses, and assume its boundary to be given by the yield function $\varphi(\mathsf{S}) = 0$: thus $\mathcal{E} = \{\mathsf{S} : \varphi(\mathsf{S}) \leq 0\}$. The flow relation may be written as the normality relation

$$\mathsf{L}^\mathsf{p} : (\mathsf{T} - \mathsf{S}) \leq 0 \quad \text{for all } \mathsf{T} \in \mathcal{E} \quad \text{or} \quad \mathsf{L}^\mathsf{p} = \lambda\frac{\partial\varphi}{\partial\mathsf{S}} \tag{56}$$

together with the complementarity conditions $\lambda \geq 0, \varphi \leq 0, \ \lambda\varphi = 0$. Equivalently, we introduce the convex, positively homogeneous dissipation function $D = \hat{D}(\mathsf{L}^\mathsf{p})$ and a flow relation

$$D(\mathsf{Q}) \geq D(\mathsf{L}^\mathsf{p}) + \boldsymbol{M} : (\boldsymbol{Q} - \boldsymbol{L}^p) + g(\zeta - \dot{\eta}), \tag{57}$$

where $\mathsf{Q} = (\boldsymbol{Q}, \zeta)$. Henceforth we make use of the Mises–Hill yield criterion with isotropic hardening relevant to large deformations, for which case the yield function and dissipation functions are

$$\varphi(\boldsymbol{M}, g) = |\operatorname{dev}\boldsymbol{M}| + g - c_0 \leq 0, \quad D(\boldsymbol{L}^p, \dot{\eta}) = \begin{cases} c_0|\boldsymbol{L}^p| \ \dot{\eta} \leq |\boldsymbol{L}^p|, \\ +\infty \text{ otherwise.} \end{cases} \tag{58}$$

The initial-boundary value problem and variational problem. The boundary Γ of the body is decomposed into nonoverlapping subsets Γ_u and Γ_t with $\Gamma_u \cup \Gamma_t = \Gamma$. The body force $\boldsymbol{b}_0 : [0, T] \times \Omega \to \mathbb{R}^d$ and surface traction $\bar{\boldsymbol{t}} : [0, T] \times \Gamma_t \to \mathbb{R}^d$ are assumed to be continuous in time. We also assume a prescribed, time-independent displacement $\bar{\boldsymbol{u}}$ on Γ_u. Then the initial-boundary value problem for large-deformation plasticity is as follows: find the displacement field $\boldsymbol{u}(\boldsymbol{X}, t)$ and $\boldsymbol{F}^p(\boldsymbol{X}, t)$ that satisfy the equation of equilibrium (50), the elastic relation $(55)_1$, the kinematic relations (46) and (49), flow relation (56) or (57), and boundary conditions. The weak form of the equilibrium equation is given by

$$\int_\Omega \boldsymbol{P} : \text{Grad}\, \boldsymbol{v}\, dx = \int_\Omega \boldsymbol{b}_0 \cdot \boldsymbol{v}\, dx + \int_{\Gamma_t} \bar{\boldsymbol{t}} \cdot \boldsymbol{v}\, ds. \tag{59}$$

The variational problem is then one of finding the displacement $\boldsymbol{u}$, plastic strain rate in the form $\boldsymbol{L}^p$, and hardening variable η that satisfy (56) or (57), and (59).

5.1 The Incremental Problem

The variational problem does not have an equivalent formulation as a minimization problem, but it is possible to formulate the corresponding *incremental* problem as an unconstrained minimization problem. As before the time interval is partitioned uniformly according to $0 = t_0 < t_1 < \cdots < t_N = T$. Then from (57) and (59) the incremental problem becomes one of finding $(\boldsymbol{u}_{n+1}, \boldsymbol{L}^p_{n+1}, \eta_{n+1})$ that satisfy

$$\int_\Omega \boldsymbol{P}_{n+1} : \text{Grad}\, \boldsymbol{v}\, dx = \int_\Omega \boldsymbol{b}_{0,n+1} \cdot \boldsymbol{v}\, dx + \int_{\Gamma_t} \bar{\boldsymbol{t}}_{n+1} \cdot \boldsymbol{v}\, ds, \tag{60a}$$

$$D(\mathsf{Q}) \geq D(\mathsf{L^p}_{n+1}) + \boldsymbol{M}_{n+1} : (\boldsymbol{Q} - \boldsymbol{L}^p_{n+1}) + \frac{1}{\Delta t} \int_\Omega g_{n+1}(\zeta - \Delta\eta). \tag{60b}$$

Here $\boldsymbol{M}_{n+1}$ is found from $(52)_2$ with (52), and $g_{n+1} = -\partial\psi^h/\partial\eta\big|_{n+1}$.

Consider the functional

$$J(\boldsymbol{v}, \boldsymbol{Q}, \zeta) = \int_\Omega \left[\psi^e(\boldsymbol{E}^e) + \psi^h(\zeta)\right] dx + \int_\Omega D(\Delta t \boldsymbol{Q}, \Delta\zeta)\, dx \\ - \int_\Omega \boldsymbol{b}_{0,n+1} \cdot \boldsymbol{v}\, dx - \int_{\Gamma_N} \bar{\boldsymbol{t}}_{n+1} \cdot \boldsymbol{v}\, ds. \tag{61}$$

Theorem 5.1 *If* $(\boldsymbol{u}_{n+1}, \boldsymbol{L}^p_{n+1}, \Delta\eta)$ *minimizes the functional (61) then* $(\boldsymbol{u}_{n+1}, \boldsymbol{L}^p_{n+1}, \Delta\eta)$ *is a solution to the variational problem (60).*

For a proof of this result see for example Reddy (2013).

As for the small-deformation problem, the appropriate algorithm for solving (60) is one of predictor–corrector type. We focus here on the corrector step, which entails finding $\boldsymbol{F}^e_{n+1}$ and η_{n+1} for given displacement $\boldsymbol{u}_{n+1}$ (determined in the predictor step) or equivalently $\boldsymbol{F}_{n+1}$. Noting that

$$\mathsf{L^p}_{n+1} = (\boldsymbol{L}^p_{n+1}, \Delta\eta/\Delta t),$$

we introduce the approximate identity (de Souza Neto et al. (2008), Appendix B)

$$\boldsymbol{F}^p_{n+1} = (\exp \boldsymbol{L}^p_{n+1})\boldsymbol{F}^p_n \tag{62}$$

in which exp is the matrix-valued exponential. Since $\boldsymbol{G}_{n+1} := [\boldsymbol{F}^p_{n+1}]^{-1} = \boldsymbol{G}_n \exp(-\boldsymbol{L}^p_{n+1})$, for a given deformation $\boldsymbol{F}_n$ at time t_{n+1} it follows that

$$\begin{aligned} \boldsymbol{F}^e_{n+1} &= \boldsymbol{F}_{n+1}\boldsymbol{G}_n \exp(-\boldsymbol{L}^p_{n+1})] \\ &= \boldsymbol{F}^{e,\mathrm{tr}} \exp(-\Delta\lambda_{n+1}\boldsymbol{N}_{n+1})]. \end{aligned} \tag{63}$$

Here we have defined the trial elastic deformation gradient $\boldsymbol{F}^{e,\mathrm{tr}}$ to be the value of $\boldsymbol{F}^e$ assuming no plastic flow in the time step $[t_n, t_{n+1}]$, that is, $\boldsymbol{F}^{e,\mathrm{tr}} = \boldsymbol{F}_{n+1}\boldsymbol{G}_n$, and we have also used $(56)_2$ with $\boldsymbol{N}_{n+1} = \mathrm{dev}\boldsymbol{M}/|\mathrm{dev}\boldsymbol{M}|$. Using the identity det[exp $\ldots = \exp\,[\mathrm{tr}\,(\ldots)]$, we find that $\det \boldsymbol{F}^e_{n+1} = \det \boldsymbol{F}^{e,\mathrm{tr}}_{n+1}$ so that $J^e_{n+1} = J^{e,\mathrm{tr}}_{n+1}$ and hence $J^e_{n+1} = J^{e,\mathrm{tr}}_{n+1}$. Thus the plastic incompressibility constraint is satisfied exactly.

The corrector step then entails the following steps:

(i) Given $\boldsymbol{F}^{e,\mathrm{tr}}$ and $\eta^{\mathrm{tr}} = \eta_n$, find $\mathsf{S}^{\mathrm{tr}} := (\boldsymbol{M}^{\mathrm{tr}}, g^{\mathrm{tr}})$ using $(55)_1$ and $(55)_2$.
(ii) If $\varphi(\mathsf{S}^{\mathrm{tr}}) < 0$ then update with $\Delta\lambda = 0$.
(iii) Otherwise, solve (63) and $\varphi(\mathsf{S}_{n+1}) = 0$ for $\boldsymbol{F}^e_{n+1}$ and $\Delta\lambda$.

The return mapping algorithm can be simplified considerably by making use of the logarithmic elastic strain $\boldsymbol{\varepsilon}^e$ whose components ε^e_A $(A = 1, 2, 3)$ in the spatial principal basis are given by $\varepsilon^e_A = \log B^e_A$ and $\boldsymbol{B}^e = [\boldsymbol{F}^e]^T\boldsymbol{F}^e$. Then with the yield function written as a function of the Kirchhoff stress $\boldsymbol{\tau} = J\boldsymbol{\sigma}$, the elastic strain update is found from

$$\boldsymbol{\varepsilon}^e = \boldsymbol{\varepsilon}^{e,\mathrm{tr}} - \Delta\lambda \left.\frac{\partial\varphi(\boldsymbol{\tau}, g)}{\partial\boldsymbol{\tau}}\right|_{n+1}.$$

Further details may be found in de Souza Neto et al. (2008), Simo and Hughes (1998).

Acknowledgments The support of the South African Department of Science and Technology and National Research Foundation through the South African Research Chair in Computational Mechanics is gratefully acknowledged.

References

de Souza Neto, E. A., Perić, D., & Owen, D. R. J. (2008). *Computational methods for plasticity: Theory and applications*. Chichester: Wiley.

Djoko, J. K., Ebobisse, F., Reddy, B. D., & McBride, A. T. (2007). A discontinuous Galerkin formulation for classical and gradient plasticity. Part 1. *Comp. Meths Appl. Mech. Eng.*, *196*, 3881–3897.

Gurtin, M. E., Fried, E., & Anand, L. (2010). *The mechanics and thermodynamics of continua*. Cambridge: Cambridge University Press.

Han, W., & Reddy, B. D. (2013). *Plasticity: Mathematical theory and numerical analysis* (2nd ed.). New York: Springer.

Lubliner, J. (1990). *Plasticity theory*. New York: MacMillan.

Reddy, B. D. (2013). Some theoretical and computational aspects of single-crystal strain-gradient plasticity. *Zeit. ang. Math. Mech. (ZAMM)*, *93*, 844–867.

Simo, J. C., & Hughes, T. J. R. (1998). *Computational inelasticity*. New York: Springer.

On the Use of Anisotropic Triangles with Mixed Finite Elements: Application to an "Immersed" Approach for Incompressible Flow Problems

Ferdinando Auricchio, Adrien Lefieux and Alessandro Reali

Abstract In this chapter, we discuss the use of some common mixed finite elements in the context of a locally anisotropic remeshing strategy, close in philosophy to "immersed" approaches for interface problems. A characteristic of the present method is the presence of highly flat triangles. Such a distinctive feature may imply stability issues for mixed elements with incompressible flow problems. First, we present a review of the literature dealing with interface problems and we illustrate these results with a simple 1D framework alongside of numerical tests. Second, we present the locally anisotropic remeshing approach for interface problems in 2D with a focus on the incompressible Stokes problem. We then present numerical tests to show stability issues of common mixed elements, as well as possible stable ones. We also deal with conditioning issues. Finally, we illustrate the results with two applications, including the fluid–structure interaction of a rotational rigid bar.

1 Introduction

Solutions to engineering problems depend more on numerical methods. One of these methods, the so-called Finite Element Method (FEM), has acquired over the years a central role in solving such problems. Its versatility relies on the capacity of the method to discretize the physical domain of the problem into simple elements: triangles, quadrilateral, tetrahedra, etc. However, for problems with complex geometries, interfaces, or involving important topological changes in time, such as fluid–structure

F. Auricchio · A. Reali
Department of Civil Engineering and Architecture, University of Pavia, Pavia, Italy
e-mail: auricchio@unipv.it

A. Lefieux (✉)
Department of Medicine, Emory University, Atlanta, GA, USA
e-mail: adrien.lefieux@unipv.it

A. Reali
Institute for Advanced Study, Technische Universitat München, Munchen, Germany
e-mail: alereali@unipv.it

J. Schröder and P. Wriggers (eds.), *Advanced Finite Element Technologies*,
CISM International Centre for Mechanical Sciences 566,
DOI 10.1007/978-3-319-31925-4_8

interaction problems, the construction of the domain partition becomes a bottleneck. The present work discusses some issues and solutions of some classes of such problems.

A key issue of the FEM is to measure how fast the numerical solutions converge to the solution of the continuous ones. The rate of convergence of the FEM depends on the degree of the polynomial, in case we use polynomials, and of the regularity of the solution of the continuous problem. For instance, if the solution is sufficiently regular, then piecewise linear elements converge quadratically in the L^2-norm. The regularity of the solution of the continuous problem may depend on the geometry of the domain, initial and boundary conditions, on the physical parameters, as well as on the load. In particular, in the case of interface problems, the physical parameters induce a singularity along the interface. However, if these singularities are not dealt with care, numerical errors are introduced, which might spread over the whole domain or might be troublesome if having an accurate solution on the interface is important, as for coupled problems.

In particular, fluid–structure interaction is a coupled problem where the common boundary between the fluid and the solid defines the interface. For such a type of problems, their solutions are expected to show singularities or discontinuities across the interface. However, the solution is likely to be regular away from the interface and thus a strategy could be to find a way to correctly capture the singularities only in the proximity of the interface. We may divide such problems into two distinctive categories: with *sharp* or *spread* interfaces. For the former, the interface is codimension one with respect to the geometry of the problem, while the latter is codimension zero. In this work, we only deal with sharp interfaces.

A common approach to solving sharp interface problems is to divide the "global" domain, i.e., the domain of definition of the problem, into sub-domains that are separated by the interfaces. Then, sub-finite element spaces are built on each sub-domain and interfacial constraints are enforced between the spaces. If we look at the problem in such a way we might see new issues, in particular for moving interfaces: how to build a mesh for the sub-domains and how to enforce the constraints at the interface.

The second issue is also an active area of research, especially regarding weak enforcement of interfacial constraints, i.e., the constraints are enforced in the weak formulation of the problem and not in the finite element spaces. In the present work, we do not get into details regarding weak strategies for enforcing interfacial constraints. Rather, the method we present relies on a strong enforcement of interfacial constraints; that is directly into the finite element spaces.

In the first section, we discuss interface problems within the context of a simple 1D framework by comparing the theory and numerical experiments. The results are not original (see, e.g., Babuška and Strouboulis 2001, where the error theory for finite elements with low regularity problems are extensively discussed); nevertheless, we believe that this section provides a simple discussion of the subject with recent results from the literature. It also serves as an inception and it motivates the strategy presented in the second section for the 2D case.

In the second section, we start by reviewing the literature on immersed methods and then we present the main strategy that consists in remeshing solely the elements of the mesh of the global domain that are cut by the interface. Indeed, we always consider an initial mesh given over the whole domain, which we remesh in a particular way. For 2D problems, such a strategy involves anisotropic triangles which may be very flat. It is well known that the FEM retains optimal convergence on anisotropic meshes (under several conditions, which we discuss and describe precisely) but the discussion, in the literature, of the present strategy for incompressible flow problems with the mixed finite element, has been limited.

The issue of stability of mixed finite element method within the presented remeshing strategy is the core of the present chapter. Indeed, there exist many velocity–pressure finite element schemes that have been formally proven to be (inf-sup) stable. However, most of the proofs require the mesh to be isotropic and the necessity of this assumption is an open issue for many mixed finite element schemes. To clarify, we first present stability results for four common mixed finite elements ($\mathbf{P}_2/P_0$, $\mathbf{P}_2/P_1$, $\mathbf{P}_2^+/P_1^d$ and $\mathbf{P}_2^+/P_1$, where $+$ designates a cubic bubble and a superscript d indicates that the pressure is discontinuous) on a set of simple tests and then we provide two applications showing stability issues, in particular for the elements with discontinuous pressures.

2 A 1D Interface Toy Problem

In this section, we provide a simple 1D presentation of the issues related to interface problems and immersed methods in order to clearly enlight what follows in higher dimensions and to justify the strategy employed in the second section.

Typically, solutions of elliptic interface problems are not smooth over the whole domain, but they are smooth away from the interface (see, e.g., Kellogg 1971; Lemrabet 1977; Nicaise 1993). Earliest error estimates can be traced back to 1970 with the work of Babŭska (see Babuška 1970) in which the author provided a method based on a penalty approach. An almost optimal order of convergence is recovered for piecewise linear elements in the H^1-norm, more precisely $\mathcal{O}(h^{3/4})$. In Barrett and Elliott (1987), the authors proposed to enrich the finite element space on elements cut by the interface and to enforce weakly the continuity constraint using a penalty approach. They proved the optimal error estimate in the H^1-norm but the estimate in the L^2-norm is still suboptimal, i.e., $\mathcal{O}(h^{3/2})$. However, these authors show that the optimal error in the L^2-norm can be recovered far away of the interface. In Hansbo and Hansbo (2002) a similar method is proposed using the Nitsche method, instead of the penalty method, and the optimal error estimate is obtained in both H^1- and L^2-norms. In Li et al. (2010) the finite element space is not enriched but a constraint on the mesh around the interface is added. The constraint consists in defining a "resolution" of the interface by the mesh, and it has to be at least of $\mathcal{O}(h^2)$ for piecewise linear elements such that the optimal rate of convergence for the L^2- and H^1-norms can be attained.

We aim at studying a simple model reproducing the typical characteristics of a fluid-structure problem, that is, a problem with continuity of the primal fields and with a possible discontinuity in the gradient at the interface. In fact, we focus on a steady Poisson problem defined in 1D domains with two different materials, where one surrounds the other; the problem under consideration requires that the continuity of the primal fields and of their fluxes is maintained at the interface.

Finally, we perform numerical tests to analyze the convergence properties in the L^2-norm, in the H^1-norm, and in the H^1-norm far away from the interface, with different material parameters.

2.1 Model Problem

We consider a Poisson problem characterized by two distinct materials, such that at the interface only continuity of the primal fields and of the corresponding fluxes have to be guaranteed.

As described in Fig. 1a, Material 1 and Material 2 are defined in Ω_1 and Ω_2, respectively, with $\Omega_1 =]A, B[\cup]C, D[$ and $\Omega_2 =]B, C[$. We denote the interface between Ω_1 and Ω_2 by Γ (i.e., $\Gamma = \{B, C\}$). The global domain Ω is the union of Ω_1, Ω_2, and Γ, that is $\Omega =]A, D[$. External boundaries (i.e., $\partial\Omega = \{A, D\}$) are denoted by Σ.

In the following, we introduce classical functional spaces that will be used in the rest of the chapter. In particular, $L^2(\Omega)$ is the space of square integrable functions on Ω, $H^1(\Omega)$ is the space of functions defined on Ω that belong to $L^2(\Omega)$ together with their first derivative, and $H^1_0(\Omega)$ the space of functions belonging to $H^1(\Omega)$ and vanishing on $\partial\Omega$.

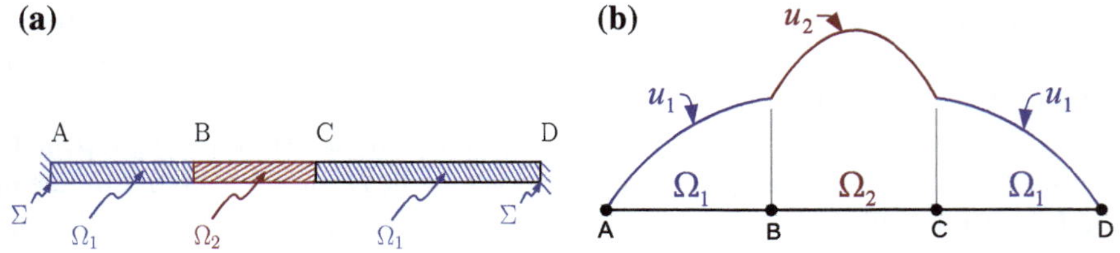

Fig. 1 A two-material 1D poisson framework. **a** Geometry of the problem. **b** Analytical solution for a uniform load, showing a kink at the interface

The strong formulation for the described problem can be written as follows:
Find *two* functions $u_1 : \Omega_1 \to \mathbb{R}$ and $u_2 : \Omega_2 \to \mathbb{R}$ smooth enough such that

$$\begin{cases} -(\alpha_1 u_1')' = f_1 & \text{on} \;\; \Omega_1, \\ -(\alpha_2 u_2')' = f_2 & \text{on} \;\; \Omega_2, \\ u_{1|\Gamma} = u_{2|\Gamma}, \\ (\alpha_1 u_1')_{|\Gamma} = (\alpha_2 u_2')_{|\Gamma}, \\ u_{1|\Sigma} = 0, \end{cases} \tag{1}$$

where $\alpha_1 \geq \bar{\alpha} > 0$, $\alpha_2 \geq \bar{\alpha} > 0$, f_1, and f_2, are given regular functions, and $u_{|\Gamma}$ is the restriction of u on Γ.

Remark 2.1 In Problem (1), we consider for simplicity homogeneous Dirichlet boundary conditions on Σ but other boundary conditions can be considered as well.

Considering the spaces

$$V := \left\{(u_1, u_2) \in H^1(\Omega_1) \times H^1(\Omega_2) \big| u_1|_\Gamma = u_2|_\Gamma\right\}$$

and

$$V_0 := \left\{(u_1, u_2) \in H^1(\Omega_1) \times H^1(\Omega_2) \big| u_1|_\Gamma = u_2|_\Gamma \text{ and } u_1|_\Sigma = 0\right\},$$

the standard weak formulation corresponding to Problem (1) can be readily obtained as:

Find $(u_1, u_2) \in V_0$ for all $(v_1, v_2) \in V_0$ such that

$$\int_{\Omega_1} \alpha_1 u_1' v_1' \mathrm{d}x + \int_{\Omega_2} \alpha_2 u_2' v_2' \mathrm{d}x = \int_{\Omega_1} f_1 v_1 \mathrm{d}x - \int_{\Omega_2} f_2 v_2 \mathrm{d}x. \tag{2}$$

2.2 A Single Field Formulation

The one field strong formulation for Problem (1) can be written as follows:
Find *one* function $u : \Omega \to \mathbb{R}$ with $u_{|\Sigma} = 0$ such that

$$\begin{cases} -(\alpha u')' - f = 0 & \text{on} \;\; \Omega, \\ [\![\alpha u']\!]_\Gamma = 0, \end{cases} \tag{3}$$

with

$$\alpha = \begin{cases} \alpha_1 & \text{on} \;\; \Omega_1 \\ \alpha_2 & \text{on} \;\; \Omega_2 \end{cases} \text{ and } f = \begin{cases} f_1 & \text{on} \;\; \Omega_1 \\ f_2 & \text{on} \;\; \Omega_2 \end{cases}. \tag{4}$$

In (3), the derivatives are in the sense of distribution and the symbol $[\![\cdot]\!]_\Gamma$ denotes the jump on Γ.

The weak formulation for Problem (3) reads: Find $u \in H_0^1(\Omega)$ such that

$$\int_\Omega \alpha u'v'\mathrm{d}x = \int_\Omega fv\mathrm{d}x \qquad \forall v \in H_0^1(\Omega). \tag{5}$$

Moreover, in Problem (5), we look for a function $u \in H_0^1(\Omega)$, thus satisfying automatically the continuity of the primal field on Γ. The continuity of the flux on Γ is instead naturally enforced in the weak formulation by the continuity of the test function; see also Carey (1982), Carey et al. (1985), van Brummelen et al. (2011).

In the next section, we provide a discretization for (5).

2.3 Discrete Formulation and Error Estimates

The discrete problem reads as follows. Given a finite-dimensional space $V_h \subset H_0^1(\Omega)$, the discrete formulation for Problem (5) can be readily obtained as:

Find $u_h \in V_h$ such that

$$\int_{\Omega_h} \alpha(u_h)'(v_h)'\mathrm{d}x = \int_{\Omega_h} fv_h\mathrm{d}x \qquad \forall v_h \in V_h. \tag{6}$$

Now, given the following approximation

$$u_h(x) = \mathbf{N}(x)\hat{\mathbf{u}},$$

with $\mathbf{N}(x)$ being the standard piecewise linear shape functions defined on Ω_h (where Ω_h describes a mesh over Ω) and $\hat{\mathbf{u}}$ the primal field nodal value vector, the algebraic formulation corresponding to Problem (6) reads

$$\mathbf{A}\hat{\mathbf{u}} = \mathbf{b}, \tag{7}$$

where

$$\begin{cases} \mathbf{A}_{|ij} = \int_{\Omega_h} \alpha N_i' N_j' \mathrm{d}x, \\ \mathbf{b}_{|i} = \int_{\Omega_h} fN_i \mathrm{d}x. \end{cases}$$

We note that in (7), the evaluation of the integrals defined over Ω_h is not an easy task since α is discontinuous on Γ. From a practical point of view, the main issue is to integrate over sub-elements which may be not trivial in higher dimensions. We do not discuss in detail such an issue in the present chapter, but the interested reader is referred to, e.g., Fries and Belytschko (2010), Chin et al. (2015) for a review on possible integration strategies.

In case, α and f are sufficiently smooth, more precisely with $\alpha \in C^1(\Omega)$ and $f \in L^2(\Omega)$, then $u \in H^2(\Omega)$ and as a consequence one can prove that

$$||u - u_h||_{0,\Omega} \leq Ch^2 \qquad ||u - u_h||_{1,\Omega} \leq Ch^1, \tag{8}$$

where $||u - u_h||_{0,\Omega}$ and $||u - u_h||_{1,\Omega}$ denote, as usual, the errors in the L^2-norm and H^1-norm, respectively, and C is a constant independent of h which might be different for each norm estimate.

However, for our application $\alpha_1 \neq \alpha_2$ on Γ and thus $\alpha \notin C^1(\Omega)$ which implies that we have now the following estimates (see, e.g., Ern and Guermond 2004)

$$||u - u_h||_{0,\Omega} \leq Ch^1 \qquad \lim_{h \to 0} ||u - u_h||_{1,\Omega} = 0, \tag{9}$$

but, as we shall see, numerical results indicate that the error in the H^1-norm is $\mathcal{O}(h^{1/2})$.

It is easy to understand why there is a loss of accuracy in the estimates in (9) with respect to the estimates in (8); while using a mesh over Ω that is defined independently of the position of the interfaces. Indeed, as it can be seen on Fig. 2, when a node does not fit the interface we actually try to approximate a discontinuous function by a continuous one, while if a node fits the interface then the finite element can exactly approximate the discontinuous function.

So, in order to solve the problem one can simply remesh at the interface, which actually consists in adding a function to the finite element space that captures the discontinuity in the gradient. This approach can be related to the Partition of Unity Method (PUM) (see, e.g., Melenk and Babuška 1996) or the eXtended Finite Element Method (XFEM) (see, e.g., Fries and Belytschko 2010).

In the next section, we perform some numerical tests for meshes that do not fit the interface.

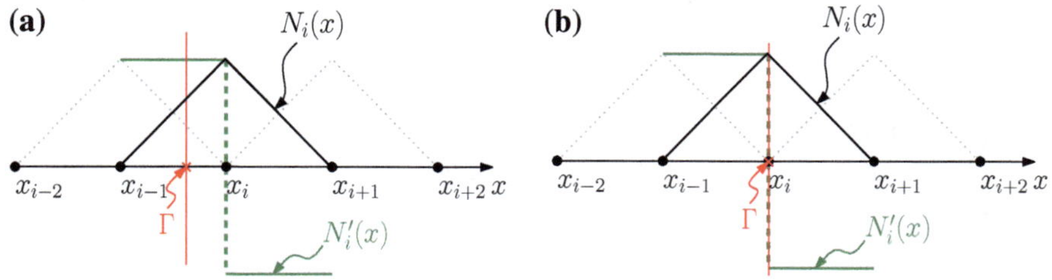

Fig. 2 Finite element basis for Problem (6) with respect to the position of the interface. **a** Unfitted mesh. **b** Fitted mesh

Table 1 Material parameters definitions

Material	Test 1	Test 2	Test 3	Test 4
α_1	1	4	1	100
α_2	4	1	100	1

2.4 Numerical Tests

We consider a *h*-refinement strategy with piecewise linear finite elements on a uniform mesh and different material parameters (see Table 1).

For all methods, we deal with the following geometry: $A = 0, B = e, C = 1 + \pi$, $D = 6$ (see Fig. 1a for a description of the geometry). Interfaces B and C are such that the problem remains unfitted for all refinement steps, and, to accomplish this goal easily, we select irrational numbers for B and C and rational numbers for A and D. The material parameters for Material 1 (α_1) and Material 2 (α_2) are chosen constant on $]A, B[\cup]C, D[$ and $]B, C[$, respectively, and we select constant loads $f_1 = 1$ on $]A, B[\cup]C, D[$ and $f_2 = 1$ on $]B, C[$.

The different sets of material parameters are given in Table 1 with the corresponding analytical solutions reported in Fig. 3.

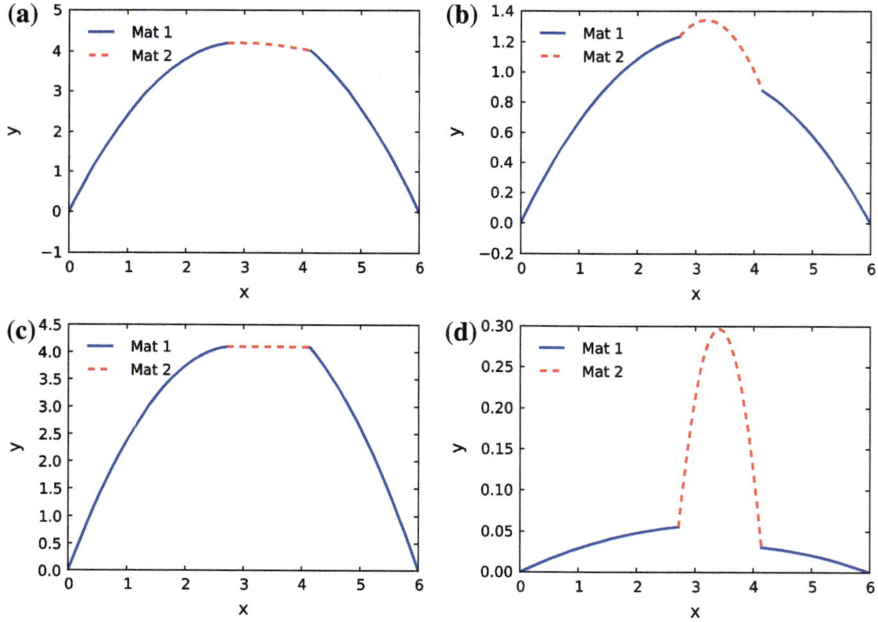

Fig. 3 Analytical solutions for the numerical test with $f = 1$ on $]A, D[$, for the different material parameters reported in Table 1. **a** Test 1. **b** Test 2. **c** Test 3. **d** Test 4

Error Measurement We use the relative error in the L^2-norm, defined by

$$E_{0,\Omega} = \frac{\left(\int_\Omega (u - u_h)^2 \mathrm{d}x\right)^{\frac{1}{2}}}{\left(\int_\Omega u^2 \mathrm{d}x\right)^{\frac{1}{2}}} = \frac{||u - u_h||_{0,\Omega}}{||u||_{0,\Omega}}, \tag{10}$$

where u is the analytical solution of the problem over Ω.

In the same fashion, we define the relative H^1-seminorm (denoted sH^1 in the figures) by

$$E_{1,\Omega} = \frac{\left(\int_\Omega (u' - u_h')^2 \mathrm{d}x\right)^{\frac{1}{2}}}{\left(\int_\Omega (u')^2 \mathrm{d}x\right)^{\frac{1}{2}}} = \frac{|u' - u_h'|_{1,\Omega}}{|u'|_{1,\Omega}}. \tag{11}$$

We note that the H^1-seminorm is equivalent to the H^1-norm in virtue of the Poincaré-Friedrichs inequality.

In Li et al. (2010), it is pointed out that when computing the H^1-norm away from the interface the optimal convergence rate in the H^1-norm can be obtained, precisely when using the error measurement:

$$E_{1,\Omega\backslash\Gamma_\epsilon} \quad \text{with} \quad \Gamma_\epsilon = \{x \in \Omega : \mathrm{dist}(x, \Gamma) < \epsilon\}. \tag{12}$$

In Li et al. (2010), a constraint for the construction of the mesh is added. It is required that the mesh is "ϵ-resolved" near the interface, i.e., there must not be an element that overlaps Γ_ϵ. In that work, it is proved that for $\epsilon = \mathcal{O}(h^2)$ the method has the optimal rate of convergence in both L^2- and H^1-norms. However, in our numerical experiments the mesh is not ϵ-resolved for $\epsilon = \mathcal{O}(h^2)$ but it is for $\epsilon = \mathcal{O}(h)$. For this reason, we choose $\epsilon = h$. In the present numerical tests, we show that we do not have the optimal rate of convergence for the $H^1(\Omega)$- and $L^2(\Omega)$-norms, but we may attain it using the $H^1(\Omega\backslash\Gamma_\epsilon)$-norm.

We observe in Fig. 4 that the rates of convergence oscillate, but averagely a convergence of order 1 is attained in L^2-norm and of an order 1/2 in H^1-norm. Instead, in the $H^1(\Omega\backslash\Gamma_\epsilon)$-seminorm a linear order of convergence is achieved.

We can observe in Fig. 5a that the error $u - u_h$ at the interface propagates to the whole domain, preventing a possible optimal convergence rate in the L^2-norm away from the interface. On the contrary, for the H^1-norm, we can observe (see Fig. 5b) that the error in the derivatives clearly converge linearly away from the interface, even showing a super-convergence property at the middle of the elements not cut by the interface. Differently, to $u - u_h$, the quantity $u' - u_h'$ does not appear to converge on the interface, but here the support of the large error values is limited to the elements crossed by the interface. It follows that the optimal rate of convergence would be obtained if the error is integrated only on elements not crossed by the interface. This example also shows that if the mesh is ϵ-resolved with $\epsilon = \mathcal{O}(h^2)$, then we may obtain the optimal rate of convergence in the H^1-norm for a smaller ϵ, i.e., $\epsilon < h$, as showed in Li et al. (2010).

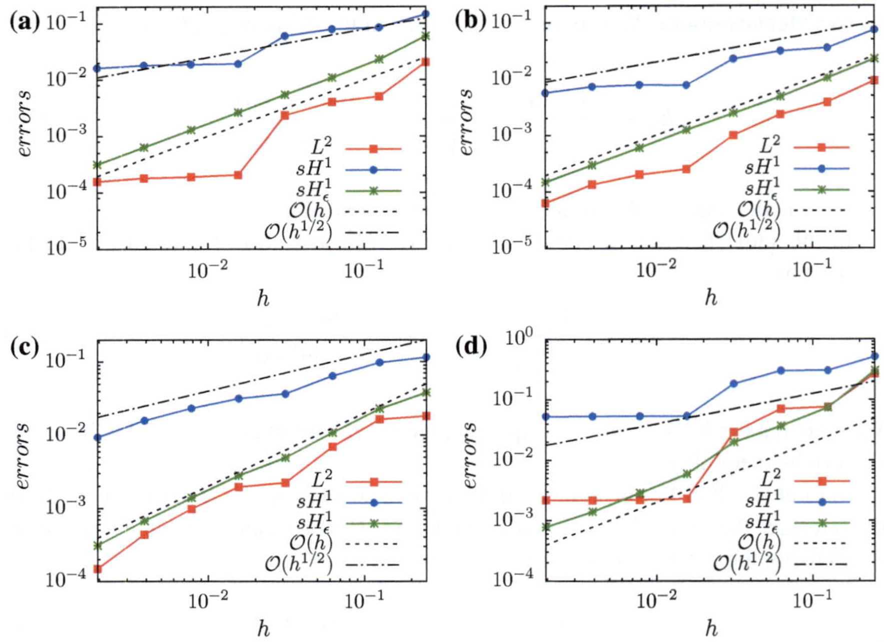

Fig. 4 Convergence plots in the various norms. **a** Test 1 ($\alpha 1/\alpha 2 = 1/4$). **b** Test 2 ($\alpha 1/\alpha 2 = 4$). **c** Test 3 ($\alpha 1/\alpha 2 = 1/100$). **d** Test 4 ($\alpha 1/\alpha 2 = 100$)

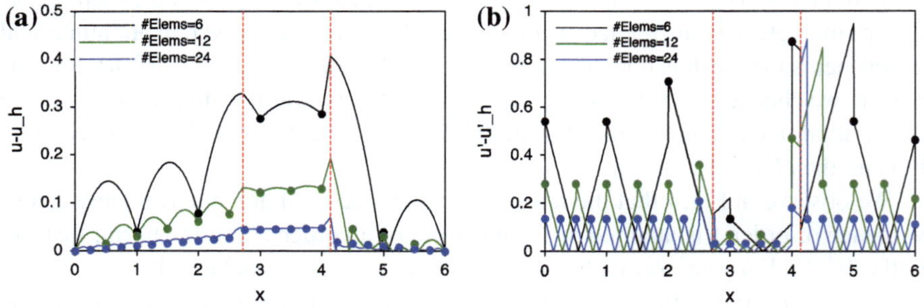

Fig. 5 Errors for test 1 ($\alpha_1/\alpha_2 = 1/4$). The *dots* symbolize the position of the nodes, while the *red lines* the position of the interface. **a** Error $u - u_h$. **b** Error $u' - u'_h$

2.5 Conclusive Remarks

Clearly, the order of accuracy of the Galerkin method is limited by the regularity of the solution and by the fact that the computational mesh does not fit the interface. The solution we envision is simply to remesh along the interface while all vertices of the initial mesh are kept fixed. Such a strategy implies the presence of triangles that are highly distorted and, in the context of the mixed finite element method for incompressible flow problems, additional considerations, specific to 2D problems, have to be made. Such considerations are described in detail in the following sections.

3 Extension to 2D: An Anisotropic Remeshing Strategy

3.1 Immersed Approaches

We consider the term *immersed approach* as a class of methods and not as a method in particular. The "immersed approach" can be seen as a concept, consisting of all methods that do not require a-priori a discrete (geometrical) representation of the geometries of the physical bodies under consideration. In the literature, this concept can be found under many names, such as *immersed boundary*, *unfitted*, *fictitious domain*, and *embedded methods*.

The main idea is as follows. On the one hand, we consider a physical domain composed of a single or multiple bodies (fluids or solids for instance) and on the other an extended domain, which is larger than the physical one. The immersed boundary class of methods aim at describing the physical problem on such an extended domain. Ultimately, a numerical method is used, not on the physical domain, but on the extended domain. Even if the concept may apply to a wide range of problems in the present document, we have fluid–structure interaction problems in mind.

Original Immersed approaches can be dated back to the 1970–1990s and they may be divided into two main approaches: the Immersed Boundary Method (IBM) originally developed by C. Peskin (2002) and the Fictitious Domain Method (FDM) originally developed by R. Glowinski (2003).

Past (1970–1990s): The Immersed Boundary Method may be traced back to the work of C. Peskin (see, e.g., Peskin 1977, or even a bit earlier if we consider Hyman 1952). The method developed by C. Peskin is based on the finite difference method, but the method was extended to the finite element method in, e.g., Boffi et al. (2008).

The main idea is to rewrite the problem as a function of a single field defined on the extended domain, which is the union of the fluid and the solid domains. In general, the fluid model is extended over the solid domain, and the solid problem acts as a constraint on the fluid extended domain. It follows that the value of the extended fluid field describes the fluid in the fluid domain and the solid in the solid region. We point out that the fluid problem is written in the Eulerian setting while the solid part is written in the Lagrangian setting which is transformed, using a specific mapping, to the Eulerian setting in order to act as a body force onto the extended fluid.

Originally, the IBM was developed for incompressible slender bodies (i.e., codimension one with respect to the fluid). It was extended to large bodies (i.e., codimension zero with respect to the fluid) with various densities and viscous coefficients (see Peskin 2002 for a finite difference setting and Boffi et al. 2011 for a finite element setting). Incompressibility of the solid is a key component of the previously cited work since, usually, incompressibility is assumed for the fluid and, therefore, conservation of volume is required also for the solid. However, more general settings have been developed both for the finite difference method (see Bhalla et al. 2013) and the finite element method (see Heltai and Costanzo 2012) to allow for compressible solids.

The Fictitious Domain Method may be traced back to Glowinski et al. (1994) (see, e.g., Glowinski 2003 for a review). The method is based on the finite element approach and, on the contrary to the IBM of C. Peskin, is a two-field method, that is both fields (the fluid in an Eulerian setting and the solid in a Lagrangian setting) are considered, but the fluid is evaluated on the extended domain and not only on the fluid physical domain. A key aspect of the method is that specific techniques are used to imposed (weakly) the essential constraints (continuity of the velocities) between the fluid and the solid. Two techniques are used to enforce interface constraints: continuity only at the interface (for instance with a boundary Lagrange multiplier) or by matching the solid domain velocity to the (extended) fluid velocity in the solid domain (for instance with a distributed Lagrange multiplier).

Originally the FDM was developed for fluid/rigid body interactions, but the method was extended to fluid/incompressible slender body interactions in Diniz dos Santos et al. (2008) and fluid/incompressible large solid interactions (see, e.g., Yu 2005). We point out that a much more detailed presentation of immersed approaches is provided in Auricchio et al. (2014) in a similar 1D setting as in the first section.

Present (2000–2010s): The methods discussed do not represent explicitly (i.e., explicitly in the finite element spaces) the jump in the fluid gradient at the interface which leads to a loss of accuracy of the methods. Many methods have been developed from the 2000s to tackle this problem. In particular, we may consider the following methods: the immersed interface method, the fat boundary method, the finite cell method, the extended finite element method, and local remeshing strategies.

The Immersed Interface Method (IIM) was originally developed using the finite difference method but the method was extended to the finite element method (see Li and Ito (2006) for a review). The IIM modifies locally (i.e., on elements crossed by the interface) the shape functions such that they represent the interface constraints, introducing physical parameters in the definition of the shape functions. A key point of the method is that a linear system of equation has to be computed on each element cut by the interface in order to recompute all shape functions such that interface constraints are correctly taken into account.

The Fat Boundary Method (FBM) has been developed in Maury (2001) and is a type of Dirichlet-Neumann domain decomposition approach and as such it is an iterative method. In this method, the representation of the interface is not "sharp" (i.e., a codimension one representation of the interface) but "spread" (i.e., a codimension zero representation of the interface, explaining the term "fat").

The Finite Cell Method (FCM) was developed in Parvizian et al. (2007) and is a high-order immersed method. The FCM takes into account the presence of the interface by integrating only in the physical domain and not over the whole domain as in the previously cited methods. In this manner, the finite cell method is similar to the eXtended Finite Element Method (see, e.g., Hansbo and Hansbo 2002; Haslinger and Renard 2009).

The XFEM was developed to tackle problems with singularities (the original XFEM can be traced back to Moës et al. (1999)) and, therefore, the method can be used to represent the jump in the fluid gradient at the interface, as done in Gerstenberger and Wall (2008). The method consists of enriching the finite element basis with tailored shape functions to represent singularities present in the solution of the problem. A key issue with the XFEM is the enforcement of essential "immersed" boundary condition, as already pointed out in Girault and Glowinski (1995) for the fictitious omain method with boundary Lagrange multipliers. For this reason, important researches have been done in order to ensure a correct enforcement of essential constraints. In particular, we may consider the Lagrange multiplier method, investigated in, e.g., Béchet et al. (2009), Hautefeuille et al. (2012), a stabilized by projection Lagrange multiplier method (see, e.g., Burman and Hansbo 2010; Amdouni et al. 2014; Barrenechea and Chouly 2012) or using the Nitsche method (see, e.g., Hansbo and Hansbo 2002; Sanders et al. 2012; Burman and Hansbo 2011a; Zunino et al. 2011). Most of the previously cited papers are for Poisson problems and few results exist for the Stokes problem, (see, e.g., Burman and Hansbo 2011b; Massing et al. 2012; Hansbo et al. 2013, Bazilevs and Hughes 2007).

An alternative approach to the XFEM has been developed in Ilinca and Hétu (2011). The method consists in a local remeshing on the elements cut by the interface. In the previously cited work, a low-order stabilized finite elements pair is used, which results in the possibility to eliminate the pressure by static condensation and because, Dirichlet boundary conditions are imposed at the interface: no nodes are added to the system. However, two remarks may be drawn: first only a staggered scheme can be used, second it is only a low order scheme. A higher order approach has been developed in Auricchio et al. (2015a) which results in added degrees of freedom. Moreover, the authors in Ilinca and Hétu (2011) claim that their approach is robust on distorted elements (i.e., on flat elements the inf-sup condition is maintained), while in Auricchio et al. (2015a) specific cares have to be taken on how to choose the finite elements pairs, due to inf-sup stability issues on distorted elements. We point out that anisotropic elements are also used for interface problems in Frei and Richter (2014).

The main aspect of the last three cited methods is that they are purely geometrical such that *almost* no assumptions are made on the fluid or on the solid. As a consequence, they are fairly general approaches and they may be used for a wide range of problems.

Of course, the presented state of the art is not exhaustive. Many other approaches may be considered, such as the methods presented in Lew and Buscaglia (2008) or in Basting and Weismann (2013) with the discontinuous Galerkin method, and in Hachem et al. (2013) by smoothing the interface and by using anisotropic meshes. Differently, in Fabrèges (2012), an iterative approach in the context of the fictitious domain method with boundary Lagrange multiplier for codimension zero "immersed" bodies is used with a specific body load extension to avoid a jump in the gradient at the interface.

In the next sections, we present the results described in detail in Auricchio et al. (2015b). First, we discuss the geometric aspects of the method and the considered model (i.e., the incompressible Stokes problem) in Sects. 3.2 and 3.3. We then provide an extensive discussion on mesh regularities assumptions in Sect. 3.4, such as the minimal and maximal angle conditions. We also provide a proof that the method satisfies the maximal angle condition. Since our work focuses on incompressible fluid materials, we discuss the inf-sup condition on distorted elements and we provide results for the two common pairs of mixed elements ($\mathbf{P}_2/P_1$ and $\mathbf{P}_2^+/P_1$, stabilized on flat triangles by a cubic bubble), and additional results for two other mixed finite elements with discontinuous pressures: $\mathbf{P}_2/P_0$ (Fortin's element) and $\mathbf{P}_2^+/P_1^d$ (i.e., Crouzeix-Raviart's element). This discussion is the core of the present work. In Sect. 3.5, using a set of numerical tests, we show that the elements with continuous pressure are much more stable than their counterpart with discontinuous pressures when elements are distorted, and that the discontinuous pressure mixed schemes should actually be avoided. We also focus on the conditioning of the various matrices involved and we compare the results with estimates in the literature.

Finally, we provide two practical applications showing how stability (or instability) of the inf-sup condition on distorted elements may adversely impact the solution of the problem we wish to solve. In particular, for the mixed elements with discontinuous pressures along the elements edges. The first application is a Stokes flow problem around a disk (extensively described in Auricchio et al. 2015a with the continuous pressure elements) and the second application is a fluid–structure interaction problem using the incompressible Navier-Stokes flow and an "immersed" rotational rigid bar (extensively described in Auricchio et al. 2015b).

3.2 Geometry

In this section, we consider the geometric aspects of the method, i.e., the problem of the construction of a mesh conveniently discretizing the considered physical domain. Two strategies are possible: "fitted" or "unfitted" (cf. Fig. 6).

In the fitted approach, the discretized domain fits the boundary of the problem while in the unfitted approach the physical domain is a subset of the discretization. More precisely, in the unfitted case, we consider a problem defined in $\Omega \subset \mathbb{R}^2$ such that a part of the boundary of $\partial\Omega$, denoted by Γ (named *immersed boundary*) is not fitted a-priori by the triangulation of $\hat{\Omega}$, with $\Omega \subset \hat{\Omega}$. Part of the boundary $\partial\Omega$ that is fitted by the triangulation of $\hat{\Omega}$ is denoted by Σ.

We illustrate the problem in Fig. 6. To avoid the difficulties and the costs connected with the generation of fitted meshes in complicated situations, we propose to start with a regular unfitted mesh $\hat{\Omega}$ and to represent Γ by a linear reconstruction on such a triangulation, as illustrated in Fig. 7. The reconstruction procedure is presented in detail in the next section.

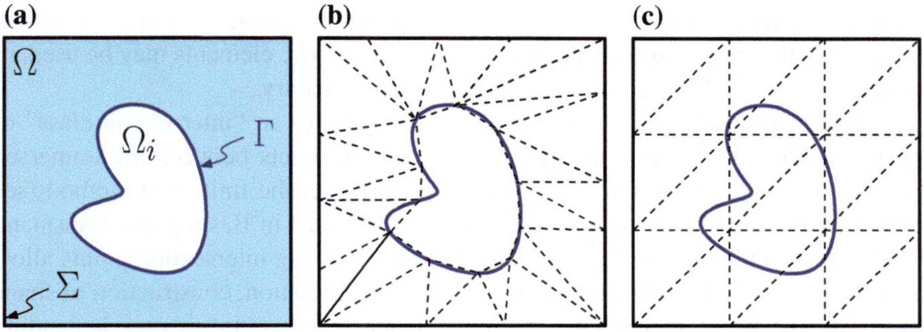

Fig. 6 Fitted and unfitted discretizations of the physical region Ω: Ω_i is the interior (nonphysical) domain, Γ is the immersed boundary, $\Sigma = \partial\hat{\Omega}$ is the external boundary, and $\hat{\Omega} := \Omega \cup \Omega_i \cup \Gamma$ is the discretized domain. **a** Physical domain. **b** Fitted grid. **c** Unfitted grid

Interface Reconstruction We assume that a regular triangulation $\hat{\mathcal{T}}$ of $\hat{\Omega}$ (named *background* mesh) and the interface Γ satisfy the conditions presented in Hansbo and Hansbo (2002), that is, the boundary Γ crosses once the two triangle edges. We note that there always exists a sufficiently fine triangulation of $\hat{\Omega}$ such that the conditions are fulfilled for any smooth immersed boundary. We point out that from a practical point of view and for dynamic problems these conditions are quite restrictive but they could be relaxed. Such issues are the focus of ongoing works. The reconstructed boundary of Γ is denoted Γ_h and it is the linear interpolation of all intersections with the background mesh edges. It follows that the reconstructed interface is a segment in each intersected element, and it defines a new domain Ω_h such that $\partial\Omega_h = \Sigma \cup \Gamma_h$ (cf. Fig. 7). The domain Ω_h is referred to as integration domain. We point out that

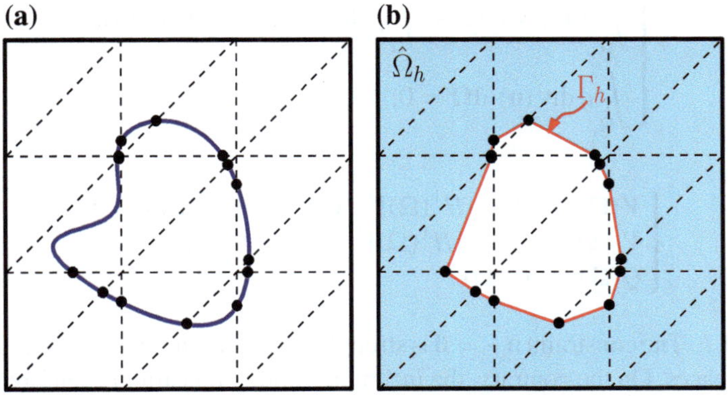

Fig. 7 Description of the interface reconstruction process. The immersed boundary is denoted by Γ and the linear reconstruction of the immersed boundary, with respect to the background mesh, is denoted by Γ_h. In the remainder of the chapter, we also consider the integration domain Ω_h (in *blue*), defined such that $\partial\Omega_h = \Sigma \cup \Gamma_h$. **a** Immersed boundary and a triangulation of $\hat{\Omega}$. **b** Interface reconstruction (in *green*) and integration domain (in *blue*)

the linear reconstruction of Γ is not a limitation of the method we propose and that, in a case with a curved immersed boundary, isoparametric elements may be used, as well as more complex algorithms, to describe the boundary.

We consider such types of methods as belonging to an "intersection class" of methods since they require to compute intersection points between the immersed boundary and the mesh. On the contrary, for instance, the finite cell method (see Parvizian et al. 2007) or the approach recently proposed in Basting and Weismann (2013) does not belong to this class of methods. Knowing intersection points allow a subdivision of the mesh, which may be used for integration, construction of shape functions, etc. We point out that computing the intersection points is very demanding in terms of computational cost, and is a fundamental part of all codes using such an approach.

3.3 The Incompressible Stokes Problem

Let $\Sigma = \Sigma_D \cup \Sigma_N$ where Σ_D denotes the part of the external boundary on which we impose a Dirichlet boundary condition and Σ_N the part on which we impose a Neumann boundary condition, whose value is assumed to be zero without loss of generality. On the other hand, on Γ, we consider homogeneous Dirichlet boundary conditions on Γ but non-homogeneous Dirichlet boundary conditions can be applied as well. Neumann boundary conditions are not considered here because they can be enforced "naturally" in the variational formulation, and as a consequence, they are easier to tackle. The model problem we consider in this chapter is given by the following standard weak form of the incompressible Stokes equation:

Find $(\mathbf{u}, p) \in V(\Omega) \times Q(\Omega)$ such that $\forall (\mathbf{v}, q) \in V_0(\Omega) \times Q(\Omega)$:

$$\begin{cases} \displaystyle\int_\Omega \nabla\mathbf{u} : \nabla\mathbf{v}\, \mathrm{d}\Omega - \int_\Omega p \,\mathrm{div}\,(\mathbf{v})\, \mathrm{d}\Omega = \int_\Omega \mathbf{f}\cdot\mathbf{v}\, \mathrm{d}\Omega, \\ \displaystyle\int_\Omega q \,\mathrm{div}\,(\mathbf{u})\, \mathrm{d}\Omega = 0, \end{cases} \tag{13}$$

where

$$\begin{cases} V(\Omega) := \{\mathbf{v} \in [H^1(\Omega)]^2;\ \mathbf{v}_{|\Sigma_D} = \mathbf{u}_D \text{ and } \mathbf{v}_{|\Gamma} = \mathbf{0}\}, \\ V_0(\Omega) := \{\mathbf{v} \in [H^1(\Omega)]^2;\ \mathbf{v}_{|\Sigma_D} = \mathbf{0} \text{ and } \mathbf{v}_{|\Gamma} = \mathbf{0}\}, \\ Q(\Omega) := L^2(\Omega). \end{cases}$$

Remark 3.1 The constraint $\mathbf{u}_{|\Gamma} = \mathbf{0}$ is strongly enforced since it is imposed in the trial and test spaces. On the contrary, the incompressibility constraint is enforced weakly in the formulation and the pressure p is the corresponding Lagrange multiplier. We note that since a weak imposition of a constraint with a Lagrange multiplier results in a saddle point problem, we have to choose a stable pair of elements for the velocity and the pressure satisfying an inf-sup condition (see Boffi et al. 2013). This issue is discussed further in Section "The Inf-Sup Condition on Anisotropic Elements". We note that in the case Σ_N is empty then $Q(\Omega) := L^2(\Omega)/\mathbb{R}$.

In this section, we present a classical unfitted method (see the example in Lew and Buscaglia 2008) which uses the triangulation $\hat{\mathcal{T}}$ to build the finite element spaces and we point out its difficulties. We present the discretized problem with classical Hood-Taylor $\mathbf{P}_2/P_1$ finite elements (but the approach may be generalized). The considered problem reads

Find $(\mathbf{u}_h, p_h) \in \boldsymbol{V}^h \times Q^h$ such that $\forall(\mathbf{v}_h, q_h) \in \boldsymbol{V}_0^h \times Q^h$

$$\left\{\begin{array}{l} \displaystyle\int_{\Omega_h} \nabla \mathbf{u}_h : \nabla \mathbf{v}_h \, \mathrm{d}\Omega_h - \int_{\Omega_h} p_h \operatorname{div}(\mathbf{v}_h) \, \mathrm{d}\Omega_h = \int_{\Omega_h} \mathbf{f} \cdot \mathbf{v}_h \, \mathrm{d}\Omega_h, \\ \displaystyle\int_{\Omega_h} q_h \operatorname{div}(\mathbf{u}_h) \, \mathrm{d}\Omega_h = 0, \end{array}\right. \tag{14}$$

where

$$\left\{\begin{array}{l} \mathbf{V}^h := \{\mathbf{v} \in \left[C^0(\bar{\hat{\Omega}})\right]^2 ; \mathbf{v}_{|T} \in [\mathcal{P}_2]^2, \mathbf{v}_{|\Sigma_D^h} = \mathbf{u}_D \text{ and } \mathbf{v}_{|\Gamma^h} = \mathbf{0}, \forall T \in \hat{\mathcal{T}}\} \\ \quad \subset \mathbf{V}(\hat{\Omega}), \\ \mathbf{V}_0^h := \{\mathbf{v} \in \left[C^0(\bar{\hat{\Omega}})\right]^2 ; \mathbf{v}_{|T} \in [\mathcal{P}_2]^2, \mathbf{v}_{|\Sigma_D^h} = \mathbf{0} \text{ and } \mathbf{v}_{|\Gamma^h} = \mathbf{0}, \forall T \in \hat{\mathcal{T}}\} \\ \quad \subset \mathbf{V}_0(\hat{\Omega}), \\ Q^h := \{q \in C^0(\bar{\hat{\Omega}}); q_{|T} \in [\mathcal{P}_1], \forall T \in \hat{\mathcal{T}}\} \subset L^2(\hat{\Omega}), \end{array}\right.$$

where $\hat{\mathcal{T}}$ is a triangulation of $\hat{\Omega}$, $\mathcal{P}_k$ is the space of polynomials of degree k, and Σ_D^h is the discrete external Dirichlet boundary.

It is important to note that in Problem (14) the integration is performed on Ω_h and not on $\hat{\Omega}$ (see Sect. 3.3 for a subdivision strategy of $\hat{\Omega}$ to perform the quadrature). Indeed, as discussed in the work of Maury (2009) one cannot hope to obtain an optimal rate of convergence if the integration is performed on $\hat{\Omega}$. This result is independent of how the constraint $\boldsymbol{u} = \mathbf{0}$ on Γ is imposed.

For the considered problem, it is not possible to obtain an optimal rate of convergence because the spaces $\boldsymbol{V}^h$ and $\boldsymbol{V}_0^h$ are not rich enough (see Lew and Buscaglia 2008 for more details). We illustrate this issue in Fig. 8. Indeed, for a general set of elements there are more constraints on the immersed boundary (i.e., at the intersection of the immersed boundary with the background mesh element edges) than nodes of the intersected elements that do not belong to the physical domain (named "free nodes"). As a consequence, the system is overconstrained and locking may occur. For example, in Béchet et al. (2009) an algorithm is presented such that two degrees of freedom are uniquely associated with an interface constraint. But, one of the drawbacks of the approach is that it weakens the imposition of the Dirichlet boundary constraint on the immersed boundary.

We point out that since it is not possible to strongly impose the condition $\boldsymbol{u} = \mathbf{0}$ on Γ_h in order to obtain optimal rates of convergence, the weak imposition of the Dirichlet condition is often used. A weak imposition can be performed, for instance, with a Lagrange multiplier (but checking the inf-sup condition for such a method is

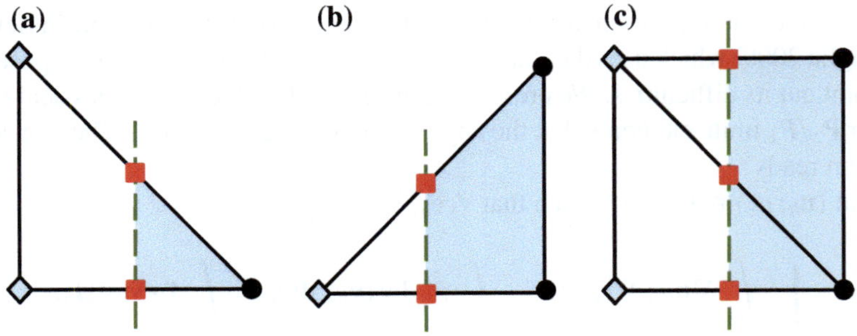

Fig. 8 In this example we consider a single field problem. The elements are P_1 and the physical domain is depicted in *blue*. It follows that the diamonds are "free" nodes (i.e., their values have no physical relevance) while the *dots* are physical nodes. We want to illustrate the difficulty of imposing the internal constraint $\mathbf{u} = \mathbf{0}$ on the *red squares*. **a** Two internal constraints are satised since two "free" nodes are associated to it. **b** The problem is over-constrained since only one "free" node is associated with the two internal constraints. **c** This generic macro-element shows that the internal constraints cannot be imposed and thus locking occurs

not an easy task, see Béchet et al. 2009 and references therein) or the Nitsche method which requires additional user parameters. As discussed in the introduction already, weak imposition of essential boundary conditions is still an active area of research (see for instance Burman and Hansbo 2011b for an example of the Nitsche method for the Stokes problem or alternative approaches in Baiges et al. 2012 and Court et al. 2014). The method we propose in the following avoids the use of complex strategies for weakly imposing essential boundary conditions. It consists in building a finite element basis that is interpolatory on the intersection points of the immersed boundary and the background mesh edges in order to impose Dirichlet boundary conditions strongly.

A Method by a Locally Anisotropic Remeshing In the following, we propose a method that considers a special local refinement using a subdivision of elements cut by the immersed boundary. The method differs from the classical one presented in Problem (14), which uses the triangulation $\hat{\mathcal{T}}$ to build the finite elements. The proposed method consists in refining all elements cut by the immersed boundary such that a locally fitting mesh may be built.

Subdivision For triangles cut by the immersed boundary, we consider the two cases depending on the sub-element belonging to Ω_h is: (a) a triangle or (b) a quadrilateral.

In the present work, we consider finite elements only in triangles and thus in case (b) we have to subdivide the quadrilateral into two triangles. As depicted in Fig. 9a, the subdivision into triangles of a quadrilateral is not unique. For instance, the Delaunay triangulation leads to the best subdivision by maximizing the minimal angles (see, e.g., Bern and Eppstein 1992 and the discussion in the next section).

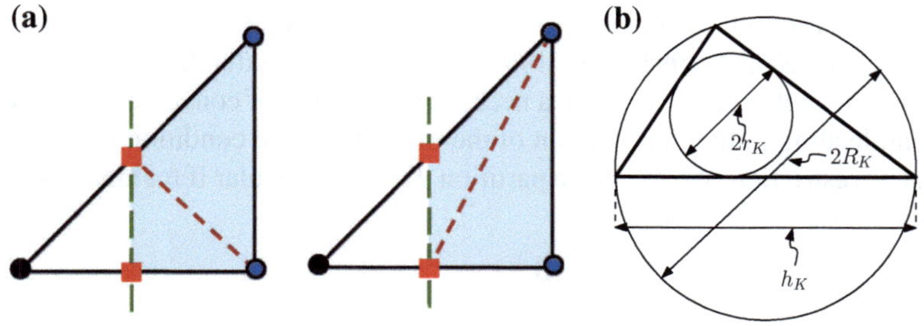

Fig. 9 Selection of the quadrilateral subdivision in subtriangles and description of the element ratio. **a** Non unicity of the quadrilateral subdivision. The pair of triangles giving the smallest element ratios is selected (i.e., the pair on the *left* in this example). **b** The diameters of the triangle, the inscribed and the circumscribed *circles* are denoted h_K and $2r_K$ and $2R_K$, respectively.

3.4 Triangular Shape Regularity Mesh Restrictions

It is clear that such a subdivision may imply flat triangles and the shape of the triangles is crucial to obtain error estimates. We now present two common restrictions on the shape of triangles. The first one is often denoted in the literature as the "regularity" restriction.

For every element K of a partition $\mathcal{T}$, we associate the number $h_K = \text{diam}(K)$. We also set h as the maximum of all h_K. We say that a partition $\mathcal{T}$ is regular if for every $K \in \mathcal{T}$ there exists a constant $C > 0$ such that

$$\frac{h_K}{r_K} = \sigma_r \leq C,$$

where r_K is the radius of the incircle of K (see Fig. 9b). The regularity restriction is related to the minimal-angle condition (i.e., that the smallest angle of any triangle in $\mathcal{T}$ be bounded away from 0). Indeed, we have that

$$r_K = \frac{2|K|}{p},$$

where $|K|$ is the area of K and p its perimeter. We have that $|K| = ch_K \sin(\alpha)/2$ where c and h_K are the length of the two sides associated to α, and α is the smallest angle of K. Let $C = c/p$ then we have

$$r_K = Ch_K \sin(\alpha).$$

As a consequence, a regular mesh cannot allow anisotropic elements. The *minimal angle condition* for triangles was introduced in two famous independent papers: Zlámal (1968) and Ženíšek (1969). The minimum angle condition is a sufficient condition to guarantee the convergence of the finite element method.

It was later found, particularly in e.g., Babuška and Aziz (1976) and Jamet (1976) (see also, e.g., Křížek 1991; Apel 1999; Rand 2009 for extended discussions), that the minimal angle condition is not a necessary condition of convergence of the finite element, leading to the introduction of the maximal angle condition or shape semi-regularity restriction. We say that a partition $\mathcal{T}$ is semi-regular if for any $K \in \mathcal{T}$ there exists a constant $C > 0$ such that

$$\frac{R_K}{h_K} = \sigma_R \leq C,$$

where R_K is the radius of circumscribed circle of K.

The semi-regularity restriction is indeed related to the maximal-angle condition (i.e., that the largest angle of any triangle in $\mathcal{T}$ is bounded away from π). We have by the Sine law that

$$R_K = \frac{h_K}{\sin(\gamma)},$$

where γ is the largest angle of the triangle K. As a consequence a semi-regular mesh allows for flat and elongated triangles. In this case, we say that the mesh contains distorted or anisotropic elements.

Again, the condition is sufficient to guarantee the optimal convergence of the finite element method. However, it has been noted in Hannukainen et al. (2012) that the maximum angle condition is not necessary and the finite element method for 2D problems may converge optimally without a maximum angle condition satisfied.

Accordingly, in the following sections, we consider a triangulation $\mathcal{T}_r$ built as follows. Given a shape regular triangulation $\hat{\mathcal{T}}$ of $\hat{\Omega}$ (i.e., the background mesh), we denote by $\mathcal{T}_\Gamma$ the triangulation of all elements that are crossed by Γ. As previously explained, it is possible to build a subtriangulation $\mathcal{T}_\Gamma'|_T$ on every $T \in \mathcal{T}_\Gamma$ such that $\mathcal{T}_\Gamma'$ fits Γ, with respect to the linear reconstruction of Γ. Then, we consider the triangulation $\mathcal{T}_r$ made of all elements in $\hat{\mathcal{T}}$ that are entirely in Ω_h and all elements of $\mathcal{T}_\Gamma'$ that are in Ω_h. The operation is illustrated in Fig. 10 for the case of an immersed disk.

We now wish to show that $\mathcal{T}_r$ is semi-regular. We briefly recall the hypothesis for Γ: Γ crosses once two triangle edges (for example, on X and Y as shown on Fig. 11). As a triangle of $\hat{\mathcal{T}}$ can be subdivided into a triangle and a quadrilateral. Furthermore, any triangles T in $\hat{\mathcal{T}}$ satisfies the minimal angle condition (since we assume that $\hat{\mathcal{T}}$ is a regular partition of $\hat{\Omega}$) and thus there exists a real $\bar{\Theta}$ such that for all $i \in \{A, B, C\}$ we have

$$\Theta_i \geq \bar{\Theta} > 0. \tag{15}$$

The proof is divided into two parts, Part 1 and Part 2, for the triangle and the quadrilateral subdivisions, respectively.

Part 1: we show that the triangle XBY satisfies the maximal angle condition. By definition we have

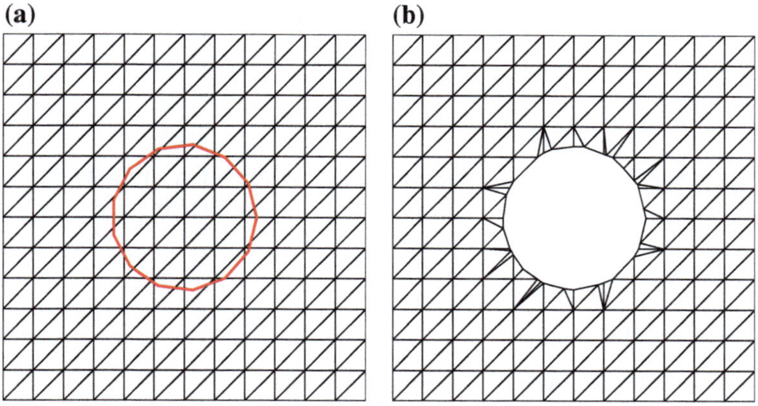

Fig. 10 Subdivision operations of $\hat{\mathcal{T}}$ into $\mathcal{T}_r$. **a** Original mesh $\hat{\mathcal{T}}$. **b** Refined mesh $\mathcal{T}_r$

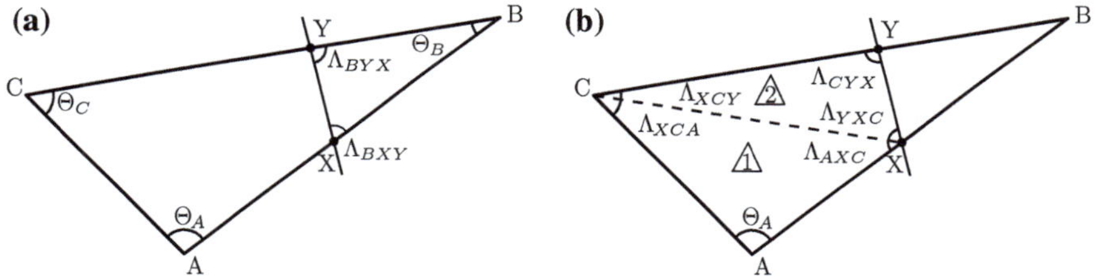

Fig. 11 Angles notation for the maximal angle condition satisfaction of partition $\mathcal{T}_r$. **a** Notation for Part 1. **b** Notation for Part 2

$$\Theta_B + \Lambda_{BYX} + \Lambda_{BXY} = \pi,$$

then since $\Theta_B > 0$ it follows that

$$\Lambda_{BYX} + \Lambda_{BXY} < \pi,$$

and therefore the triangle XBY satisfies the maximal angle condition.

Part 2: we now show that there exists a subdivision into two triangles of the quadrilateral AXYC that satisfies the maximal angle condition. By definition we have

$$\Theta_A + \Lambda_{AXC} + \Lambda_{XCA} = \pi$$

but since $\Theta_A > 0$ then

$$\Lambda_{AXC} + \Lambda_{XCA} < \pi,$$

and thus the triangle AXC satisfies the maximal angle condition. Again, by

$$\Lambda_{XCY} + \Lambda_{CYX} + \Lambda_{YXC} = \pi \tag{16}$$

and

$$\Lambda_{CYX} + \Lambda_{BYX} = \pi \tag{17}$$

but because triangle XBY satisfies the minimal angle condition it follows that $\Lambda_{BYX} < \pi$ thus by (17) we obtain that $0 < \Lambda_{CYX}$ and by (16) we have that

$$\Lambda_{XCY} + \Lambda_{YXC} < \pi,$$

which concludes the proof.

Application to the Incompressible Stokes Problem

In the following, we give an example of the discretized Stokes problem using the locally remeshing strategy with the $\mathbf{P}_2/P_1$ finite element scheme:

Find $(\mathbf{u}_h, p_h) \in \boldsymbol{W}^h \times R^h$ such that $\forall(\mathbf{v}_h, q_h) \in \boldsymbol{W}_0^h \times R^h$:

$$\begin{cases} \displaystyle\int_{\Omega_h} \nabla \mathbf{u}_h : \nabla \mathbf{v}_h \, \mathrm{d}\Omega_h - \int_{\Omega_h} p_h \, \mathrm{div}\,(\mathbf{v}_h) \, \mathrm{d}\Omega_h = \int_{\Omega_h} \mathbf{f} \cdot \mathbf{v}_h \, \mathrm{d}\Omega_h, \\ \displaystyle\int_{\Omega_h} q_h \, \mathrm{div}\,(\mathbf{u}_h) \, \mathrm{d}\Omega_h = 0. \end{cases} \tag{18}$$

In the present work, we consider four mixed approximation schemes. For simplicity, the spaces $\mathbf{W}_0^h$, that is, the spaces of tests functions having null trace on Σ_D^h are not explicitly written (in such a case we would build $\mathbf{W}_0^h$ by taking the intersection with respect to $\{\mathbf{V}_0(\Omega_h) \cap [C^0(\bar{\Omega}_h)]^2\}$ instead of $\{\mathbf{V}(\Omega_h) \cap [C^0(\bar{\Omega}_h)]^2\}$).

- $\mathbf{P}_2/P_0$: (Fortin) continuous piecewise quadratic velocity and piecewise constant pressure:

$$\begin{cases} \mathbf{W}^h &= \{\mathbf{v} : \mathbf{v}_{|T} \in (\mathcal{P}_2)^2, \forall T \in \mathcal{T}_r\} \cap \{\mathbf{V}(\Omega_h) \cap [C^0(\bar{\Omega}_h)]^2\}, \\ R^h &= \{q : q_{|T} \in \mathcal{P}_0, \forall T \in \mathcal{T}_r\} \cap Q(\Omega_h). \end{cases}$$

  Here, $\mathcal{P}_k$ denotes the space of polynomials of order k.
- $\mathbf{P}_2^+/P_1^d$: (low order conforming Crouzeix-Raviart) continuous piecewise quadratic with a cubic bubble velocity and discontinuous piecewise linear pressure

$$\begin{cases} \mathbf{W}^h &= \{\mathbf{v} : \mathbf{v}_{|T} = \mathbf{v}_{|T}^2 + \mathbf{v}_{|T}^+;\ \mathbf{v}_{|T}^1 \in (\mathcal{P}_2)^2, \\ & \qquad \mathbf{v}_{|T}^+ \in (\mathcal{P}_3)^2, \mathbf{v}_{|\partial T}^+ = \mathbf{0}, \forall T \in \mathcal{T}_r\} \cap \{\mathbf{V}(\Omega_h) \cap [C^0(\bar{\Omega}_h)]^2\} \\ R^h &= \{q : q_{|T} \in \mathcal{P}_1, \forall T \in \mathcal{T}_r\} \cap Q^2(\Omega_h), \end{cases}$$

- $\mathbf{P}_2/P_1$: (Hood-Taylor) continuous piecewise quadratic velocity and continuous piecewise linear pressure but discontinuous across the structure

$$\begin{cases} \mathbf{W}^h &= \{\mathbf{v} : \mathbf{v}_{|T} \in (\mathcal{P}_2)^2, \forall T \in \mathcal{T}^n\} \cap \{\mathbf{V}(\Omega_h) \cap [C^0(\bar{\Omega}_h)]^2\} \\ R^h &= \{q : q_{|T} \in \mathcal{P}_1, \forall T \in \mathcal{T}_r\} \cap \{Q(\Omega_h) \cap C^0(\bar{\Omega}_h)\}. \end{cases}$$

- $\mathbf{P}_2^+/P_1$: continuous piecewise quadratic with a cubic bubble velocity and continuous piecewise linear pressure but discontinuous across the structure

$$\begin{cases} \mathbf{W}^h &= \{\mathbf{v} : \mathbf{v}_{|T} = \mathbf{v}^2_{|T} + \mathbf{v}^+_{|T};\ \mathbf{v}^1_{|T} \in (\mathcal{P}_2)^2, \\ &\quad \mathbf{v}^+_{|T} \in (\mathcal{P}_3)^2, \mathbf{v}^+_{|\partial T} = \mathbf{0}, \forall T \in \mathcal{T}_r\} \cap \left\{\mathbf{V}(\Omega_h) \cap [C^0(\bar{\Omega}_h)]^2\right\} \\ R^h &= \{q : q_{|T} \in \mathcal{P}_1, \forall T \in \mathcal{T}_r\} \cap \{Q(\Omega_h) \cap C^0(\bar{\Omega}_h)\}. \end{cases}$$

See e.g., Boffi et al. (2013) for more details.

Remark 3.2 For the $\mathbf{P}_2^+/P_1^d$ element, we use for the pressure basis $N_i(x, y) = a_i + b_i x + c_i y$ evaluated at the "actual" elements, and not on the reference element. Indeed, such a choice is possible because $p \in L^2(\Omega_h)$.

Remark 3.3 As presented, the bubbles are used on all elements of the mesh $\mathcal{T}_r$. In practice, we add the bubble only on subtriangles.

The Inf-Sup Condition on Anisotropic Elements

We first provide a brief presentation of the inf-sup condition.

Given the approximations $\mathbf{u}_h = \sum_{i=1}^n \mathbf{N}_i \hat{\mathbf{u}}_i$ and $p_h = \sum_{i=1}^m M_i \hat{p}_i$, where $\mathbf{N}_i$ and M_i are the finite element bases for $\mathbf{W}^h$ and R^h (with n and m the number of degrees of freedom, respectively) the discrete incompressible Stokes problem in matrix form reads

$$\begin{bmatrix} \mathbf{A} & \mathbf{B}^T \\ \mathbf{B} & \mathbf{0} \end{bmatrix} \begin{Bmatrix} \hat{\mathbf{u}} \\ \hat{\mathbf{p}} \end{Bmatrix} = \begin{Bmatrix} \hat{\mathbf{f}} \\ \hat{\mathbf{g}} \end{Bmatrix}, \tag{19}$$

where

$$\begin{cases} \mathbf{A}|_{ij} = \displaystyle\int_{\Omega_h} \nabla \mathbf{N}_i : \nabla \mathbf{N}_j \, \mathrm{d}\Omega_h & \forall (i,j) \in \{1, 2, \ldots, n\} \times \{1, 2, \ldots, n\}, \\ \mathbf{B}|_{ij} = -\displaystyle\int_{\Omega_h} M_i \, \mathrm{div}(\mathbf{N}_j) \, \mathrm{d}\Omega_h & \forall (i,j) \in \{1, 2, \ldots, m\} \times \{1, 2, \ldots, n\}. \end{cases}$$

Let $n+1, \ldots, n+n_D$ be the eliminated degrees of freedom laying on Σ_D, the right hand side reads

$$\begin{cases} \hat{\mathbf{f}}|_i = \displaystyle\int_{\Omega_h} \mathbf{f}_h \cdot \mathbf{N}_i \, \mathrm{d}\Omega_h - (\bar{\mathbf{A}} \hat{\mathbf{u}}_D)|_i & \forall i \in \{1, 2, \ldots, n\}, \\ \hat{\mathbf{g}}|_i = -(\bar{\mathbf{B}} \hat{\mathbf{u}}_D)|_i & \forall i \in \{1, 2, \ldots, m\}, \end{cases}$$

where

$$\begin{cases} \bar{\mathbf{A}}|_{ij} = \int_{\Omega_h} \nabla \mathbf{N}_i : \nabla \mathbf{N}_j \, \mathrm{d}\Omega_h, \\ \qquad \forall (i,j) \in \{1,2,\ldots,n\} \times \{n+1, n+2, \ldots, n+n_D\}, \\ \bar{\mathbf{B}}|_{ij} = -\int_{\Omega_h} M_i \, \mathrm{div}(\mathbf{N}_j) \, \mathrm{d}\Omega_h, \\ \qquad \forall (i,j) \in \{1,2,\ldots,m\} \times \{n+1, n+2, \ldots, n+n_D\}, \end{cases}$$

and $\hat{\mathbf{u}}_D$ are the nodal boundary values of $\mathbf{u}_D$.

In the following, we also use the pressure mass matrix defined by

$$\mathbf{M}_{|ij} = \int_{\Omega_h} M_i M_j \mathrm{d}\Omega_h \qquad \forall (i,j) \in \{1,2,\ldots,m\} \times \{1,2,\ldots,m\}. \tag{20}$$

The Euclidean norm is given by $||\hat{\mathbf{v}}||_0^2 = \hat{\mathbf{v}}^T \hat{\mathbf{v}}$ with $\hat{\mathbf{v}} \in \mathbb{R}^n$. We also consider the norm defined by the stiffness matrix $\mathbf{A}$, that is $||\hat{\mathbf{v}}||_A^2 = \hat{\mathbf{v}}^T \mathbf{A}^T \hat{\mathbf{v}}$ and its associated dual norm given by $||\hat{\mathbf{v}}||_{A'}^2 = \hat{\mathbf{v}}^T \mathbf{A}^{-T} \hat{\mathbf{v}}$. Let $\hat{\mathbf{q}} \in \mathbb{R}^m$, then the norm used for the pressure field is given by $||\hat{\mathbf{q}}||_M^2 = \hat{\mathbf{q}}^T \mathbf{M}^T \hat{\mathbf{q}}$ and its associated dual norm by $||\hat{\mathbf{q}}||_{M'}^2 = \hat{\mathbf{q}}^T \mathbf{M}^{-T} \hat{\mathbf{q}}$, where $\mathbf{M}$ is defined in (20).

It is well known that a key component for (19) to have a unique solution is the satisfaction of the following condition (see Boffi et al. 2013):

Inf-sup: $\exists \beta_h > 0$ (independent of h) such that

$$\max_{\hat{\mathbf{v}} \in \mathbb{R}^n \setminus \{\mathbf{0}\}} \frac{\hat{\mathbf{v}}^T \mathbf{B}^T \hat{\mathbf{q}}}{||\hat{\mathbf{v}}||_A} \geq \beta_h ||\hat{\mathbf{q}}||_M \qquad \forall \hat{\mathbf{q}} \in \mathbb{R}^m. \tag{21}$$

Being $\hat{\mathbf{u}}^I$ and $\hat{\mathbf{p}}^I$ the vectors of analytical solutions at the nodes for the velocity and the pressure, respectively, an error estimate is given by (see, e.g., Boffi et al. 2013):

$$||\hat{\mathbf{u}}^I - \hat{\mathbf{u}}||_A \leq C \left(||\hat{\mathbf{f}}||_{A'} + \beta_h^{-1} ||\hat{\mathbf{g}}||_{M'} \right), \tag{22}$$

$$||\hat{\mathbf{p}}^I - \hat{\mathbf{p}}||_M \leq C \left(\beta_h^{-1} ||\hat{\mathbf{f}}||_{A'} + \beta_h^{-2} ||\hat{\mathbf{g}}||_{M'} \right), \tag{23}$$

where C denotes a general constant independent of h and β_h.

We clearly can see from (22) and (23) that if $\beta_h \to 0$ as $\sigma_r \to \infty$ then the error for the pressure may not be bounded, and it depends on $1/\beta_h^2$, while the velocity field may also not be bounded but it depends only on $1/\beta_h$.

We equip the space $\mathbf{V}$ and Q (see (13)) with the norms

$$\begin{cases} ||\mathbf{v}||_V^2 &= \int_\Omega \nabla \mathbf{v} : \nabla \mathbf{v} \mathrm{d}\Omega, \\ ||q||_Q^2 &= \int_\Omega q^2 \mathrm{d}\Omega, \end{cases}$$

where $\mathbf{v} \in \mathbf{V}(\Omega)$ and $q \in Q(\Omega)$. Given that $\mathbf{u}^I = \sum_i \hat{\mathbf{u}}_i^I \mathbf{N}_i$ and $q^I = \sum_i \hat{q}_i^I M_i$ are the interpolant of the analytical solution using the finite element basis, it can be shown that (see, e.g., Auricchio et al. 2004)

$$||\hat{\mathbf{u}}^I - \hat{\mathbf{u}}||_A \leq C\left(\beta_h^{-1}||\mathbf{u}^I - \mathbf{u}_h||_V + ||p^I - p_h||_Q\right), \quad (24)$$

$$||\hat{\mathbf{p}}^I - \hat{\mathbf{p}}||_M \leq C\left(\beta_h^{-2}||\mathbf{u}^I - \mathbf{u}_h||_V + \beta_h^{-1}||p^I - p_h||_Q\right). \quad (25)$$

To conclude, it is very important that for the chosen finite element choice β_h remains bounded from below as σ_r increases. In other words, we would like to have β_h to be independent of σ_r.

The literature on inf-sup stability of mixed finite elements on anisotropic meshes is scarce, particularly on triangles. In Apel and Randrianarivony (2003), it is showed numerically that the $\mathbf{P}_2/P_1$ is stable on triangular edge meshes but that it fails on triangular corner meshes. More importantly, it is shown that $\mathbf{P}_2^+/P_1$ passed all proposed tests. No proof of the stability of the $\mathbf{P}_2^+/P_1$ is known, to the best of the authors' knowledge. In Apel and Nicaise (2004) a proof of the stability of the $\mathbf{P}_2/P_0$ element for both edge and corner meshes is given but with some restriction on corner meshes and, as we shall see later, this element is actually unstable for the problems we are dealing inhere. In Micheletti et al. (2004), a residual-free bubble stabilized formulation for the $\mathbf{P}_1/P_1$ is proposed on general triangular meshes. The element is proven stable but under some restrictions on the orientation of the mesh with respect to the solution of the problem. In Liao and Silvester (2013) a proof of the stability of the $\mathbf{Q}_1/P_0$ element (continuous bilinear quadrilateral velocity and piecewise constant pressure) stabilized with a pressure jump strategy is given. It is proved that the inf-sup constant is independent of the element mesh ratio on both edge and corner meshes. The authors claim that their results can be extended to triangles, but no formal proof is provided. See also Ainsworth et al. (2014) for a discussion on this topic regarding the $\mathbf{P}_2/P_0$ element.

Numerical Methods to Measure the Inf-Sup Condition (A Smallest Generalized Eigenvalue Test)

In order to test if our finite element scheme choice remains stable as σ_r increases, we compute numerically the inf-sup constant.

It can be proven that (see, e.g., Elman and Wathen 2005) the inf-sup constant β_h is given by the square root of the lowest positive eigenvalue of the following generalized eigensystem:

$$\mathbf{B}\mathbf{A}^{-1}\mathbf{B}^T\mathbf{q} = \lambda\mathbf{M}\mathbf{q}, \quad (26)$$

where $\mathbf{B}\mathbf{A}^{-1}\mathbf{B}^T$ is called the Schur complement.

Remark 3.4 In the case of an enclosed flow, the first eigenvalue is zero, since it represents the constant pressure mode. In such a case β_h is estimated by the square root of the second lowest eigenvalue. On the contrary, if the problem admits a Neumann boundary condition then all eigenvalues are strictly positive.

Conditioning of the Schur Complement

The satisfaction of the inf-sup condition is not only important for an accurate resolution of the velocity and the pressure but also for a better conditioning of the systems at hand. Indeed, one can show that (see, e.g., Ern and Guermond 2004)

$$\kappa(\mathbf{B}\mathbf{A}^{-1}\mathbf{B}^T) \leq C\beta_h^{-2}\kappa(\mathbf{M}), \tag{27}$$

where κ designates the condition number. The Schur complement is later also denoted by $\mathbf{S}$.

3.5 *Numerical Tests*

We perform the tests with the four mixed finite elements described in Section "Application to the Incompressible Stokes Problem", that is: $\mathbf{P}_2/P_0$, $\mathbf{P}_2/P_1$, $\mathbf{P}_2^+/P_1$, and $\mathbf{P}_2^+/P_1^d$, which we recall are known to be inf-sup stable on isotropic meshes.

We also provide and discuss with some representative tests the condition number, denoted by κ, of the Schur complement, see (26).

In all presented tests, integration was performed exactly. However, further numerical experiments showed that the use for $\mathbf{P}_2^+/P_1$ of the integration rule exacts on $\mathbf{P}_2/P_1$ (clearly leading to an under-integration of the terms involving bubble shape functions) leads to practically identical results. This is in agreement with what is expected from a theoretical point of view. It thus follows that $\mathbf{P}_2^+/P_1$ at a cost similar to $\mathbf{P}_2/P_1$ (Table 2).

Regarding, the reduced-integration of $\mathbf{P}_2^+/P_1$, numerical evidence showed that it has an impact on the convergence rate of the method but not on the stability of

Table 2 Impact of the reduced-integration on the convergence rates of $\mathbf{P}_2^+/P_1$

(a) Convergence rates for $\mathbf{P}_2^+/P_1$ using exact integration

h	$\|\|\mathbf{u}-\mathbf{u}_h^e\|\|_0$	Rate	$\|\mathbf{u}-\mathbf{u}_h^e\|_1$	Rate	$\|\|p-p_h^e\|\|_0$	Rate
0.24	8.21E–4		4.35E–2		1.40E–2	
0.12	1.00E–4	3.03	1.08E–2	2.01	3.40E–3	2.04
0.06	1.24E–5	3.01	2.70E–3	2.00	8.57E–4	1.99
0.03	1.55E–6	3.00	6.71E–4	2.01	2.14E–4	2.00

(b) Convergence rates for $\mathbf{P}_2^+/P_1$ using reduced-integration

h	$\|\|\mathbf{u}-\mathbf{u}_h^r\|\|_0$	Rate	$\|\mathbf{u}-\mathbf{u}_h^r\|_1$	Rate	$\|\|p-p_h^r\|\|_0$	Rate
0.24	1.60E–3		9.96E–2		1.42E–2	
0.12	2.60E–4	2.62	4.66E–2	1.10	3.50E–3	2.02
0.06	5.33E–5	2.29	2.29E–2	1.02	8.58E–4	2.03
0.03	1.25E–5	2.09	1.14E–2	1.01	2.14E–4	2.00

The errors are actually relative errors

the element, as previously checked in Auricchio et al. (2015a). Indeed, considering the Stokes problem in the domain $[-1, 1]^2$ with the analytical solution provided in (28) and performing a h-refinement on a uniform mesh, the results show a loss of an order of accuracy if reduced-integration is performed on the velocity field but not on the pressure. However, further analysis showed that if all polynomials of order 2 are correctly integrated (as performed here) the element passes the path-test. This issue was not observed in Auricchio et al. (2015a) due to the anisotropy of the elements in the SGE-tests.

$$\begin{cases} u_x(x, y) &= 20x^3 y, \\ u_y(x, y) &= 5x^4 - 5y, \\ p(x, y) &= 60x^3 y - 2xy^3. \end{cases} \tag{28}$$

Smallest Generalized Eigenvalue Test Problems

We propose the constant flow problem (see Eq. 29) (Fig. 12).

$$\begin{cases} u_x(x, y) &= 1, \\ u_y(x, y) &= 0, \\ p(x, y) &= 0. \end{cases} \tag{29}$$

More importantly, we consider three meshes for the inf-sup test problems (see Fig. 13).

We first present a summary of the results in Table 3 that reports for each mesh and each test (i.e., $(-1 + \alpha) \to -1$ and $-\alpha \to 0$) if it passes the test (i.e., if the numerical inf-sup constant remains bounded below) or, if it fails, the number of spurious modes associated. An extensive presentation of the results is provided in Lefieux (2014). We provide also additional figures to analyze the behavior of the conditioning of the

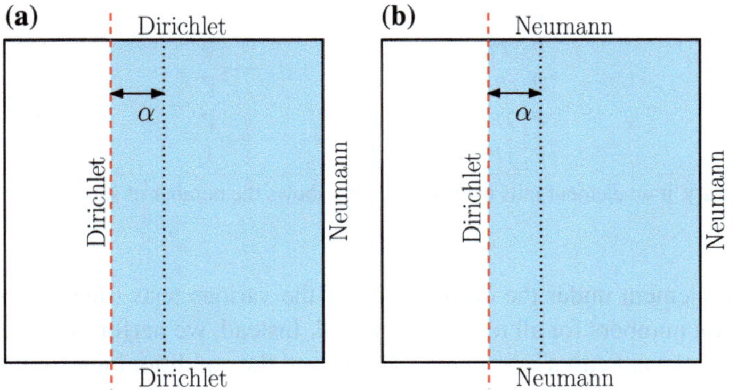

Fig. 12 Boundary value problems under consideration for the inf-sup eigenproblem. **a** Test 1 (T1): with Dirichlet boundary conditions on $y = \pm 1$. **b** Test 2 (T2): with Neumann boundary conditions on $y = \pm 1$

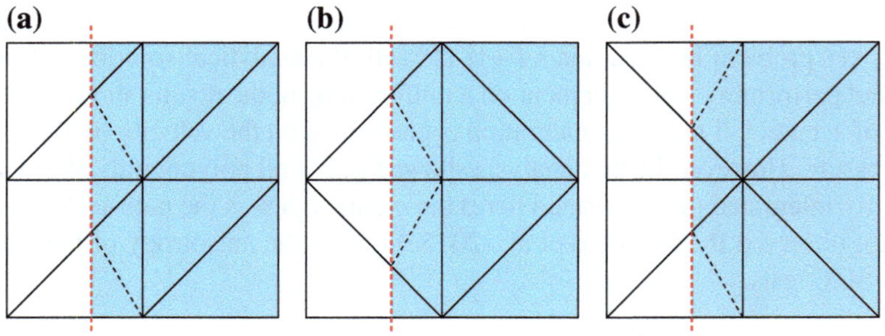

Fig. 13 The three meshes used for the generalized eigenproblem with different immersed boundary positions. The background domain is defined on $[-1, 1]^2$. **a** Mesh 1 (M1). **b** Mesh 2 (M2). **c** Mesh 3 (M3)

Table 3 Summary of the results: if an element passes the test it is denoted by P

(a) Test 1 (T1): with Dirichlet boundary conditions on $y = \pm 1$				
Mesh		1	2	3
$(-1+\alpha) \to -1$	$\mathbf{P}_2/P_0$	P	P	P
	$\mathbf{P}_2/P_1$	P	P	P
	$\mathbf{P}_2^+/P_1$	P	P	P
	$\mathbf{P}_2^+/P_1^d$	2	2	2
$-\alpha \to 0$	$\mathbf{P}_2/P_0$	2	2	1
	$\mathbf{P}_2/P_1$	1	2	P
	$\mathbf{P}_2^+/P_1$	P	P	P
	$\mathbf{P}_2^+/P_1^d$	5	5	4
(b) Test 2 (T2): with Neumann boundary conditions on $y = \pm 1$				
Mesh		1	2	3
$-1+\alpha \to -1$	$\mathbf{P}_2/P_0$	P	P	P
	$\mathbf{P}_2/P_1$	P	P	P
	$\mathbf{P}_2^+/P_1$	P	P	P
	$\mathbf{P}_2^+/P_1^d$	2	2	2
$-\alpha \to 0$	$\mathbf{P}_2/P_0$	1	P	1
	$\mathbf{P}_2/P_1$	P	P	P
	$\mathbf{P}_2^+/P_1$	P	P	P
	$\mathbf{P}_2^+/P_1^d$	2	1	2

On the contrary, if an element fails the test, the table shows the number of spurious modes

Schur complement under the deformation of the various tests (denoted by **S**). Not all condition numbers for all tests are reported. Instead, we performed a selection to discuss the relevant cases in view of stability and the conditioning estimate in (27).

The instability of the $\mathbf{P}_2/P_0$ element comes from a single pressure mode on elements with the area behaving as $\mathcal{O}(\alpha^2)$. This element fails for both problems. It implies that the spurious modes are not concentrated only in corners in which

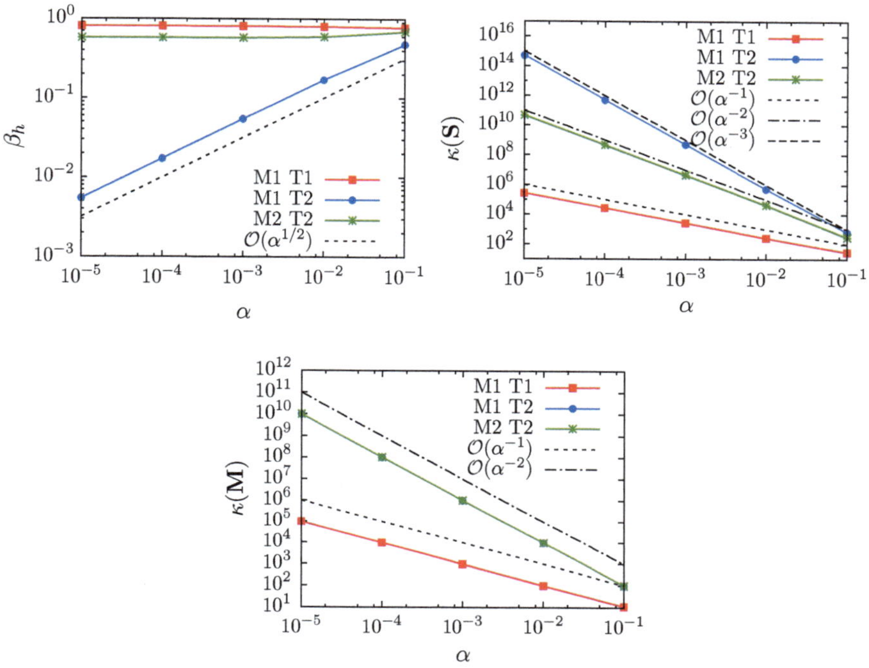

Fig. 14 Schur complement conditioning of $\mathbf{P}_2/P_0$, with $M_1/T_1(-1+\alpha) \to -1$; M_1/T_2 and $M_2/T_2 - \alpha \to 0$

Dirichlet boundary conditions are imposed. Such a result has important implications in practice, as we shall see in Section "Applications". Also, all spurious modes have a behavior $\mathcal{O}(\sqrt{\alpha})$.

Regarding the conditioning of the Schur complement, it clearly appears from Fig. 14 that it satisfies the estimate (27). Actually, the conditioning of $\mathbf{M}$ is the one we would expect on such meshes since $\mathbf{M}$ is a diagonal matrix in which the smallest eigenvalue behaves as the smallest area of the mesh, that is, with $(-1+\alpha) \to -1$ as $\mathcal{O}(\alpha^2)$ and $(-\alpha \to 0)$. The largest eigenvalue remains constant in our test problems.

The instability of $\mathbf{P}_2/P_1$ comes from a single rogue pressure mode on corners with Dirichlet boundary conditions. Indeed, for Test 2 and for Test 1 with Mesh 3, the pair is stable for all tests, even when the smallest element area is $\mathcal{O}(\alpha^2)$, in the test $(-\alpha) \to 0$. Convergence rates of the spurious modes have a dependence on $\sqrt{\alpha}$. Regarding the conditioning of the Schur complement, again, we obtain results in accordance with the estimate in (27). Standard estimate on the conditioning of $\mathbf{M}$ states that its lowest eigenvalue behaves as the area of the smallest element, while the largest one as the largest element area (see also Kamenski et al. 2014).

The $\mathbf{P}_2^+/P_1$ element passes all tests. Numerical tests showed that the conditioning of the Schur complement behaves as the estimate in (27) (Fig. 15).

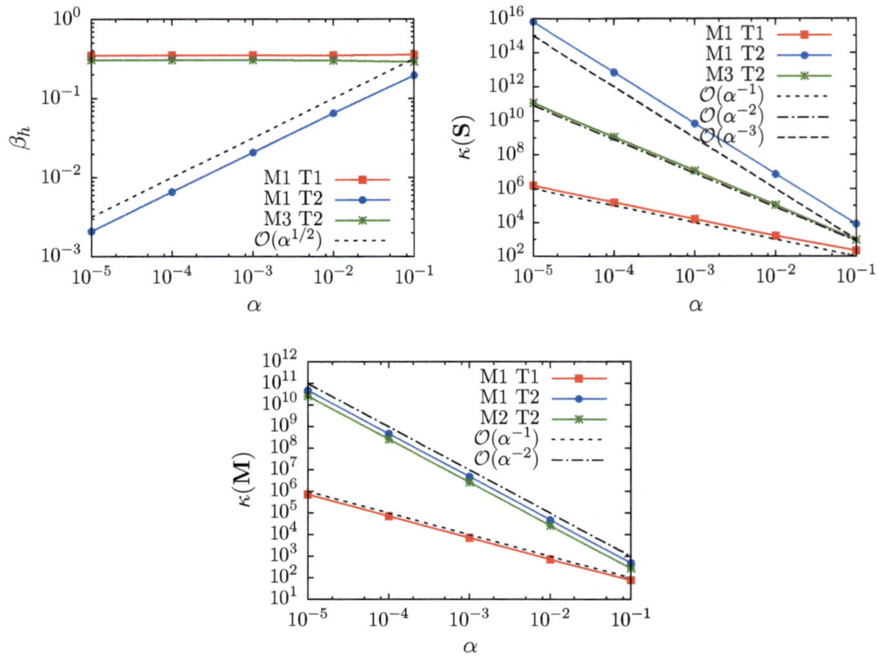

Fig. 15 Schur complement conditioning of $\mathbf{P}_2/P_1$, with $M_1/T_1(-1+\alpha) \to -1$; M_1/T_2 and $M_3/T_2 - \alpha \to 0$

The $\mathbf{P}_2^+/P_1^d$ element is the only one that fails for all tests. Furthermore, this element does not only have spurious modes of behavior $\mathcal{O}(\sqrt{\alpha})$ but also $\mathcal{O}(\alpha)$. In particular, such a result implies that the condition number of the Schur complement would be much larger than for the other finite elements. We also observe that the number of spurious modes with $\mathcal{O}(\alpha^{1/2})$ corresponds to the number of spurious modes of the $\mathbf{P}_2/P_0$ element (see also Lefieux 2014), which is expected. Regarding the conditioning of the Schur complement for $\mathbf{P}_2^+/P_1^d$ our results presented in Fig. 16 do not violate the estimate in (27) but we were expecting a poorer conditioning of the Schur complement. Indeed, when the inf-sup constant behaves as $\mathcal{O}(\alpha^{1/2})$ and the mass matrix as $\mathcal{O}(\alpha^{-3})$ for $(-1+\alpha) \to -1$ and the conditioning of the Schur complement behaves as $\mathcal{O}(\alpha^{-3})$, while we were expecting $\mathcal{O}(\alpha^{-4})$. The same behavior has been observed with the other tests. We point out that the very bad behavior of the mass matrix $\mathbf{M}$ on distorted elements was not known to the authors and further work is under consideration (see Remark 3.2 for a description of the basis used for P_1^d.)

Applications

In this section, we present two applications showing issues regarding inf-sup stability for the two elements with discontinuous pressures. The first one is a Stokes flow problem around a disk while the second is a fluid-structure interaction problem in

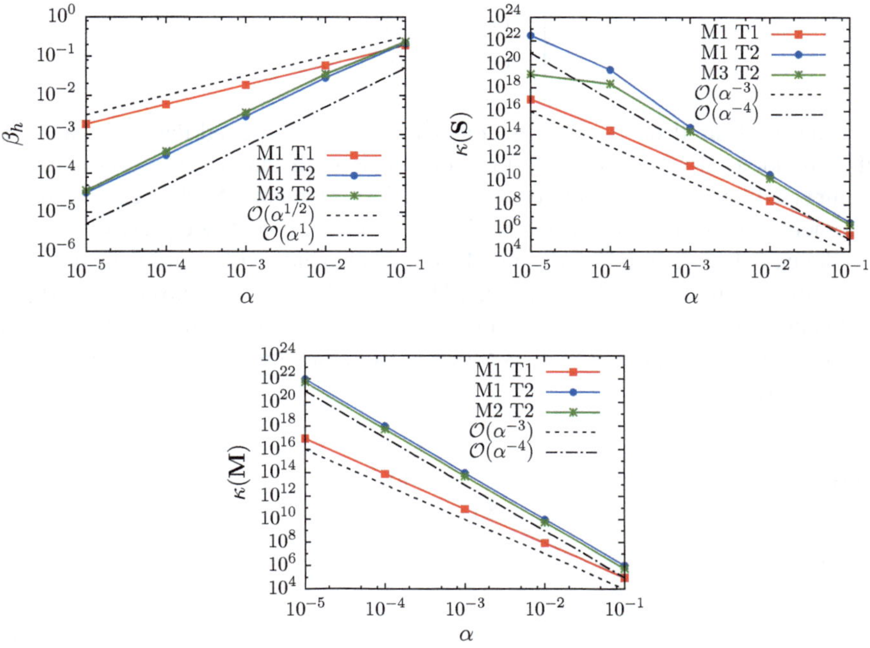

Fig. 16 Schur complement conditioning for $\mathbf{P}_2^+/P_1^d$, with $M_1/T_1(-1+\alpha) \rightarrow -1$; M_1/T_2 and $M_3/T_2 - \alpha \rightarrow 0$

which the fluid is modeled by the incompressible Navier-Stokes equation and the structure is a rigid bar with one rotating degree of freedom at one end.

We thus investigate cases with very distorted elements that can occur during simulations. We show that the $\mathbf{P}_2/P_1$ and the $\mathbf{P}_2^+/P_1$ are actually stable inhere (examples of failure for $\mathbf{P}_2/P_1$ are presented in Auricchio et al. 2015a), as expected from the SGE-tests, but the elements with discontinuous pressures fail. For all tests, we do not present the results for the velocity field but the solution is in accordance with those obtained with the SGE-tests, i.e., the accuracy of the velocity field remains very good even when highly distorted elements are present (which is not what would be expecting from the estimate in (24)).

Flow Around a Disk We now consider a problem consisting of a flow around a cylinder between two plates. By symmetry, the problem reduces to a 2D flow around a disk, whose boundary is defined as an immersed boundary. The fluid domain is defined on $[-1, 1] \times [-1, 1]$. The inflow condition is a Poiseuille flow and is given by (30), no-slip boundary conditions are prescribed on $y = \pm 1$ and a do-nothing boundary condition is applied on $x = 1$. The disk has a radius of 0.312 and a no-slip boundary condition is applied on its surface.

$$\begin{cases} u_x(x, y) &= (1 - y^2), \\ u_y(x, y) &= 0. \end{cases} \tag{30}$$

The mesh used on $[-1, 1]^2$ is composed of 33×33 quadrilaterals which are then divided into triangles with their diagonals such that $x - y =$ constant. The "immersed" boundary is defined by a set of 89 segments. Such a choice for the geometry has been picked such that highly distorted elements are present.

We can observe (see Fig. 17) that the mixed elements with continuous pressures are stable (with $\beta_h \approx 0.18$), as suggested by the SGE-tests, since no corner Dirichlet boundary conditions are enforced in this problem. On the contrary, the elements with discontinuous pressure fail as expected from the SGE-tests (Test 2) showing a $\beta_h \approx 0.07$.

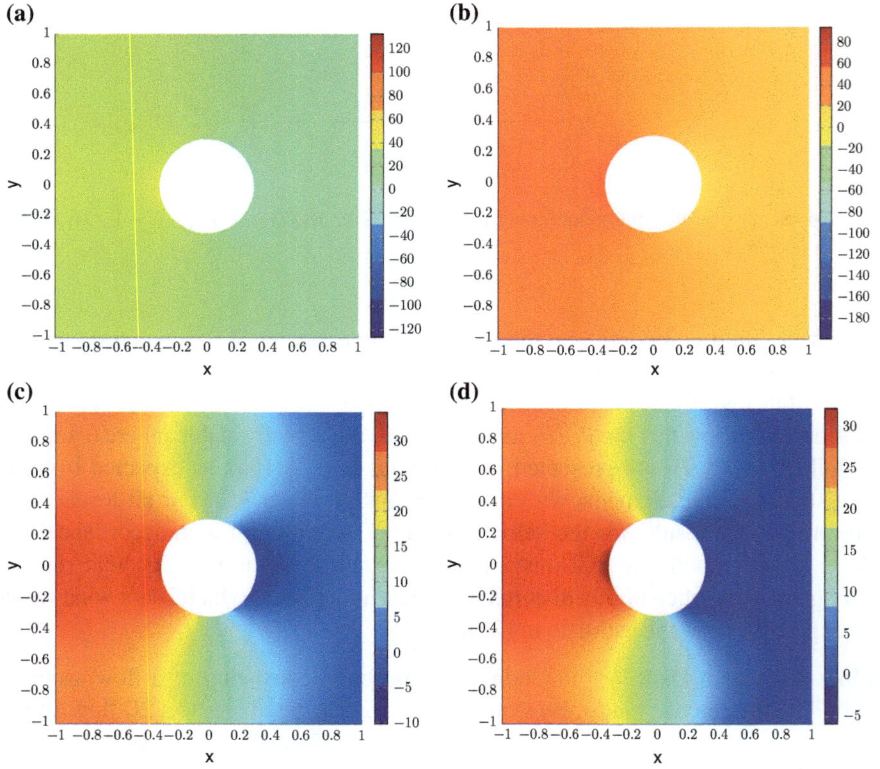

Fig. 17 Pressure field for the Stokes flow problem around a disk application. Results for this test show that the mixed elements with discontinuous pressure show instability while the elements with continuous ones are stable. **a** $\mathbf{P}_2/P_0$, $\beta_h = 0.0675$. **b** $\mathbf{P}_2^+/P_1^d$, $\beta_h = 0.0665$. **c** $\mathbf{P}_2/P_1$, $\beta_h = 0.1780$. **d** $\mathbf{P}_2^+/P_1$, $\beta_h = 0.1780$

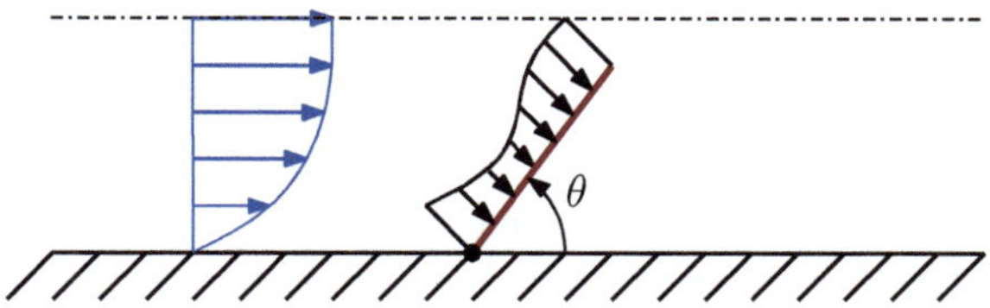

Fig. 18 An example of a hinged rigid bar sets in motion by an unsteady fluid

A Fluid—Structure Interaction Problem In this problem, we consider the fluid-structure interaction of a fluid, modeled by the incompressible Navier-Stokes equation (see (31)), and a rigid bar with a single rotational degree of freedom at one end (see Fig. 18). Here, we do not discuss in details the numerical schemes and the results regarding the fluid/structure motion but, rather, we focus on the anisotropic remeshing strategy and its implications for the various mixed finite elements considered in this work. For the interested reader, however, a detailed description of the numerical scheme is presented in Auricchio et al. (2015b).

For the fluid part, the system of equation of the problem is given by the classical incompressible Navier-Stokes equations

$$\begin{cases} \rho_f \left(\frac{\partial \mathbf{u}}{\partial t} + \mathbf{u} \cdot \nabla \mathbf{u}\right) - \mathbf{div}(\mu \nabla^s \mathbf{u}) + \nabla p - \mathbf{f} = \mathbf{0}, \\ \operatorname{div}(\mathbf{u}) = 0, \end{cases} \tag{31}$$

where ρ_f designates the density, μ the dynamic viscosity, $\mathbf{u}$ the velocity, p the pressure and $\nabla^s \mathbf{u}$ is defined as $\nabla^s \mathbf{u} = \nabla \mathbf{u} + (\nabla \mathbf{u})^T$.

We consider (31) in a fixed domain Ω with $\partial\Omega = \Sigma_D \cup \Sigma_N$ such that $\Sigma_D \cap \Sigma_N = \emptyset$, completed by the following boundary and initial conditions:

$$\begin{cases} \mathbf{u} = \mathbf{b}_D & \text{on } \Sigma_D, \\ -p\mathbf{n} + \mu(\nabla^s \mathbf{u})\mathbf{n} = \mathbf{b}_N & \text{on } \Sigma_N, \end{cases} \tag{32}$$

$$\mathbf{u}(\mathbf{x}, 0) = \mathbf{u}_i(\mathbf{x}) \qquad \text{on } \Omega, \tag{33}$$

where $\mathbf{n}$ designates the outward normal of $\partial\Omega$. We assume that Σ_N is not empty such that the pressure is uniquely defined.

For the solid part, we choose a hinged rigid bar of length L with negligible width, only one rotational degree of freedom, and a rotational spring to the bar (see Fig. 18).

For describing the motion of the bar, we use a polar coordinate system with pole $\mathbf{R} \in \mathbb{R}^2$, the point of rotation of the bar. We denote by θ and r the angular and radial coordinates, respectively.

The solid equation to solve is given by

$$\tau(t) = 0, \tag{34}$$

where

$$\tau(t) = \int_{\Gamma} r f_s(r,t)\mathrm{d}r, \tag{35}$$

with f_s the net balance of the stress acting on the two sides of the bar (see Fig. 18).

That problem has been studied in Pedrizzetti (2005), where an asymptotic analysis of the valve opening without vortex shedding is presented.

The Navier-Stokes equations are linearized using a Picard approach and a first order backward Euler scheme is used for time integration. The coupling is performed in a monolithic fashion (see Auricchio et al. 2015b for further details).

Regarding the definition of the test case. The computational domain is the rectangle $[-1\text{ cm}, 6\text{ cm}] \times [0, 1\text{ cm}]$. At the inflow $x = -1$, the velocity is given by (36).

$$\mathbf{u}(x,y,t) = \{(1-\cos(\pi t/T))/2, 0\}^T. \tag{36}$$

The length of the bar is $L = 0.999$ cm. The no-slip boundary condition on $y = 0$ is applied. The do-nothing (or "stress-free") boundary condition is applied on $x = 6$ (that is $p\mathbf{n} - \mu(\nabla^s \mathbf{u})\mathbf{n} = \mathbf{0}$). A symmetry boundary condition is imposed on $y = 1$, i.e., only normal velocity components are set to zero and tangential ones are set to do-nothing (see, e.g., Demkowicz 1991). The initial condition is $\mathbf{u}_i = \mathbf{0}$. The time period is set to $T = 10$ s. The viscosity is set to $\mu = 0.001$ g.cm^{-2}. We use a 127×19 discretization of the fluid domain, the time step is set to $\delta t = 10/128$ s and the simulation is performed from 0 to 10 seconds.

An approximation of the asymptotic solution for the motion of the rigid bar problem assuming that no vortex are generated behind the bar is given by (see Pedrizzetti 2005, which is the case here: see Auricchio et al. 2015b):

$$\frac{d\theta}{dt} = \frac{2u_x(t)\sin(\theta)}{(\sin(\theta) - 2)}, \tag{37}$$

with $u_x(t) = (1 - \cos(\pi t/T))/2$.

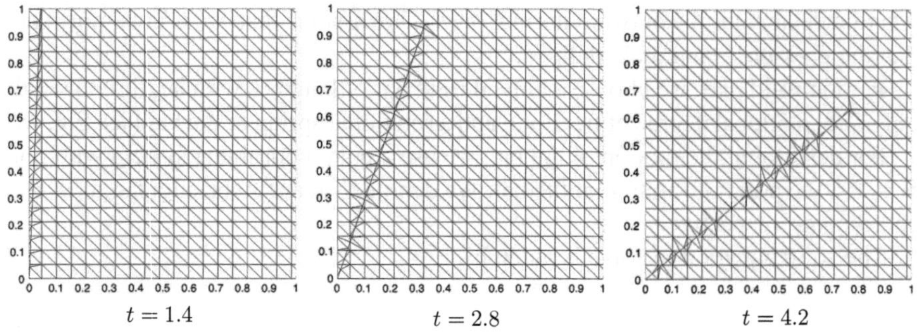

Fig. 19 Examples of the distortion of the mesh with the $\mathbf{P}_2^+/P_1$ element

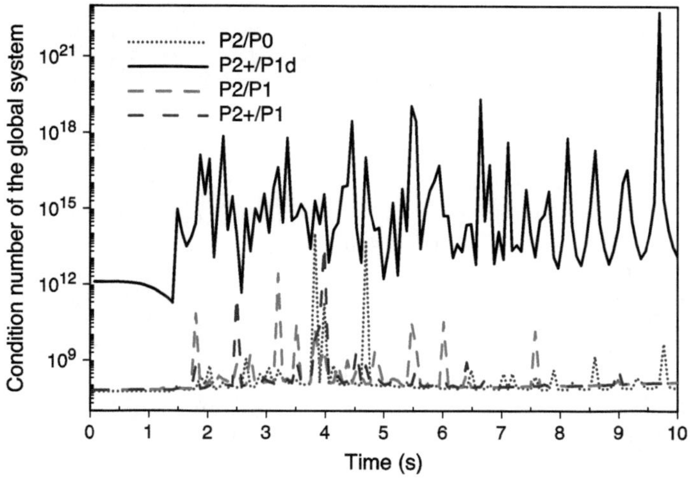

Fig. 20 Condition number of the linear system for all four mixed elements

As we pointed out in Sect. 3.3, the mesh is not isotropic, as it can be seen in Fig. 19, leading to possible issues with the inf-sup stability and condition number issues. In Fig. 20, we report the condition number of the complete linear system (see Auricchio et al. 2015b or Lefieux 2014) for the four finite element schemes. The condition number of the linear system is affected by the inf-sup constant β_h. Since β_h is affected by the distortion of the elements, the condition number of the linear system will also be affected (see Schur complement estimate in (27) and the previous section for numerical tests). Inf-sup unstable finite elements are expected to show its conditioning worsen with respect stable elements. This is precisely what we observe. We also recall that the conditioning of the Schur complement is also affected by the conditioning of the pressure mass matrix (see (27) and Section "Smallest Generalized Eigenvalue Test Problems").

1. For the $\mathbf{P}_2/P_0$ element, we see in Fig. 20 two peaks indicating ill-conditioning, of one order of magnitude higher than the highest peaks using $\mathbf{P}_2/P_1$ and $\mathbf{P}_2^+/P_1$. Indeed, we can observe in Fig. 21, which represents the pressure field at the time of the first peak, spurious modes (a zoom on the culprit is shown in Fig. 22).
2. For the $\mathbf{P}_2^+/P_1^d$ element, the associated linear system is very ill-conditioned with respect to the other elements, indicating that the inf-sup constant is much more sensitive to mesh distortion (or the conditioning of the associated pressure mass matrix). In Fig. 21, we can observe some spurious modes on the pressure field at the time of one of the peaks present in Fig. 20. A zoom on the spurious modes is presented in Fig. 22.
3. For the $\mathbf{P}_2/P_1$ element, as pointed out in Section "Smallest Generalized Eigenvalue Test Problems", the inf-sup constant may be very small on small elements in recessed corners with Dirichlet boundary conditions enforced on both bound-

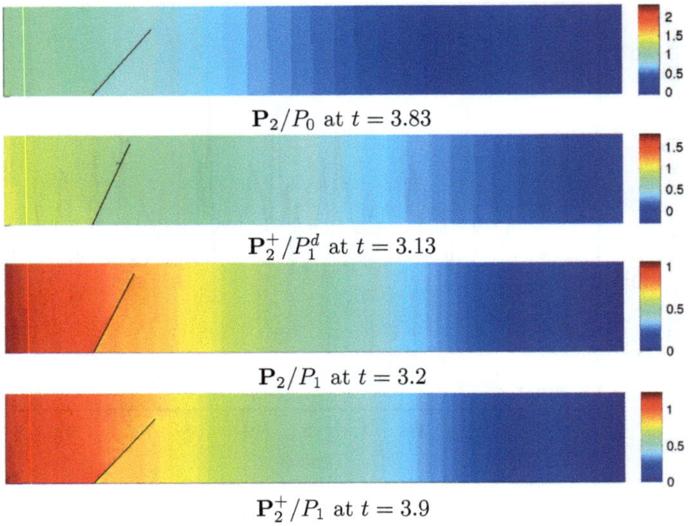

Fig. 21 Pressure field at times corresponding to ill conditioned linear systems. It shows the inf-sup stability issue for the $\mathbf{P}_2^+/P_0$ and $\mathbf{P}_2/P_1^d$ elements

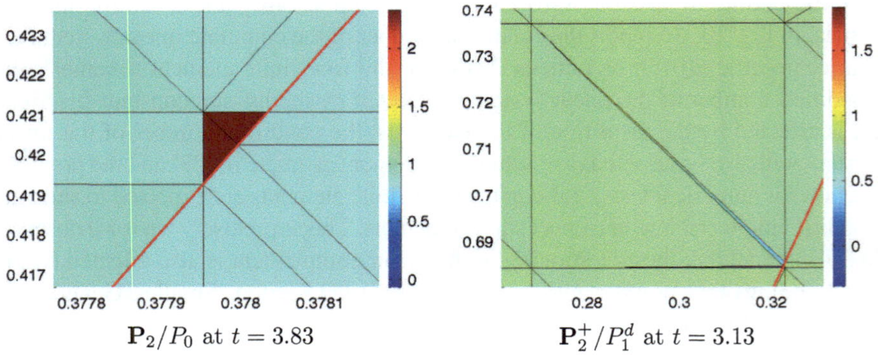

Fig. 22 Zoom on effects of elements distortion: presence of spurious modes. The rigid bar is depicted in *red*

ary edges, but such a situation is very unlikely to occur here, leading to a stable scheme. Indeed, no spurious modes are visible in Fig. 21.

4. The $\mathbf{P}_2^+/P_1$ element is stable, as discussed previously and no spurious modes are visible in Fig. 21. Globally the conditioning is of the same order as with $\mathbf{P}_2/P_1$, as expected from the results in Table 3.

The stable schemes $\mathbf{P}_2/P_1$ and $\mathbf{P}_2^+/P_1$ show a much better conditioning than the inf-sup unstable schemes pointing out the importance of having inf-sup stable elements on distorted meshes.

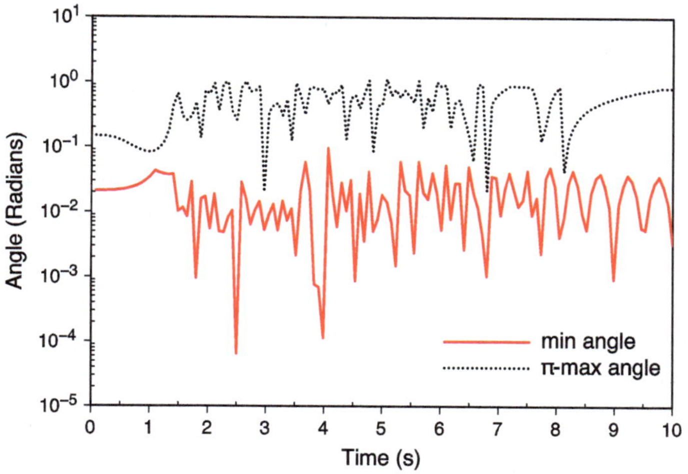

Fig. 23 Min and supplementary maximal (π-max) angles using $\mathbf{P}_2^+/P_1$

Remark 3.5 In van Loon et al. (2006) the method employed uses the $\mathbf{P}_2^+/P_1^d$ element with a similar approach for the local refinement as performed here but maintaining good element ratios by a smoothing procedure. However, our results show that extending the method presented in van Loon et al. (2006) to very stretched elements is not straightforward since inf-sup stability issues occur with $\mathbf{P}_2^+/P_1^d$. We thus show the necessity of the smoothing procedure with $\mathbf{P}_2^+/P_1^d$.

In Sect. 3.4, we discussed implications of the minimal and maximal angle conditions with the finite element method. In Fig. 23, we report the min angle and the supplementary maximal angle (denoted π-max angle) during the fluid–structure interaction simulation using the $\mathbf{P}_2^+/P_1$ element. It clearly appears that the largest angle has a much larger bound away from π than the smallest angle away from 0, which we can observe if we compare the minimal and the supplementary maximal angles of the remeshed triangles (see Fig. 23). The difference is of an order of magnitude. The largest difference between the minimal and supplementary maximal angles is of two orders of magnitude. By comparing the conditioning of the linear system associated to $\mathbf{P}_2^+/P_1$ (in Fig. 20) and the minimal and maximal angles we observe that only the minimal angle has a severe impact on the conditioning of the system, as we can observe that the two highest condition numbers correspond to two very small angles, at time $t = 2.5$ s and $t = 3.98$ s.

4 Conclusive Remarks

In this chapter, we first presented detailed results on interface problems in a 1D setting. Even though the error in the energy norm has an optimal behavior away from the interface, many coupled problems require an accurate evaluation of derivatives on the interface. For this reason, in higher dimensions, we envisioned a local remeshing strategy which maintains the vertexes of the meshes provided independently of the location interface. The presented approach leads to highly distorted elements near the interface.

For what concerns mixed elements, the issue of inf-sup stability on such distorted elements is largely open and very few proofs exist, in particular for triangles. Herein, we investigated four common mixed element schemes known to be inf-sup stable on isotropic meshes: two with continuous pressures and two with discontinuous pressures. We presented a set of simple tests to investigate the behavior of the inf-sup constant as the triangles are distorted. In particular, we showed that the elements with discontinuous pressures are more subjected to inf-sup issues within the presented local remeshing strategy and that they should not be used in this context (at least without stabilization).

Even if conceptually the extension of the present approach to 3D is straightforward, the analysis and the implementation of the method is not. Indeed, in the context of meshes with well shaped elements, the error analysis is easily extended from 2D to 3D, in particular in the affine case. However, for anisotropic transformations, from the reference to the physical element error, analysis has to be performed case by case (see, e.g., Apel 1999). We must expect that inf-sup stability of such elements in 3D may not be inherited from 2D stable mixed elements and very few proofs exist (see Acosta and Durán (1999) for an example of such a proof for the low-order non-conforming Crouzeix–Raviart mixed element: $\mathbf{P}_1^{NC}/P_0$). Furthermore, the intersection of tetrahedra by planes leads to much more complex subdivisions (not only triangles and quadrilaterals, as with the 2D case) but arbitrary polyhedra which would lead to many sliver tetrahedra using a Delaunay triangulation. For this reason, we believe that a recently developed numerical method named the Virtual Element Method (VEM) (see, e.g., Beirão da Veiga et al. 2013 and Beirão da Veiga et al. 2014) is a possible solution. This approach has two important properties that could be employed in the framework discussed in this article: it allows elements to be arbitrary polytopes (and thus we can avoid the subdivision process) and it is robust when elements are highly anisotropic. Further works are required to assess the viability of the method in the present context.

Regarding the conditioning of the various systems, inf-sup stability directly impacts the conditioning of the system and thus having an inf-sup stable element is important, as shown in the present work. Nevertheless, conditioning on anisotropic elements is a major issue for solving large problems and new techniques are required to overcome them, as performed, for example, in a 2D setting in Frei and Richter (2014).

Acknowledgments This work is partially funded by iCardioCloud project by Cariplo Foundation (No. 2013-1779)) and Lumbardy Region (No. 42938382; No. 46554874); ERC Starting Grant through the Project ISOBIO: Isogeometric Methods for Biomechanics (No. 259229); The authors would also like to acknowledge the support of Franco Brezzi and Alessandro Veneziani in the realization of this work.

References

Acosta, G., & Durán, R. G. (1999). The maximum angle condition for mixed and nonconforming elements: Application to the stokes equations. *SIAM Journal on Numerical Analysis*, *37*, 18–36.

Ainsworth, M., Barrenechea, G. R., & Wachtel, A. (2014). *Stabilisation of high aspect ratio mixed finite elements for incompressible flow*. Brown: Technical report.

Amdouni, S., Moakher, M., & Renard, Y. (2014). A local projection stabilization of fictitious domain method for elliptic boundary value problems. *Applied Numerical Mathematics*, *76*, 60–75.

Apel, T. (1999). *Anisotropic finite elements: Local estimates and applications*. Teubner.

Apel, T., & Nicaise, S. (2004). The inf-sup condition for low order elements on anisotropic meshes. *Calcolo*, *41*, 89–113.

Apel, T., & Randrianarivony, H. M. (2003). Stability of discretizations of the stokes problem on anisotropic meshes. *Journal Mathematics and Computers in Simulation*, *61*, 437–447.

Auricchio, F., Brezzi, F., & Lovadina, C. (2004). Mixed finite element methods. *Encyclopedia of computational mechanics, Chapter 9*. New York: Wiley.

Auricchio, F., Boffi, D., Gastaldi, L., Lefieux, A., & Reali, A. (2014). A study on unfitted 1d finite element methods. *Computers and Mathematics with Applications*, *68*, 2080–2102.

Auricchio, F., Brezzi, F., Lefieux, A., & Reali, A. (2015a). An "immersed" finite element method based on a locally anisotropic remeshing for the incompressible stokes problem. *Computer Methods In Applied Mechanics and Engineering*, *294*, 428–448.

Auricchio, F., Lefieux, A., Reali, A., Veneziani, A. (2015b). A locally anisotropic fluid-structure interaction remeshing strategy for thin structures with application to a hinged rigid leaflet. *International Journal For Numerical Methods In Engineering*. doi:10.1002/nme.5159.

Babuška, I. (1970). The finite element method for elliptic equations with discontinuous coefficients. *Computing*, *5*, 207–213.

Babuška, I., & Aziz, A. K. (1976). On the angle condition in the finite element method. *SIAM Journal on Numerical Analysis*, *13*, 214–226.

Babuška, I., & Strouboulis, T. (2001) *The finite element method and its reliability*. Oxford: Oxford University Press.

Baiges, J., Codina, R., Henke, F., Shahmiri, S., & Wall, W. A. (2012). A symmetric method for weakly imposing dirichlet boundary conditions in embedded finite element meshes. *International Journal for Numerical Methods in Engineering*, *90*, 636–658.

Barrenechea, G. R., & Chouly, F. (2012). A local projection stabilized method for fictitious domains. *Applied Mathematics Letters*, *25*(12), 2071–2076.

Barrett, J. W., & Elliott, C. M. (1987). Fitted and unfitted finite-element methods for elliptic equations with smooth interfaces. *IMA Journal of Numerical Analysis*, *7*, 283–300.

Basting, S., & Weismann, M. (2013). A hybrid level set-front tracking finite element approach for fluid-structure interaction and two-phase flow applications. *Journal of Computational Physics*, *255*, 228–244.

Bazilevs, Y., & Hughes, T. J. R. (2007). Weak imposition of dirichlet boundary conditions in fluid mechanics. *Comput. Methods Appl. Mech. Engrg.*, *36*, 12–26.

Béchet, É., Moës, N., & Wohlmuth, B. (2009). A stable Lagrange multiplier space for stiff interface conditions within the extended finite element method. *International Journal for Numerical Methods in Engineering*, *78*, 931–954.

Beirão da Veiga, L., Brezzi, F., Cangiani, A., Manzini, G., Marini, L. D., & Russo, A. (2013). Basic principles of virtual element methods. *Mathematical Models and Methods in Applied Sciences*, *23*, 199–214.

Beirão da Veiga, L., Brezzi, F., Marini, L. D., & Russo, A. (2014). The hitchhiker's guide to the virtual element method. *Mathematical Models and Methods in Applied Sciences*, *24*, 1541–1573.

Bern, M., Eppstein, D. (1992) Mesh generation and optimal triangulation. In *Computing in Euclidian geometry, Lecture Notes Series on Computing*, (vol. 1, pp. 23–90). Singapore, World Scientific.

Bhalla, A. P. S., Bale, R., Griffith, B. E., & Patankar, N. A. (2013). A unified mathematical framework and an adaptive numerical method for fluid-structure interaction with rigid, deforming, and elastic bodies. *Journal of Computational Physics*, *250*, 446–476.

Boffi, D., Gastaldi, L., Heltai, L., & Peskin, C. S. (2008). On the hyper-elastic formulation of the immersed boundary method. *Computer Methods in Applied Mechanics and Engineering*, *197*, 2210–2231.

Boffi, D., Cavallini, N., & Gastaldi, L. (2011). Finite element approach to immersed boundary method with different fluid and solid densities. *Mathematical Models and Methods in Applied Sciences*, *21*, 2523–2550.

Boffi, D., Brezzi, F., & Fortin, M. (2013). *Mixed finite element methods*. Heidelberg: Springer.

Burman, E., & Hansbo, P. (2010). Fictitious domain finite element methods using cut elements: I. A stabilized Lagrange multiplier method. *Computer Methods in Applied Mechanics and Engineering*, *199*, 2680–2686.

Burman, E., & Hansbo, P. (2011a). Fictitious domain finite element methods using cut elements: II a stabilized Nitsche method. *Applied Numerical Mathematics*, *62*(4), 328–341.

Burman, E., Hansbo, P. (2011b). Fictitious domain methods using cut elements: III. a stabilized nitsche method for Stokes' problem. Technical report, Jönköping University.

Carey, G. F. (1982). Derivative calculation from finite element solutions. *Computer Methods in Applied Mechanics and Engineering*, *35*, 1–14.

Carey, G. F., Chow, S. S., & Seager, M. K. (1985). Approximate boundary-flux calculations. *Computer Methods in Applied Mechanics and Engineering*, *50*, 107–120.

Chin, E. B., Lasserre, J. B., & Sukumar, N. (2015). Numerical integration of homogeneous functions on convex and nonconvex polygons and polyhedra. *Computational Mechanics*.

Court, S., Fournier, M., & Lozinski, A. (2014). A fictitious domain approach for the Stokes problem based on the extended finite element method. *International Journal for Numerical Methods in Fluids*, *74*, 73–99.

Demkowicz, L. (1991). A note on symmetry boundary conditions in finite element methods. *Applied Mathematics Letters*, *4*(5), 27–30.

Diniz dos Santos, N., Gerbeau, J.-F., & Bourgat, J.-F. (2008). A partitioned fluid-structure algorithm for elastic thin valves with contact. *Computer Methods in Applied Mechanics and Engineering*, *197*, 1750–1761.

Elman, D., Silvester, H., & Wathen, A. (2005). *Finite elements and fast iterative solvers with applications in incompressible fluid dynamics*. Oxford: Oxford Press.

Ern, A., & Guermond, J.-L. (2004). *Theory and practice of finite elements*. Heidelberg: Springer.

Fabrèges, B. (2012). *Une méthode de prolongement régulier pour la simulation d'écoulements fluide/particules*. PhD thesis, Université Paris-Sud.

Frei, S., & Richter, R. (2014). A locally modified parametric finite element method for interface problems. *SIAM Journal on Numerical Analysis*, *52*(5), 2315–2334.

Fries, T.-P., & Belytschko, T. (2010). The extended/generalized finite element method: An overview of the method and its applications. *International Journal for Numerical Methods in Engineering*, *84*, 253–304.

Gerstenberger, A., & Wall, W. A. (2008). An eXtended Finite Element Method/Lagrange multiplier based approach for fluid-structure interaction. *Computer Methods in Applied Mechanics and Engineering*, *197*, 1699–1714.

Girault, V., & Glowinski, R. (1995). Error analysis of fictitious domain method applied to a dirichlet problem. *Japan Journal of Industrial and Applied Mathematics*, *12*, 487–514.

Glowinski, R. (2003). *Handbook of numerical analysis: Numerical methods for fluids (Part 3)* (vol. 9), chapter VIII. North-Holland.

Glowinski, R., Pan, T., & Périeux, J. (1994). A fictitious domain method for external incompressible viscous flow modeled by navier-stokes equations. *Computer Methods In Applied Mechanics and Engineering*, *112*, 133–148.

Hachem, E., Feghali, S., Codina, R., & Coupez, T. (2013). Immersed stress method for fluid-structure interaction using anisotropic mesh adaptation. *International Journal for Numerical Methods in Engineering*, *94*(9), 805–825.

Hannukainen, A., Korotov, S., & Křížek, M. (2012). The maximum angle condition is not necessary for convergence of the finite element method. *Numerical Mathematics*, *120*, 79–88.

Hansbo, A., & Hansbo, P. (2002). An unfitted finite element method, based on Nitsche's method, for elliptic interface problems. *Computer Methods in Applied Mechanics and Engineering*, *191*, 5537–5552.

Hansbo, P., Larson, M. G., & Zahedi, S. (2013). A cut finite element method for a Stokes interface problem. *Applied Numerical Mathematics*, *85*, 90–114.

Haslinger, J., & Renard, Y. (2009). A new fictitious domain approach inspired by the extended finite element method. *SIAM Journal on Numerical Analysis*, *47*, 1474–1499.

Hautefeuille, M., Annavrapu, C., & Dolbow, J. E. (2012). Robust imposition of dirichlet boundary conditions on embedded surfaces. *International Journal for Numerical Methods in Engineering*, *90*, 40–64.

Heltai, L., & Costanzo, F. (2012). Variational implementation of immersed finite element methods. *Computer Methods in Applied Mechanics and Engineering*, *229*, 110–127.

Hyman, M. A. (1952). Non-iterative numerical solution of boundary-value problems. *Applied Scientific Research, Section B*, *2*, 325–351.

Ilinca, F., & Hétu, J.-F. (2011). A finite element immersed boundary method for fluid flow around rigid objects. *International Journal for Numerical Methods in Fluids*, *65*, 856–875.

Jamet, P. (1976). Estimations d'erreur pour des éléments finis droits presque dégénérés. *RAIRO. Analyse Numérique*, *10*, 43–60.

Kamenski, L., Huang, W., & Xu, H. (2014). Conditioning of finite element equations with arbitrary anistropic meshes. *Mathematics of Computations*, *83*(289), 2187–2211.

Kellogg, R. B. (1971). Singularities in interface problems. *Numerical solution of partial differential equations* (pp. 351–400). Cambridge: Academic Press.

Křížek, M. (1991). On semiregular families of triangulations and linear interpolation. *Applications of Mathematics*, *36*(3), 223–232.

Lefieux, A. (2014). *On the use of anisotropic triangles in an immersed finite element approach with application to fluid-structure interaction problems*. PhD thesis, Istituto Universitario degli Studi Superiori di Pavia.

Lemrabet, K. (1977). Régularité de la solution d'un problème de transmission. *Journal de Mathèmatiques Pures et Appliquèes*, *9*(56): 1–38.

Lew, A. J., & Buscaglia, G. C. (2008). A discontinuous-Galerkin-based immersed boundary method. *International Journal for Numerical Methods in Engineering*, *76*, 427–454.

Li, J., Melenk, J. M., Wohlmuth, B., & Zou, J. (2010). Optimal a priori estimates for higher order finite elements for elliptic interface problems. *Applied Numerical Mathematics*, *60*, 19–37.

Li, Z., & Ito, K. (2006). *The immersed interface method*. SIAM.

Liao, Q., & Silvester, D. (2013). Robust stabilized stokes approximation methods for highly stretched grids. *IMA Journal of Numerical Analysis*, *33*, 413–431.

Massing, A., Larson, M. G., Logg, A., & Rognes, M. E. (2012). A stabilized Nitsche fictitious domain method for the stokes problem. *Journal of Scientific Computing*, *61*(3), 604–628.

Maury, B. (2001). A fat boundary method for the poisson problem in a domain with holes. *Journal of Scientific Computing*, *16*, 319–339.

Maury, B. (2009). Numerical analysis of a finite element/volume penalty method. *SIAM Journal on Numerical Analysis*, *47*(2), 1126–1148.

Melenk, J. M., & Babuška, I. (1996). The partition of unity finite element method: basic theory and applications. *Computer Methods in Applied Mechanics and Engineering*, *139*, 289–314.

Micheletti, S., Perotto, S., & Picasso, M. (2004). Stabilized finite elements on anisotropic meshes: A priori error estimates for the advection-diffusion and the stokes problems. *SIAM Journal on Numerical Analysis*, *41*(3), 1131–1162.

Moës, N., Dolbow, J., & Belytschko, T. (1999). A finite element method for crack growth without remeshing. *International Journal for Numerical Methods in Engineering*, *46*, 131–150.
Nicaise, S. (1993). *Polygonal interface problems*. Switzerland: Peter Lang.
Parvizian, J., Düster, A., & Rank, E. (2007). Finite cell method. *Computational Mechanics*, *41*, 121–133.
Pedrizzetti, G. (2005). Kinematic characterization of valvular opening. *Physical Review Letters*, *95*, 194502.
Peskin, C. (1977). Numerical analysis of blood flow in the heart. *Journal of Computational Physics*, *25*, 220–252.
Peskin, C. S. (2002). The immersed boundary method. *Acta Numerica*, *11*, 1–39.
Rand, A. (2009). *Delaunay refinement algorithms for numerical methods*. PhD thesis, Carnegie Mellon University.
Sanders, J. D., Laursen, T. A., & Puso, M. A. (2012). A Nitsche embedded mesh method. *Computational Mechanics*, *49*, 243–257.
van Brummelen, E. H., van der Zee, K. G., Garg, V. V., & Prudhomme, S. (2011). Flux evaluation in primal and dual boundary-coupled problems. *Journal of Applied Mechanics*, *79*, 010904.
van Loon, R., Anderson, P. D., & van de Vosse, F. N. (2006). A fluid-structure interaction method with solid-rigid contact for heart valve dynamics. *Journal of Computational Physics*, *217*, 806–823.
Yu, Z. (2005). A DLM/FD method for fluid/flexible-body interactions. *Journal of Computational Physics*, *207*, 1–27.
Ženíšek, A. (1969). The convergence of the finite element method for boundary value problems of a system of elliptic equations (in czech). *APL Materials*, *14*, 355–377.
Zlámal, M. (1968). On the finite element method. *Numerical Mathematics*, *12*, 394–409.
Zunino, P., Cattaneo, L., & Colciago, C. M. (2011). An unfitted interface penalty method for the numerical approximation of contrast problems. *Applied Numerical Mathematics*, *61*, 1059–1076.

MIX
Papier aus verantwortungsvollen Quellen
Paper from responsible sources
FSC® C105338

If you have any concerns about our products,
you can contact us on
ProductSafety@springernature.com

In case Publisher is established outside the EU,
the EU authorized representative is:
Springer Nature Customer Service Center GmbH
Europaplatz 3, 69115 Heidelberg, Germany

Printed by Libri Plureos GmbH
in Hamburg, Germany